Praise for *Sex and* [War]

D0562488

"When I first observed the dark side of chimp[anzees]
they could be brutal just like us, especially th[e males.] [I]
do believe we have inherited aggressive tendencies from our ancient primate past—but also traits of compassion and altruism that we observe in chimpanzees as well. And with our highly developed intellect and sophisticated communication, I believe we are capable of controlling our violent behavior. Potts and Hayden make an important contribution as they explore our evolutionary origins and make suggestions as to how human society might reduce warfare in the future."

—JANE GOODALL, PhD, DBE,
Founder of the Jane Goodall Institute, UN Messenger of Peace

———

"In this fresh and fascinating examination of the causes of war and misery around the world, Malcolm Potts and Thomas Hayden have met the enemy, and he is us. *Sex and War* is a penetrating and far-ranging analysis of the long, bloody history of our species and the primitive behaviors and blinkered policies that drive nations to disaster. Tired of stale rhetoric and baseless ideology? Turn off the TV and the talking heads and read this book! It will transform your outlook on war, peace, and what needs to be done to secure a safer world."

—SEAN B. CARROLL, PhD
Author of *Endless Forms Most Beautiful* and *The Making of the Fittest*

———

"In this impressively comprehensive treatment, Potts and Hayden step as far back as possible from the human race to assess the root causes of social upheaval. The case they make for the genetic basis of aggression is both compelling and frightening."

—RANDY OLSON, PhD
Writer/Director, *Flock of Dodos: The Evolution-Intelligent Design Circus*

"*Sex and War* offers a compelling discussion of the biological roots of human warfare....*Sex and War* integrates information from multiple disciplines into a captivating read. It challenges us to ask deep questions about why the pattern of coalitional violence is so widespread in our species and why it is almost exclusively men who are involved."

—*SCIENCE*

———

"Malcolm Potts, an obstetrician-scientist, is a hero of the international movement to improve women's health. Here, with journalist Thomas Hayden, he offers a lively and highly readable account of the evolution of war and terrorism.... *Sex and War* is an important effort to raise our species' consciousness of its ugliest behaviors."

—*THE WASHINGTON POST*

———

"...Potts and...Hayden take a wide-ranging look at the many places that biology intersects with war. But the most fascinating parts of the book look at how modern technology has interacted with our Stone Age brains' risk calculators to produce the brutality and aggression of the world today."

—*WIRED*

———

"...one of the year's more thought-provoking books. *Sex and War* is worth reading, and arguing about. Nonviolently, of course."

—*THE TORONTO STAR*

———

"...highly readable must-read."

—*PITTSBURGH TRIBUNE-REVIEW*

ALSO BY MALCOLM POTTS

Ever Since Adam and Eve: The Evolution of Human Sexuality (1999)

Queen Victoria's Gene (1997)

Textbook of Contraception (1969)

ALSO BY THOMAS HAYDEN

On Call in Hell: A Doctor's Iraq War Story (2007)

How Biology Explains
Warfare and Terrorism and
Offers a Path to a Safer World

Malcolm Potts

AND

Thomas Hayden

BENBELLA BOOKS, INC.
Dallas, Texas

Copyright © 2008 by Malcolm Potts and Thomas Hayden
First BenBella Books Paperback Edition 2010

All rights reserved. No part of this book may be used or reproduced in any manner
whatsoever without written permission except in the case of brief quotations
embodied in critical articles or reviews.

BenBella Books, Inc.
6440 N. central Expressway, Suite 503
Dallas, TX 75206
www.benbellabooks.com
Send feedback to feedback@benbellabooks.com

Printed in the United States of America
10 9 8 7 6 5 4 3 2 1

Library of congress cataloging-in-Publication Data is available for this title.
ISBN 978-1935251-70-5

Proofreading by Emily Brown and Stacia Seaman
cover art by Jacques-Louis David © corbis
cover design by Laura Watkins
Text design and composition by John Reinhardt Book Design
Index by Shoshana Hurwitz
Printed by Bang Printing

Distributed by Perseus Distribution
perseusdistribution.com

To place orders through Perseus Distribution:
Tel: 800-343-4499
Fax: 800-351-5073
E-mail: orderentry@perseusbooks.com

Significant discounts for bulk sales are available.
Please contact Glenn Yeffeth at glenn@benbellabooks.com or (214) 750-3628.

Table of Contents

1

SEX AND VIOLENCE

Nature, Mr. Allnutt, is what we were put on this earth to rise above.

—KATHARINE HEPBURN as Rose Sayer, *The African Queen*, 1951

THE MINISTER OF HEALTH threw the keys across the table with the words, "And second gear doesn't work." I had not requested that an ambulance be parked outside the house we had turned into a nursing home for the raped women. I merely wanted to know how I would find one if a surgical complication arose. But this was Bangladesh in 1972, after the War of Liberation, and I was an honored visitor doing a difficult task, so I was given a converted Land Rover as an ambulance, along with a driver. The Minister was right about second gear.

My most harrowing confrontation with warfare, in a life too filled with exposure to its consequences, was to witness the result of what may well have been the largest systematic rape of women in the history of the world. Pakistan had begun in 1947 as a single nation in two parts, West and East, separated by one thousand miles of largely hostile Indian territory. The two parts shared the same Muslim religion but were deeply divided by language and culture. When East Pakistan sought independence as Bangladesh, the ethnically mixed West moved troops to the mostly Bengali East to prevent secession, and a particularly bitter war followed. The Pakistani soldiers targeted civilians, shooting the men and raping the women. India entered the war on the side of East Pakistan and an independent Bangladesh came into being in December 1971 when the West Pakistan army finally surrendered. Shortly afterward, I led a team of doctors from India, Bangladesh, Australia, the U.S.A., and England who sought to provide what help we could to those women who had survived

1

the mass rape and were pregnant as a result. We offered them abortions, and performed hundreds of the operations over several months. The pain, humiliation, and sheer human suffering I witnessed during that time have stayed with me ever since, and continue to help shape my thinking about sex, war, and human nature.

This book is about war. It is about terror, and cruelty, and the biological origins and long, brutal, vicious, and destructive history of organized aggression. Perhaps most importantly, it is about not just the depths to which human beings can sink, but also how we came to be this way and what we can do about it. We will show that killing other members of our own species—a rarity in the animal kingdom—is a male behavior that evolved early in our history, because those individuals who manifested such a predisposition were more likely to transmit their genes to the next generation than those who didn't. War and violence, then, are indelibly linked to sex and reproduction. This does not mean that human beings are inherently murderous, however, nor that war is inevitably as much a part of our future as it is a fixture of our past. In fact, we will show that humanity can take control of its most destructive impulses to build a safer, more secure world. Indeed, we have already begun to do so; despite our fiendishly effective weaponry and the very real and vicious presence of war and terrorism in the modern world, we are actually more at peace today than we have ever been.

In the pages that come, we will show that for most of history and prehistory, small groups of men who were prepared to attack their neighbors and steal their resources, and who could seduce or coerce women for sex, ended up having more offspring. Women, meanwhile, were more likely to improve their reproductive success—to have more children survive to reproduce themselves—by aligning themselves with successfully violent men rather than by joining raids and risking death themselves.* Women will certainly fight courageously to protect themselves, their children, and their communities, but unlike men, they do not seem to band together spontaneously, go out, and attack and kill other people. Fortunately, humans have culture as well as biology. Our genes may provide violent impulses, but our minds, our hearts, and our laws and social standards

---

* This is apparently still the case. In a 2008 study of twenty-eight populations over 300 years, British researchers Rebecca Sear and Ruth Mace found that the early death of a father often had "surprisingly little effect" on their offspring's survival, while a mother's death significantly increased childhood mortality. In evolutionary terms, fathers are more expendable than mothers, and thus can afford to take more risks.

are often quite good at tempering them. We will argue that while evolution has linked sex and violence over millions of years, civilization has given us the tools to separate the two again, and that this opens an important pathway toward making the world a safer place. To understand how, we must look closely at war and its impacts from a biological perspective. What we see when we do so can be distressingly ugly.

Of my foreign colleagues in Bangladesh, Dr. Geof Davis from Australia stayed longest after the war. He visited every hospital in the country, and estimated that 100,000 women had been raped during the nine-month conflict.[1] Many of them were girls, hardly past puberty. They were malnourished and their pregnancies were well advanced by the time we arrived to try to help them, and unfortunately we did have the sort of medical complication I had feared would require an ambulance. Sometimes when a fetus dies in the womb, either spontaneously or during an induced abortion, the protein that makes blood clot accumulates around the dead tissue. Doctors call the condition afibrinogenemia, and it was the reason one of the women we were treating began to bleed copiously as we were trying to deliver her dead fetus.

We needed the ambulance—fast. But the driver had chosen that precise moment for lunch and had wandered off, taking the ambulance keys with him. After a harried search and much lost time we found the driver and the keys and set off for the hospital, now in desperate need of a blood transfusion for our patient. The young girl died as I cradled her head on my lap, traveling the potholed road from our nursing home to the hospital.

In a highly conservative Muslim society such as Bangladesh, to rape a woman is to destroy her. As we will see, there are biological reasons why men often strive to control women's reproduction, and rape during wartime—a disturbingly common phenomenon—may be an extreme example of that tendency. Regardless, the systematic campaign of rape tore Bangladeshi society apart from within. If a woman had been a virgin when she was attacked, she became unmarriageable; if already married, she was abandoned by her husband as unclean. The women brought to us came from various parts of Dhaka, the capital, and they lay motionless on the iron beds we had found to furnish the little nursing home. They never talked to us, or to one another, and even our Bengali colleagues could discover very little about them. Sometimes we did not even know their names. It took some time just to find out who our dead girl was and where she came from.

The Bengali doctors were deeply worried when they heard she had been raped not by a Pakistani but by a member of the Bihari ethnic group living in Bangladesh. The Biharis, originally from the neighboring part of India, were especially hated because they had sided with the West Pakistanis in their efforts to prevent the emergence of an independent Bangladesh. Moreover, while the Pakistani military had gone home after surrendering on December 16, 1971, the Biharis had been rounded up and put in a camp for their own protection, out of fears that their neighbors might seek revenge. My Bangladeshi friends feared that news of the girl's death would ignite a retaliatory attack on the Bihari camp. "Would you come to tell the girl's parents she has died?" they asked, thinking that if the news came from me, there might be less chance of community violence. I was not sure at all how a young English doctor could help quell a riot, but we all piled into the Land Rover for a second journey.

We ran out of fuel on the way to the girl's house, and by chance we were exactly opposite the Bihari camp. Some sort of disturbance was already taking place and a fire had started. By the time our driver plodded back with a can of diesel it was getting dark. We finally found the girl's home by asking the mullah at a nearby mosque as he was closing up after evening prayers. To reduce the risk of a riot, my colleagues had decided to tell the girl's mother that her daughter was very ill, and to bring her back to the clinic. The lane to her home was too narrow to take the vehicle, so we were parked at the end, where she met us. The girl's mother explained she needed to run back to the hut where she lived to get some things. We waited in the dark for her to come back.

The dead girl's mother never returned—she knew her daughter's condition, even if she did not know she was dead. If her daughter had died from any cause other than a pregnancy resulting from rape, my colleagues' fears of a riot most likely would have been realized. But the disgrace of rape was compounded by the stigma of abortion, which at that time in Bangladesh was regarded as so shameful and secretive that the girl's mother, likely assuming her daughter had had one, simply ran away rather than confront the situation. Women suffered thrice over in this war: they were raped, they became pregnant, and they endured shame and anguish over abortion. The despair and pain of sexual violence are all too common in war. But unlike the rest of the shock and awe of warfare, these torments are hidden from public view. They must be borne by each woman alone, one by one.

Systematic rape is one of the most hideous, and most explicitly male, expressions of warfare, but it is hardly the only one. All wars are extraordinarily costly in material terms and grotesquely painful in human terms. Yet wars are so much a part of the human experience that we don't always pause to realize that one of the most astonishing aspects of war is the very fact that we so regularly go out and deliberately kill members of our own species.

Even in the now-peaceful West, we do not have to go back very far in most families to find the marks of war. Often, we celebrate the courage of those who fought. My wife's uncle Douglas Campbell was the first American-trained flying ace in World War I. He shot down eight German planes in the spring of 1918, before he himself was wounded.[2] My own father was in the British Royal Air Force the day it was founded in 1919, and again in the early days of World War II. My elder brother was a professional soldier who fought in Korea. I was a child in Cambridge, England, during the Second World War. I remember an occasional bombing raid, and my mother would sometimes invite clean-shaven American flyers from the nearby bomber base to dinner. I was too young to understand what it meant for tens of millions of men to be involved in cataclysmic global conflict, but I can still remember the unanimity of the British commitment to the war effort—and the universal hatred of Germans. Today, among my closest friends are Germans living in cities which the Allies bombed repeatedly in World War II. They confirm that the animus ran deeply in both directions, and I hate to think of the results had our families somehow met during the war. What drives such strong divisions between "us" and "them," and makes us so ready to kill our fellow human beings?

North Americans and Europeans today have lived through several decades of unprecedented peace in their own lands. People in the West no longer experience bombs falling on their houses or tanks rolling through their streets, and war has become something most of us see only on television. But that certainly doesn't mean we've lost interest. Many men especially never tire of reading war stories, and spend their time visiting battlefields or collecting memorabilia from conflicts long past. (Most women I have known have a strong sense that much of history is a catalogue of male destructiveness, and they tend not to want to spend as much time reading about tactics or heroic acts.) Wars and violence invade our fiction. The fantastic well-loved landscapes of J. R. R. Tolkien's *The Lord of*

To Douglas Dler
with love, from his uncle.
Douglas Campbell

Courtesy Martha Campbell

Douglas Campbell beside his Nieuport fighter in 1918, at age 21. The first U.S. trained ace in World War I, Campbell was wounded in action six weeks later. He lived to be 94. Relationships on the battlefield can be complex. The same chivalric rules that linked medieval knights to one another, even when fighting on opposing sides, often applied to fighter pilots in the first world war as well.

the Rings and brilliant cyber images of George Lucas's Star Wars series are not really about hobbits or intergalactic empires. They are about us: the human sense of justice and injustice, and our ready hatred of the "other" and blind love for loyal kin. There are reasons that we thrill to the spectacle of handsome young warriors defending some symbol, or rescuing nubile women—"nubile" meaning, literally, "suitable for marriage"—from an evil enemy. These stereotypes tap into readily identified volcanoes of human emotion and human biology. And they provide strong hints as to why we spend so much time fighting, often with immense courage, and trying to kill other members of our own species—sometimes in the most deliberately painful ways possible.

What I saw in Bangladesh shouldn't have been a surprise to me, but it was. I had traveled through West Pakistan extensively, studying family planning, and as a result I knew something about the towns and villages from which the Pakistani soldiers came. These rapists were not some subset of pathologically evil men—they were proud fathers who would gladly have laid down their lives to protect their own wives and daughters, and they came from a society with strict rules forbidding premarital and extramarital sex. Neither was the cruel link they forged between sex and violence an anomaly; their behavior was not a sign of a sick culture. Indeed, when a sample of American men was asked what would they do if they found themselves in a situation where they could rape a woman and know they would never be caught, over one-third said they would rape[3]—and it's hard to imagine that as many men or more weren't so honest as to admit that truth. It does seem that human males—and this must include the male authors of this book, had they been dropped into a different set of circumstances—have an intrinsically nasty side.

Social scientists have studied all this, of course, and have come up with expansive descriptions of human violence in all its varieties. But *explanations* for that behavior seem to be in short supply. As a doctor, I am trained to believe that a list of symptoms simply isn't enough. We have to look for root causes if we are to make a proper diagnosis and find an effective cure. Fortunately, we do have many of the intellectual tools and scientific data we need to begin solving the conundrum of organized human violence; pulling them together is one of the key purposes of this book.

Sex and War might just as easily have been called *Biology and War*—at its heart, this book is about the biological roots of warfare and terrorism. We take the view that human behavior can only be truly understood through the lens of human biology, including the several million years of human evolution. And so it makes sense that one promising route to understanding ourselves is to look at other mammals, our biological relatives. We test our drugs on mice and dogs, and tens of thousands of people are alive because they have had a pig's valve sewn into their hearts. But despite the obvious connections in our biology, when it comes to human behavior many people are strangely reluctant to look at animal models. And yet we can learn a great deal about ourselves by taking a closer look at our mammalian cousins, through both our similarities and our differences. Dogs, pigs, and monkeys are all just as interested in sex as we are, and they too manifest violent behaviors; understanding why they generally do not set about obliterating their own kind, for example, may be a key to understanding why *we* do.

Competition

Ultimately, the evolution of every living thing has been driven by competition. It is a simple, universal fact that all living things, from bacteria to giant redwoods, reproduce more rapidly than the resources in their environments can support. And thus all living things must compete against not just other species, but also their own kind in order to survive. Competition was there, at the molecular level, when life first began literally billions of years ago, and it has continued to shape the evolution of life in increasingly complex ways ever since. Evolution as a result is a painful, callous process of separating winners from losers, driven by what Charles Darwin called the "war of nature."

In complex animals, such as birds or mammals, this competition for survival is frequently associated with violent behaviors. This occurs between species, within species, and even between males and females of the same species. Nowhere is this war of nature made more explicit than in the battles males of many species must win in order to mate. The male deer or bull elephant seal that competes successfully against his brethren and impregnates many females will pass his genes on to future generations; his losing competitors will not. Human behavior is considerably more complex than that of deer or seals, but for our species, too, it appears that more competitive, aggressive males have often made outsized contributions to the gene pool—a contribution that inevitably reinforces our warlike tendencies.

Take Genghis Khan. In 2003, an international team of geneticists published a DNA analysis of central Asians.[4] Remarkably, the authors found, 8 percent of the men in central Asia have virtually identical Y chromosomes. Much more than a curious biological footnote, this finding offered a profound insight into our collective past. The Y chromosome defines maleness, and men who share a Y chromosome are all descended from the same man.* The fact that one in twelve men in Central Asia has the same Y chromosome can mean only one thing: At some point in history, probably within the last thousand years, one man had a vast number of offspring. And Genghis Khan, Mongol Emperor from 1206 A.D. until his death in 1227, is the one historical figure who fits this role.

Still revered as a great leader in parts of Central Asia and reviled elsewhere as the ultimate barbarian, Genghis Khan presided over one of the greatest military expansions of all time. In just over twenty years, he united the tribes of the Central Asian Plateau, and extended the reach of the Mongol Empire from the Sea of Japan in the east to the Caspian Sea some 6,000 kilometers to the west. He reinforced loyalty by sharing out the spoils of war—including vanquished women—after every battle, but the Great Khan saved the most beautiful women for himself, and he boasted of the pleasure of violating other men's wives and daughters. Just a century after his birth, a Chinese historian claimed that Genghis Khan already had 20,000 descendants, and whether that was an exaggeration or not, modern science suggests that today the number of his descendents has grown to 16 million worldwide. Put another way, fully one in 200 of all men now alive on this planet are likely descendants of Genghis Khan.

If Genghis Khan is the most dramatic example of this "founder effect," he is hardly the only one. Another recent Y chromosome study, conducted on almost 800 men in Ireland, found that 20 percent shared a common genetic signature thought to show descent from a Middle Age king called Niall. In all, the study's authors estimated that as many as one in twelve people in Ireland—and perhaps two to three million people worldwide—could be descended from King Niall.[5] And a similar analysis of the seventeenth-century Manchu ruler Nurhaci suggests that he—or perhaps his

---

* Human chromosomes come in twenty-two matched pairs, plus one potentially mixed pair of X and Y "sex" chromosomes. A person who has two X chromosomes is female and a person with one X and one Y is male. In the case of the matched chromosome pairs, the genetic material from each parent is mixed together during reproduction, but the single Y chromosome is passed virtually unaltered from father to son. Male lineages can be tracked reliably as a result. Another type of genetic material, mitochondrial DNA, is passed only from mothers to both sons and daughters, revealing maternal lineages with even less susceptibility to change over time.

grandfather Giocangga—has 1.6 million descendents alive today.* If, as we will see in the next chapter, evolution judges success by the number of an individual's progeny, then Genghis Khan, King Neill, Nurhaci, and other, lesser warriors and rapists are among the most "successful" men in history. (None of them lived long enough ago to be credited with the evolutionary invention of aggression or war, however. They, like all of us, were the products of evolution occurring over millions of years before their time on Earth.) Fortunately, human beings have also evolved as social animals capable of empathy, altruism, and love. But even if most of us would prefer to live in a world where the Genghis Khans do not win, we can't begin to understand the raids and wars that still permeate our present world without understanding why they so often won in the past.

Throughout history and across cultures, rich and powerful men typically have had more sexual partners, and thus more offspring, than those lower in the social hierarchy.** The Bible tells us that King Solomon had 700 wives, and 300 concubines. Large harems were the order of the day for the Egyptian Pharaohs, the Aztec Kings, the Turkish Sultans, the African Kings, and the Chinese Emperors.[6] Idi Amin, the Ugandan despot of the 1970s, had four wives and thirty children. Queen Victoria's son, later King Edward VII, had several mistresses and many short affairs.[7] From his box in a theater he would survey the women in the audience and send his equerry to invite the most attractive to join him—one hundred years later heavy metal bands echo the strategy, with roadies taking the place of squires. President John F. Kennedy had several long-term extramarital sexual relationships and many one-night stands, aided by a kind of cultural acceptance that Bill Clinton must surely have envied.[8] Impeachment would not have been a problem among the Incas of Peru, who provide perhaps the most organized example of how power has been used to allocate sexual opportunities to men. Incan law allowed (male) aristocrats fifty women apiece; the leaders of vassal nations were allotted thirty, the heads of 100,000 men received twenty women, and so on down to leaders of ten men, who were apportioned three. The great Inca at the top of

---

* Because the Y chromosome is shared by close male relatives, this sort of genetic analysis doesn't distinguish between the descendents of brothers, paternal uncles, and male paternal cousins, for example. And of course the descendents of any man also are descended from his grandfather. Whether a Y lineage is founded by one man only, or by a group of male relatives, the results remain the same, as do the reasons—in these cases, warfare, conquest, and rape.

** The relatively recent availability of reliable contraception has of course made sex without pregnancy quite easily achievable. Judging by the frequency of celebrity paternity suits, however, this fact seems to have escaped many of today's rich and powerful men.

the social pyramid had the pick of the empire and the men at the bottom often went unmarried.

Even if not all generals rape as did Genghis Khan, and not all kings and presidents are promiscuous—nor all philanderers violent—we still need to explain why in every society, across time and space, men on average are much more violent than women. This book will argue that our evolutionary history is the key, and that while the different behaviors of men and women in relation to sex and violence are modified by our upbringing, they are also written in our genes. The standard social science paradigm tells us that the differences between the sexes are almost entirely determined by culture. An evolutionary paradigm, in contrast, recognizes biological differences in male and female behavior, put in place over millions of years of evolution. This is not an argument that nature necessarily triumphs over nurture, nor that culture is unimportant and genes omnipotent: It is simply a recognition that humans are subject to the same biology as every other living species.*[9] History demonstrates that a behavioral predisposition is not predestination, because we can choose to express the impulses and tendencies we have inherited in many ways, some of them peaceful or creative. Our genes certainly help to set our personalities and temperaments, but our behavior can be molded by our families, environment, and cultures—and by our own decisions—in a rich variety of ways.

Team Aggression

Human warfare and terrorism require a special sort of violence in men, which we will call *team aggression*. This behavior is not limited to humans and we will document how in a handful of social mammals, a highly specialized behavior has evolved in which teams of adults—almost always males—attack and kill individuals of the same species. In the process they enlarge their territory and increase the resources available to the group to which they belong, with the side benefit of eliminating potential sexual competitors. In the case of chimpanzees and human beings, the chief practitioners, males are the primary beneficiaries of this kind of team aggression.

---

* Most parents who have a boy and a girl notice differences in behavior between the two sexes, and humans are not alone. In a 2002 study, psychologists Gerianne Alexander and Melissa Hines gave toys such as a police car and a doll to vervet monkeys. The boy monkeys spent more time pushing the car around and the girl monkeys more time cuddling the doll.

It is no accident that modern armies are built on small squads, grouped together into platoons—nor that terrorists almost always organize themselves into cells. Young men in all cultures have a common willingness to work together in small, intensely loyal teams. The majority of men awarded medals for courage on the battlefield are in their twenties—or even younger. And most will tell you that they fought not primarily for king and country, but for their comrades, and because they feared letting this most intimate of teams down. In chapter 7, we will show how modern warfare, despite its high-tech façade, is based on the same set of behaviors that drives everything from a minutes-long raid conducted by a small posse of illiterate men in the Brazilian rainforest, to the million-plus warriors locked in deadly combat for months on end at Stalingrad in the winter of 1942–1943. The nineteen young men who killed over 3,000 people on September 11, 2001 also fit into this seamless continuum of violent raids.

Evolution has tailored our bodies to digest meat, and our hands to manipulate small objects. In the same vein we have also inherited certain universal behavioral frameworks, such as sexual jealousy, which are expressed across virtually all cultures and all periods of history. We will argue that the predisposition to engage in team aggression against our own species grows out of just these sorts of inherited frameworks, including aggression, group cohesion, and a strong sense of loyalty to our own group backed by hostility toward outsiders. These impulses stretch back five to seven million years or more, to the ancestors we shared with chimpanzees. The evidence suggests that a series of genetic mutations occurred in our ape-like, forest-living forebears, which predisposed adult males to band together with their brothers and cousins to raid and kill their neighbors, and that those who manifested such a trait acquired more territory. A larger territory meant more resources, more resources meant more females, more females meant the opportunity for more sex, and more sex meant more offspring carrying the male's genes, aggressive tendencies and all, to the next generation. Those males who coordinated their violence in teams became the winners in the ruthless war of nature.

Tempting as it is to think of wars or terrorist attacks as temporary aberrations interrupting an intrinsically peaceful world, biology and history show that this simply is not the case. A careful examination of evolutionary clues, the archaeological record, and modern behavior proves that raids and wars are fundamental aspects of our nature, human behaviors going back to our pre-human history, and ones to which men return with remarkable ease. The face of war has changed dramatically as our societies have become more complex and our technologies more lethal, but there is no war without willing warriors, and those have never been in short supply. During the two World Wars, whole generations of young men in more than a dozen countries were recruited to fight, and surprisingly few opted out. Even during the current American war in Iraq, when public awareness of the physical, economic, and psychiatric burdens borne by warriors and their families has arguably never been higher, military recruiters are still able to find considerable numbers of young men not just willing to serve, but eager to fight. A central argument in this book is that violent conflict, terrorism, and war—and the emotions that they are built on—derive from an evolved behavior that once benefited the males who expressed it.

Women

It is indeed difficult—perhaps impossible—for one sex to get into the mind of the other, which is why I appreciate the contributions of my wife to this book. As an obstetrician I have delivered hundreds of babies, and it always has a powerful impact—the first time I saw a birth as a medical student, I nearly fainted and had to sit down. But as a man I will never truly understand what it means to give birth. If we could live in the mind of the opposite sex for just one day, I suspect we would be amazed at the differences that internal hormones (themselves an expression of genes) and the external environment have wrought on the ways the two sexes look at the world.

The evolutionary perspective we take here highlights the fact that team aggression is predominately an activity of young males, and although it affects women it very often does not benefit them in a lasting way, as we will show. Women compete for resources for themselves and their offspring, and may even benefit from associations with the most aggressive, successful male leaders or groups, but their competition is not usually expressed with such violent destructiveness as among men. There are women gangsters, warriors, and terrorists, of course, just as there have been great female military leaders. But despite some fanciful misinterpretations of prehistory and anthropology, women have never shared men's propensity to band together spontaneously and sally forth to viciously attack their neighbors.

The same can be said for older males, who may well gain more from war as generals than the young do as foot soldiers, but who also generally have much more to lose. We will show that societies where the proportion of young males to older males is high are often particularly prone to conflict—and that one way to reduce the risk of violence is to empower women and maximize their role in society.

This is perhaps the most profound insight to come from taking an evolutionary perspective on war: empowering women reduces the risk of violent conflict. Far from being a politically correct notion of feminist philosophy, women's role in reducing the risk of war is born out by rigorous study and historical experience. We will argue that contemporary Western nations have a great opportunity to make the world more secure and reduce terrorism by doing everything they can to empower women who live in countries where they currently enjoy few choices and wield little or no political power.

There is a clear and important evolutionary link between this biological perspective on war and Martha's and my work in family planning. For most women, the first step toward autonomy and equality is the ability to choose when to have children, and how many to have. From Italy to Iran, it is an empirical fact that wherever women have access to modern family planning, family size always falls. This is of crucial interest at a time when rapidly growing human population stresses our infrastructure and social systems, and threatens what remains of the natural environment. Reducing birth rates can also have a dramatic effect on the prospects for peace by reducing competition for resources and lowering the ratio of testosterone-filled young men to older men and women in the population. By understanding this and other biological dynamics underlying war, we can begin to see our way toward taking real, effective action to reduce war's likelihood and limit its destructive potential.

Civilization at Its Best

Homo sapiens has been a distinct species for more than 200,000 years. This is a mere blink of the eye in the billions of years life has existed on Earth, but even so, 95 percent of human history was spent surviving and reproducing—and fighting—in small clans of hunter-gatherers.* In

* Nineteenth-century writers called such groups "savages." Today the neutral term "preliterate" is used.

other words, the behavioral frameworks underlying our sexual behavior and driving violence evolved to suit Stone Age conditions,* rather than a globalizing community where cooperation can benefit many and nuclear bombs and biological weapons can harm us all, even those who would use them. Culture evolves more rapidly than biology does, however, which lends hope to the challenge before us: to understand and rein in our Stone Age behaviors.

An evolutionary perspective not only suggests strategies to help peace break out, it also helps explain why men still try to limit women's freedoms in so many societies. In the 1970s I worked for a while in Beirut, Lebanon. Quite often newspapers would report the case of a man who had murdered his sister or niece and received a light prison sentence, or had been acquitted altogether. These were "honor killings" in which a woman who had fallen in love or had a sexual relationship before marriage had her throat cut by a male relative in order to protect the family's reputation.[10] (In 2004, for example, UNICEF recorded twenty-three honor killings in Jordan, fifty-two in Egypt, 300 in Pakistan, and 400 in Yemen.) In the West, the same desire to control women manifests itself in ways less obvious or violent, but equally real. In the United States, considerable numbers of people want to restrict access to contraception, especially to the unmarried. Moral and religious reasons, which I obviously believe to be mistaken, are offered, but the effect is the same—to limit the ability of women to control how many children they have and when. Overseas, the U.S. preaches democracy and free markets, but is slow to challenge the traditional restraints so cruelly heaped on women in many developing countries—restraints that keep women from participating as equals in political and economic life. Western science and technology have provided the cell phones and the jumbo jets, the high-yield seeds and the vaccines, but we are ambivalent and stingy when it comes to helping poor women in developing countries gain control over their own reproduction. Unfortunately, the current disputes over international assistance for family planning, especially in U.S. political circles, prevent us from taking steps that are not only humane and wanted by women (and many men) of the global South, but which would also do much to reduce conflict and forestall terrorism in the next generation.

* Archaeologists define the Stone Age as beginning two to three million years ago with the first use of stone tools by early hominids. The Stone Age is divided into the Paleolithic or Old Stone Age, the Mesolithic or Middle Stone Age, and the Neolithic or New Stone Age, which began six to ten thousand years ago with the adoption of agriculture, development of pottery, and settled living.

Our willingness to understand the evolution of our behavior could actually help determine whether we survive as a species or destroy ourselves through some combination of warfare and ecological disaster. The biological analysis we present here is not pessimistic but realistic, and thus ultimately hopeful. It does not imply that war is inevitable, but it does emphasize how deeply warring behaviors are embedded in our nature. I suspect that nearly every young man, placed in the appropriate environment, could find himself acting as a heroic warrior, or a torturer, or perhaps even a rapist—all for the same evolutionary reasons. In 1972 I found myself in Bangladesh as an educated, privileged young man, an honored guest of the government and able to provide at least some aid to women who had been brutalized as part of warfare. It is uncomfortable for me to consider that I was separated from their rapists not by my evolutionary nature, but simply by the circumstances of my birth and upbringing. But it was only once I accepted that frankly appalling reality that I could begin to make sense of what I saw in Bangladesh and to see that there was at least a glimmer of hope, even there.

The remark quoted at the beginning of this chapter comes from the classic 1951 film, *The African Queen*. It is part of an exchange between the Christian missionary Rose Sayer, played by Katharine Hepburn, and Charlie Allnut, played by Humphrey Bogart, the hard-drinking owner of the battered but reliable riverboat on which the couple find themselves. The idea that civilization at its best can rise above our warring nature is not a purely religious one, and it will be a recurring theme here. This book suggests that there is real hope for humankind to build a more peaceful future, but that hope shines with an uncertain light at the end of a long, exceedingly dark and dangerous tunnel. If we are ever to tame our Stone Age behaviors, we must begin by understanding their origins.

2

THE WAR OF NATURE

Thus from the war of nature, from famine and death, the most exalted object which we are capable of conceiving, namely, the production of higher animals follows.

—CHARLES DARWIN, *On the Origin of Species*, 1859

Y WIFE AND I have both been fortunate enough to watch chimpanzees in Gombe Stream National Park in Tanzania, where Jane Goodall has studied our closest living relatives for forty years. Martha has visited twice, and I have returned a half-dozen times, often while working on family planning programs in the area. We have not spent the thousands of hours primatologists devote to researching chimpanzee behavior but we have been able to glimpse their complex social life and the parallels between chimpanzees and ourselves. Human beings are not descended from chimpanzees, but we had a common ancestor five to seven million years ago, which makes us something like evolutionary cousins. There are obvious differences between chimps and humans but also many similarities, which are the legacy of our shared biological lineage. There is something very moving about observing such intelligent animals, and I suspect they also find us quite interesting to watch. Once I sat on the edge of a narrow plank bridge across the Gombe stream as two adult males pulled apart a small palm tree to eat the pith. Once finished, they crossed the bridge just behind me, and as they did so the biggest animal brushed against my back softly but deliberately, his very large scrotum swinging a foot from my face. The message I got was, "I know I am a male like you, I could crush you if I wanted to, but with balls this size I don't need to worry about you." I agreed with him.

The scientific observation of chimpanzee behavior goes far beyond simple curiosity. The idea that we can learn something about our own nature by teasing apart the complexities of chimpanzee society may seem strange to some readers, but there are these good reasons to do so. Watch female chimps and you see immediately the meaning of being a good mother. The energy, happiness, and occasional frustration of young chimps playing are identical to human play. We will explore what chimpanzees have to teach us about war and terrorism soon. But first, we'll need to understand something of the history of thought about warfare and human nature.

Einstein believed that "the theory determines what we observe."[11] This seems to be particularly true when it comes to thinking about human behavior, especially in its more violent forms. Explanations of wars and terrorism tend to fall into two categories. One intellectual tradition is associated with the Enlightenment-era French philosopher Jean-Jacques Rousseau and frames the world as innately good. Our ancestors are seen as living in a peaceful—even idyllic—world where "men were innocent and virtuous."[12]

Rousseau's perspective took on a new academic popularity after World War I, with anthropologists such as Franz Boas, Margaret Mead, and Ashley Montagu holding that war was solely the outcome of culture, uninfluenced by biological inheritance. "Evil," wrote Montagu, "is not inherited in human nature, it is learned."[13] The Dutch historian Johan Huizinga went further, claiming, "All fighting that is bound by rules bears the formal characteristics of play."[14] A year after he made this assertion that war, in essence, was some sort of game that got out of hand but had no real intention of killing people, Hitler invaded the Netherlands and Huizinga was held under detention until his death in 1945. Still, Rousseau's influence persisted, and the impulse to distance human nature from the horrors of war and genocide is certainly understandable. After World War II, Marxist archeologists argued that Stone Age societies were economically self-sufficient and therefore incapable of warfare[15]—that is until they excavated skeletons with flint arrows embedded in them.[16]

The opposing perspective predates that of Rousseau. It is mostly associated with the English philosopher Thomas Hobbes, who wrote in *Leviathan* in 1651, "every man is enemy to every man." In Rousseau's interpretation, people become violent because of some defect in their upbringing or injustice in the environment, which makes peace the natural human condition, and casts wars or acts of terrorism as social pathologies.[17] To a follower of Hobbes, however, warfare is not an aberration but the norm to which society returns when other influences are removed. This has been interpreted as an inherently pessimistic model of human behavior, but that need not be the case. This chapter builds the stage on which we will argue that warfare is indeed a partly inherited behavior that predates the first modern humans, and thus has been with us since our very beginnings as a species. We will show that as a result, we can learn a great deal about war by looking at the behavior of other animals, particularly chimpanzees. And we will demonstrate that this more realistic, biological perspective is anything but pessimistic—indeed, it offers the best chance we have to recognize and rein in our most destructive tendencies.

In examining the nature of warfare, an evolutionary approach builds bridges between biology, anthropology, and history. This type of synthesis is outlined in E. O. Wilson's landmark 1998 book *Consilience: The Unity of Knowledge*,[18] and we are hardly the first to attempt it. The first thinker to frame war as an instinctive biological phenomenon was an early twentieth-century polymath, Wilfred Trotter, who had been a surgeon at University College Hospital where I trained. His book, *Instincts of the Herd in Peace and War*, was published in 1916, at the height of World War I. In it, Trotter argued that human beings evolved as an instinctively gregarious species that reacts strongly to external threats.[19] He saw very clearly that this behavior was both inherited and unconscious. "Irrational belief forms a large bulk of the furniture of the mind," he wrote, "and is indistinguishable by the subject from rational verifiable knowledge." *Instincts of the Herd in Peace and War* had gone through seven printings by 1922 but it was then almost totally forgotten. In the 1960s Robert Ardrey in *The Territorial Imperative* also posited that human beings had evolved a tendency to defend a territory.[20] At about the same time, the pioneering animal behaviorist Konrad Lorenz, a Nobel laureate, suggested in his book *On Aggression* that human beings are innately aggressive.[21] Ardrey and Lorenz moved the discussion forward but they stopped short of explaining why human beings, unlike nearly all other animals, systematically kill their own species—something that is difficult, and crucially important, to explain.

This is where the chimpanzees of Gombe Stream National Park,[22] including the one who so clearly let me know where I stood in the social order, come to our intellectual rescue. The observations of chimpanzee violence in Gombe and elsewhere were vividly summarized by Richard Wrangham of Harvard in his 1996 book with Dale Peterson, *Demonic Males: Apes and the Origins of Human Violence*.[23] Our book takes the insights of *Demonic Males* as a starting point, and tracks the development of human raids, warfare, and terrorism from our common ancestry with chimpanzees through the fossil, archaeological, and historical record to the present day. The essence of history is to put ourselves in the shoes of historical persons while carrying a minimum of our contemporary baggage on the journey back through time. In the same way, the essence of observing other animals is to place ourselves in their paws, so to speak, leaving behind as much of our human perspective as we can while traveling into the world of another species. These can be challenging journeys, but they offer willing travelers a uniquely compelling and important story as a reward.

Explaining Evolution

To understand what chimpanzees can tell us about ourselves, we have first to understand exactly what it means to be related—to share a common ancestor. And that means starting with Charles Darwin. His *On the Origin of Species* (1859) is in many ways the most important book in the history of human thought. It is also one of the most challenging. The Darwinian vision replaced a static world, in which a Divine Creator had fashioned each species of living organism separately, with a dynamic world where all living things are related and slowly changing. Design and purpose were replaced by random, directionless, accidental changes that we call mutations.

At its core, Darwinian evolution is a remarkably simple process. All living things resemble their parents, but can differ in ways both subtle and dramatic. And all living things are, whether they recognize it or not, in a race to reproduce before they die. How well they manage that trick depends on many factors, and chance is one of the most important— just ask the dinosaurs, done in by a falling asteroid, of all the lousy luck.

But there are other factors in survival, including an organism's physical strength and intelligence, its susceptibility to disease, its age at sexual maturity, and its tendency to run or fight when confronted, to name but a few. All of these factors, to greater or lesser degrees, are influenced by genes, and all genes are susceptible to change, or mutation. When a mutation makes a gene more useful, the individuals that inherit the new form have a better chance of avoiding death before they pass their genes, mutation and all, along to future generations. The new therefore gains a reproductive advantage over the old, and that is the very heart of evolution.

Darwin did no complicated experiments, he used no statistics, and although he did own a microscope he could have created his theory without it. This naturalist deduced that evolution occurred from two simple observations. First, that all plants and animals produce more offspring than their environment can support and individuals must therefore compete to survive. Second, that individual animals and plants vary, that these variations are inherited, and that they lead to what Darwin called "the preservation of favoured races in the struggle for life." By carefully observing the natural world, while circumnavigating the world in the HMS *Beagle* between 1831 and 1836 and for many years afterward in his garden at Down, Kent, Darwin produced overwhelming evidence that species change, compete, and evolve over time.

It's important to keep in mind that evolution is a "therefore" not a "because." That is, giraffes weren't designed to have long necks *because* they could reach more treetop food to eat. Rather, those giraffes that happened to inherit a longer neck had a better rate of survival and reproduction, and *therefore* the genes for longer-necked giraffes became more common in succeeding generations. Darwin speculated that the several species of finch he saw on the Galapagos Islands had evolved from a few individuals of a single species blown to the remote location eons ago. Today there are fourteen species of finches there, mainly defined by small variations in the shape and size of their beaks. They did not develop a particular shape of beak so they could crack open a particular seed; rather, those finches with the optimum shape of beak to take advantage of particular food sources were more successful in obtaining food and therefore were more likely to survive and pass their genes to the next generation. Since 1973 Peter and Rosemary Grant of Princeton University have measured meticulously the inherited variation in the size of the beak in twenty generations of these same birds.[24] They have documented evolution in action and shown that even a millimeter difference in beak size can mean the difference between life and death during a time of drought. Evolution, like Newton's theory of gravity, has been proved by observation.

Variation is the fuel of Darwinian evolution, and that variability lives in the genes.* Our genetic program comes ready with several mechanisms which virtually guarantee that no child is exactly similar to either of its parents, and only rarely so to any of its siblings, in the case of identical twins. Each new child is a new experiment, in effect, blending the package of genes that each of its parents brings to the union and quite often introducing new arrangements and mutations of its own. Most mutations and many gene rearrangements are neutral so far as survival is concerned. Some, like so-called "disease genes" (which are actually genes that fail to perform some crucial purpose, rather than having no purpose other than to contribute to the organism's downfall), are harmful. And some—usually in concert with specific environmental conditions—give the endowed individual a leg up when it comes to surviving long enough to reproduce.

The Role of Genes

It was part of Darwin's genius to come up with his remarkable insights while lacking the understanding of genetic inheritance we enjoy today. Unbeknownst to Darwin, in the 1860s a monk called Gregor Mendel in Brno, in the present-day Czech Republic, demonstrated that many individual characteristics—in his case of the pea plants he grew in the monastery garden—were inherited according to simple mathematical rules.

By a quirk of history, at the time Mendel was conducting his studies, a key biochemical discovery was being made as a result of the destructiveness of warfare. During the Seven Weeks War in 1866, Prussia decisively defeated Austria at the battle of Tauberbischofsheim, close to Stuttgart, Germany. A Swiss doctor, Friedrich Miescher, did a chemical analysis of the large quantities of pus that accumulated in the infected wounds of sol-

---

* Variability can also be found in external factors, including nutrition, exposure to disease, and parental care—and genetic variations can be enhanced or obscured by everything from steroids for athletes to fetal exposure to alcohol, mercury, or other compounds. Until recently, scientists believed that unless such environmental factors caused genetic mutations, the effects were not passed along to offspring. But in one of the most dramatic examples of how much we still have to learn about biology, recent experiments have shown that environmental factors can cause changes in the cellular machinery that processes DNA—and that these changes can in fact be passed on to the next generation. The extent and significance of these "epigenetic" modifications are still uncertain, but it seems clear that they can be significant in susceptibility to cancer, as one example.

diers. Pus is mainly composed of white blood cells and bacteria, both containing a great deal of DNA. Miescher collected the sticky yellow material and was the first to study its chemical makeup, but he didn't know the large molecule he found would turn out to be the key to biological inheritance.[25] In 1944, it was shown that DNA transmitted genetic information and in February, 1953, Francis Crick and James Watson,[26] using superb x-ray crystallography studies conducted by Rosalind Franklin, worked out the now-famous double helix structure of DNA.

DNA consists of tens of thousands of atoms arranged in a small number of orderly molecular building blocks that form the letters of the genetic code. Miescher had speculated that "just as the words and concepts of all languages can find expression in the 24–30 letters of the alphabet," so chemicals might transmit information. It was a perspicacious insight. The DNA code is written with only four different chemical "letters" set out in a limited number of combinations but repeated many million times. We call the sequences of information carried by DNA *genes*, and genes control the manufacture of proteins. These large molecules in turn are the stuff of life; they make up the structure of our muscles, are the basis of our nerves, and carry oxygen throughout the body in red blood cells.

Genes are not static blueprints, like the plans of a house. They interact with one another and with the environment, and they are turned on and off as they sculpt the embryo during development, giving its various parts different characteristics. Some DNA sequences, such as the HOX genes that tell animal embryos how to divide into the thorax and abdomen and similar segments, have remained constant across billions of generations. Other genes, such as one dubbed FOXP2, which appears to be related to speech and language in humans, spread through our species as recently as 120,000 to 200,000 years ago.[27] In nearly all cases, many genes work together to produce the anatomical or behavioral characteristics we identify in one another.

Modern molecular biology spells out unequivocally and in atomic detail our place in nature and it confirms beyond doubt what Darwin discovered in 1859—that all living things are related, and that different species often use common anatomical structures and behavioral pathways. Ultimately, we share a common ancestor with the tuberculosis bacterium, with frogs, and with oak trees. Prior to the sequencing of the human genome—our full set of genes—in 2001, scientists assumed that human DNA differed from that of other species much more than it really does. Our separation from mice occurred seventy million years ago and from chimpanzees five to seven million years ago, and we now know that we continue to share 93 percent of our genes with mice and over 98 percent with chimpanzees. An understanding of genetics also unites races: all indications are that we differ more within the broadly defined racial groups than those groups vary from each other. Evolutionary understanding makes it crystal clear that all humans are one as a people—which may help explain why Daniel François Malan, the Dutch Reformed Church minister and architect of apartheid who became South Africa's prime minister in 1948, found Darwinian evolution to be a deeply distasteful idea.

Given that evolution is the product of random accidents, it is not surprising that biology is replete with what might well be called "ridiculous design" rather than "intelligent design." The larynx evolved to separate the food and air passages, for example, but as it migrated nearer to the base of the tongue the larynx was able to play a key role in creating speech. The price of this beneficial evolutionary step, however, was that sometimes human beings choke to death when food goes down the trachea to the lungs instead of to the stomach. A freshman engineering student could come up with a more intelligent design than that. Perhaps 400 million years ago in one of our fish ancestors a duplication occurred in the tube carrying urine from the kidneys to the outside world. In males, this second tube served as a convenient conduit to carry sperm to the outside world, and in females it evolved into the Fallopian tubes, uterus, and vagina. The exit to this system was behind the pelvic fin of a fish, but when fish evolved into land animals they needed a bony ring, or pelvis, to support their hind limbs. This is why a woman in labor has to push her baby through the pelvis amidst a great deal of pain and danger for her and her infant. The six year old who thinks babies are delivered through their mother's belly button reflects exactly what an intelligent designer would have done reroute the birth canal through the abdominal wall. But evolution does not permit redesign. It simply waits for another accidental mutation to build on that which already exists.*

---

* There are innumerable arguments against the supposed theory of "intelligent design"—the claim that biological structures and systems are too complex or too perfect to have arisen by chance and thus must be the product of an intelligent (and presumably divine) designer. Having seen young women in Africa who pushed for long pain-filled hours to deliver an infant, only to have that child die within and still not be deliverable until it began to soften and decay, and having seen the obstetric fistulas such trauma can cause—tears between the vagina and the bladder that leave the woman leaking urine for the rest of her life—it is difficult indeed for me to believe that there is much intelligence in human design.

Traits that enhance survival and reproduction, whether physical or behavioral, are said to be "adaptive" in that they make an organism better suited to its circumstances and thus more likely to reproduce. But especially when it comes to the subjects of this book—sex, war, and the underlying connections between the two—it is important to point out that not all adaptive traits are beneficial in all circumstances. Lungs were certainly an important evolutionary innovation when fish first left the sea and started evolving into the four-legged animals more than 350 million years ago. And they were evidently no great disadvantage when the ancestors of whales returned to the ocean 50 million years later. But then circumstances changed, and commercial whaling all but eliminated the great whales precisely because they had to surface in order to breathe air—the same trait went from being highly adaptive, to neutral, to wildly maladaptive, solely because the great whales' environment changed.

We propose that, like the lungs of a whale, the human male predispositions to violence, team aggression, and war are, in essence, an evolutionary hangover. The impulses behind the behavior must once have given a reproductive advantage to those who expressed them—and in some cases may still. But culture and conditions have changed while the genes have not, and many of our Stone Age impulses are decidedly undesirable in an Information Age man. The simple fact is that biology has painted us into a behavioral corner, and it's up to us to find a way back out of it. That's where we have the advantage over whales—they can't stop breathing, but humans could choose to stop fighting wars.

Sexual competition, which pits males of a single species against each other in a battle for available females, is another product of evolution that even a modestly intelligent—to say nothing of benign—designer would have chosen to avoid. Sometimes, sexual competition is highly destructive, as when one bull elephant seal fights another and in the process squashes to death a pup he sired previously. And it seems inevitable that the male propensity to rape during wartime is also a product of sexual competition. Evolution is silent on the morality of whether men should kill each other, but it will much more readily reward the man who rapes rather than murders women. This does not mean that men have evolved to rape, but it does suggest that many if not all men have at least the biological potential to experience the aggression, loose emotional control, and dissociation from empathy that presumably underlie the act. We will argue that warfare, terrorism, and their attendant horrors are based on just this sort of inherited predisposition for team aggression which, whatever its origins, has become a horribly costly and counterproductive behavior in the modern world.

Behavioral Predispositions

Darwin knew that his theory of evolution by natural selection would be controversial in Victorian society, and he avoided the issue of human evolution in *On the Origin of Species*, noting only that "light will be thrown on the origin of man and his history." It wasn't until 1871, in *The Descent of Man and Selection in Relation to Sex*, that he finally laid out clearly how natural selection and sexual competition had shaped human behavior.

We recognize instincts in other animals, as when a cat or dog cares for its young, but the word instinct is rather too mechanical to capture the subtlety of human behavior. E. O. Wilson calls the behavioral frameworks we have inherited *predispositions*, a term we have already used. Sometimes, a single gene can control these tendencies or proclivities. It has been shown, for example, that the lack of a particular gene in a mouse (and possibly a person) produces a self-destructive fearlessness. Mice lacking the gene run about in the open and climb on prominent objects exposing themselves to cats and other predators instead of hiding. However, most behaviors, such as our desire for sex or our ability to hate our neighbors, depend on complex, shifting interactions between large sets of genes and are influenced by the environment around us.

Most people appreciate that genes can determine the size or shape of a body part, say, or contribute to a disease. It is more challenging for some to see that genes also shape behaviors, such as aggression. Yet even humans, in all our complexity, display a variety of inherited impulses and behaviors, all evolved toward the single goal of making sure the individual carrying them survives long enough to reproduce successfully. Many behavioral predispositions can be found across geographical regions and historical epochs. Others are so strong and universal that they persist without the usual cues—even blind people who have never seen a smile know how and when to do so. Ultimately, an unfolding pattern provided by an animal's DNA must specify a way to build an interlocking system of neurons influenced by their internal chemical environment and responsive to the world outside—as represented by both past and present experiences—so as to underpin our behavioral predispositions.

Violence, aggression, and the other warlike predispositions are apparently found in all human males, likely owing to the behaviors' origin in our pre-human ancestors. Individuals do vary significantly in their levels of expression, but there is no evidence that particular ethnic or racial groups have consistently higher or lower levels of innate aggression, for example. Cultural and ethnic groups certainly have shown warlike or peaceful tendencies over time, but one need only look to the peaceful modern Scandinavian descendants of the once universally feared Vikings to appreciate that the differences are largely due to culture and circumstance, not genes. It makes no sense to speak of inherently violent or aggressive peoples then—we are all inherently violent and aggressive, unfortunately. But so too are we all capable of a Viking-like turnaround.

Jane Goodall observed that male chimpanzees make mistakes and fall out of trees twice as often as female chimpanzees, and that all the falls over ten meters involved males.[28] Human accident statistics also show that young men are twice as likely as young women to break a leg or an arm. Obviously, culture plays an important role (a little girl who is taught to stay at home is unlikely to be run over in the road), but the chimpanzee data does suggest that the predisposition of men to take more risks than women probably also has a biological basis. At the same time, the interaction of genes and the environment is complex and ever changing. Brothers and sisters inherit their genes from the same parents, but birth order can affect behavior. The optimum strategy for first-born children is to uphold their own authority and use their greater strength to defend their position against younger siblings, but it is in the interest of later-born children to question the status quo and to devise innovative strategies to compete with older, stronger siblings. Darwin was a later-born child who devised a revolutionary idea.*[29] Evolution gives us behavioral frameworks to succeed in the struggle to survive, but how these predispositions are expressed can depend on the environment.

Evolutionary psychology, the study of inherited behavioral predispo-

---

* Frank Sulloway of the Massachusetts Institute of Technology analyzed the voting records of U.S. Supreme Court justices and found that first-born justices were twice as likely to render conservative opinions than later-born justices. Biological insights into human behavior, such as the impact of birth order, can be fascinating and seductive, but also difficult to validate. Sulloway's analysis of the effects of birth order has been challenged by other evolutionary psychologists. For what it's worth, the authors of this book are the last-born in their respective families.

sitions, can provide rational explanations for seemingly irrational behaviors, including killing our own species. Evolutionary psychology is predicated on two important principles. First, our inherited predispositions are largely unconscious, built-in shortcuts to survival. Many people, like apes and monkeys, have an innate fear of snakes. We do not need to work out that a snake may be poisonous and kill us; it is sufficient to be programmed to be afraid of long slithering things and get away from them as quickly as possible.*[30] Second, as already pointed out, evolution is a slow process and our genes are a record of past success; the behaviors we inherit are mainly those that adapted us to a world long past. For millions of years our human, and before them, hominid ancestors lived in small bands of a few hundred persons wherein the women contributed most of the calories by gathering edible plants and the men provided much of the protein through hunting. Most of our behavioral predispositions were evolved to adapt us to this type of life and not to our very different, contemporary world of computers, cars, and concrete.

Gombe Stream National Park

Perhaps it is precisely because we have always been warriors that scientists, philosophers, and statesmen have rarely asked why it is that we kill our own species while so few other animals do so. And perhaps it is not surprising that the key evolutionary insight into same-species killing did not come from a knowledgeable anthropologist, a great commander, or an erudite sociologist—and certainly not from a man. It came instead from a young female secretary with no formal academic training.** In 1960, Jane Goodall and her mother Vanne set up a tent beside a tiny creek called Gombe Stream on the edge of Lake Tanganyika. Her inspiration, and Dian Fossey's to study gorillas[31] in Rwanda (just north of Jane's work in Tanzania), and Birute Galdikas's to study orangutans halfway across the world in Borneo, was the distinguished Kenyan anthropologist and paleontologist Louis Leakey.

---

* Further evidence of our residual "hardwired" behavior can be seen when our fear of surprise causes us to break out in goose bumps. When a chimpanzee is frightened it uses tiny muscles in each hair follicle to make its fur erect in order to look bigger. Human beings retain the same response—we just happen to have lost the fur that made it effective.

** Goodall did eventually complete a Ph.D., and indeed, we (Malcolm) had the same mentor at Cambridge—Robert Hinde, an authority on monkey behavior and a pioneer in the study of the causes of war.

Credit: Cyril Ruoso/JH Editorial/Minden Pictures

Like humans, female chimpanzees are usually caring mothers. They first mate at about ten years old, pregnancy lasts eight months, and they breastfeed for two-and-half to five years. For the first few years, mother and child are rarely more than a few meters apart. Male chimpanzees are indifferent fathers, but if a child dies its mother will grieve. This baby bonobo chimpanzee in the Democratic Republic of Congo is just a few hours old.

Initially Jane Goodall and Dian Fossey saw the apes they assiduously followed as gentle giants of the forest. However, in January, 1974, Hilali Matama, a Tanzanian whom Goodall had trained, provided the first report of an attack by a group of chimpanzees from one troop on an individual of another troop. After witnessing these attacks herself, Jane Goodall wrote, "I could never have imagined when I first knew the chimps, the series of brutal attacks made by the males of one social group against the individuals of a smaller neighboring community: attacks that led to the death of the victims, male and female alike."[32] In a horrible way, these killing episodes make apes even more like human beings than we originally suspected. In the next chapter we will explore this same-species killing in much more detail.

Even a brief acquaintance with Gombe has taught Martha and me about the tenacity and patience of those dedicated to studying animal behavior. Once when we stayed in Goodall's house in the park the only other person there was a young Polish primatologist. We became quite worried when it got pitch dark and she had not returned from the field. But she had been following one chimp until it made a nest just before sunset and then plodded back several miles along dark slippery paths. She ate a cold meal, bathed in the lake, slept briefly, and got up early enough to return to the nest site before dawn. Studying primate behavior is hard work, but the results have been invaluable. Jane Goodall and those who followed her have changed the way we look at the world and our place within it.

The Role of Hormones

So how does evolution affect behavior? One important way is through the action of hormones—molecules which our cells use to communicate with each other. Some hormones are proteins, and as with all proteins they are encoded in our genes and produced by our cells in response to developmental programs and environmental circumstances. But unlike other proteins, which provide structural support to skin and connective tissue, or facilitate the myriad chemical reactions that our metabolism is built upon, hormones regulate our moods and our actions. And even those hormones that are not proteins are made by proteins, and thus are also subject to evolution.

Among the behaviors we share with chimps is the tendency for males to be more violent and aggressive than females. I don't consider myself a particularly violent person, but the symptoms of maleness need not be as obvious as bar fights or commando raids. I shout obscenities at the computer when it does something I don't expect, while Martha would never do anything so pointless. And she has to calm me down every time I go through airport security because the forced passivity and implied submission of the screening process tweaks my competitive impulses in a way she never experiences. Some of the difference may be a matter of personality, but it is also driven by circulating testosterone, which is twenty times higher in adult men than in women.[33] At puberty, rising testosterone levels in boys give the young man his greater muscle strength, deep voice, beard, and sex drive. Testosterone also provides much of the male drive for status and aggression, and accounts for our poor impulse control. As with most things in biology, hormonal differences between the sexes are more often a matter of degree rather than absolute differences. Women too produce testosterone, and it is probably the hormone that drives female libido. In women the "male" sex hormone is made in the adrenal glands as well as the ovaries but levels are low and relatively static. In men testosterone output from the testes fluctuates according to both age and immediate circumstances.

Even relatively low levels of testosterone have a "permissive" effect on aggression, but there is also a dose response, such that men with above-average testosterone levels tend to be verbally and physically aggressive.[34, 35] Testosterone levels are lower in married men than in bachelors (perhaps because high testosterone levels are inherently risky, and the endocrine system can afford to relax levels a little in the less-competitive context of a long-term relationship), and predictably, unmarried men aged twenty to twenty-four are three times as likely to murder someone as married men in the same age group. Some studies find high levels of testosterone in men convicted of violent offenses and domestic violence.[36] Athletes who use testosterone and other steroids illicitly to enhance their performance and build body mass also boost their susceptibility to "roid rage"—an enhanced tendency to be short-tempered and to get into fights.

The effects of testosterone can be subtle, but they are pervasive. Levels rise in male soccer players prior to a competitive match and remain high during the game. They fall in the losers but remain elevated in the winners, sometimes for days. Remarkably, the same changes take place in the fans—physiological evidence of just how strongly we can identify with our side, even in a proxy war. And that "us versus them" dynamic is powerful. When a team of domino players was studied on a Caribbean island it was found that testosterone levels were higher when they competed against a team from the next village than from their own village. Changes in testosterone occur even in male chess players and, interestingly, if a competitive game looks as if it will be won easily, then testosterone levels do not rise as much as when the competition is challenging or prolonged.[37] Oliver Schultheiss and Kenneth Campbell measured the desire of Harvard University undergraduate men to have an impact on other people by asking them to write a story about a set of pictures illustrating either manipulating another or offering unsolicited help. Those whose stories demonstrated the highest desire to gain power by manipulating others had the highest testosterone levels. It seems that one role of testosterone is to reinforce assertive behavior.[38] In rhesus monkeys testosterone levels track social status and are low in those animals that lose battles for dominance.[39] On average, testosterone levels are higher in trial lawyers than religious ministers.[40]

Other hormones and enzymes also influence behavior. Adrenaline levels rise in both sexes in stressful situations.[41] Serotonin is another chemical messenger which among other things influences an animal's social status: Low-rank vervet monkeys have low levels of serotonin, but if given drugs that increase serotonin levels, their behavior changes in ways that help them climb the social ladder.[42] Conversely, changes in the diet can reduce serotonin levels in the brain, with significant results. In one study, the addition of micronutrients to the diet of young men in a British jail reduced aggression toward guards and other prisoners by a measurable 35 percent.[43] The genetic predispositions evolution has given us are constantly manipulated by the hormones our internal glands produce. They are also enhanced, even reversed, and nearly always strongly influenced by the external environment.

The Environment's Influence

Perhaps the most readily understood example of interaction between genes and the environment is furnished by the acquisition of language. All human languages have certain common rules of grammar and syntax, but they vary from each other in ways that are often baffling and frustrating to the adult brain. Children, however, are like sponges, able to absorb and acquire the structure and subtleties of whatever language it is they hear around them, whether English or Japanese or any other. It is genes that grant children this remarkable ability and their environment which shapes it into a particular set of words and rules that we recognize as a language.* Children do this very early in life and without any schooling because the human brain has evolved a predisposition to handle languages in a consistent way. The exact way the genetic code specifies particular neuronal pathways in order to analyze words and structure them so they can be used to transmit meaningful information remains a baffling mystery. Yet, profound ignorance of the exact details of a biological mechanism is not unusual and need not halt our broader understanding of evolutionary processes. As we

---

* If people speaking several different languages come together in one place they soon develop a "pidgin" language, mixing and melding words from each separate tongue. When the next generation comes along it is the children who add a grammar and turn a pidgin into a genuine Creole language. During the 1970s and 1980s a group of deaf children in Nicaragua developed their own set of hand signals, which after a single generation had developed complex grammar, in effect becoming a "Creole."

saw earlier, Darwin had only a shadowy idea of the rules of genetic inheritance, and no knowledge whatsoever of the molecular structure of DNA, yet he was able to deduce the grand overall structure of the evolutionary process.

Even in ants, whose behavior is easy to misinterpret as a set of mechanical behavioral responses, the environment and genes interact in subtle and complex ways. For example, one species of desert ant is extremely aggressive toward members of its own species when foraging near its nest, but one hundred meters from its home becomes much more tolerant.[44] The relationship between genes and the environment in human beings is well illustrated by the story of an enzyme called monoamine oxidase. Monoamine oxidase influences the rate at which chemical transmitters are removed from the brain once they have performed their function. Like other enzymes, which are all proteins, the level of monoamine oxidase is determined by our genes. Low levels are associated with above-average aggression, and higher levels appear to have a protective effect against environmental influences that would normally result in aggressive impulses. Both animal and human studies also show that maltreatment during childhood is associated with increased levels of violence in adults. Researchers from the U.K., the U.S., and New Zealand tracked levels of violence in 537 adult men.[45] A remarkable eight out of ten of those men were unlucky enough to have both a low level of enzyme *and* to have been maltreated as kids. As one researcher put it, "They're doing four times their share of rape, robbery, and assault," thanks to the interaction of genes and environment.

Childhood abuse can bring about life-long changes in the brain; for example, a region called the hippocampus is smaller in adults who were physically or sexually abused as children. After studying the long-term impact of child abuse, Martin Teicher and colleagues at Harvard came to understand these changes not as damage but as an adaptation: "exposure to early stress generates molecular and neurological effects that alter neural development in an adaptive way that prepares the adult to survive and reproduce in a dangerous world." However, a predisposition to react aggressively to any challenge is costly and individuals abused as children have more type II diabetes, a higher risk of suicide, and a plethora of psychiatric problems.[46]

The drive for status seems to be a virtually universal behavioral predisposition among primates, including human beings.* High social rank in males is associated with more mating opportunities and high rank among females provides access to more secure food supplies. However, the predisposition to strive for social status is modified by the social and physical environment in which the individual lives. Animals kept in a zoo, like men in prisons, cannot escape from one another. Throughout the late 1970s the Dutch primatologist Frans de Waal studied the large colony of chimpanzees at Arnhem Zoo in the Netherlands. One chimpanzee de Waal called Yeroen was the alpha male for three years, but then his rule was challenged by a coalition of two lower ranking males called Luit and Nikkie. In the wild, Yeroen might have fled to another part of the forest, but one day in Arnhem zoo, when the keepers opened the cage they found that

> Luit had many deep gashes on his head, flanks, back, around the anus, and in the scrotum. His feet in particular were badly injured [and he] had also sustained bites on this hands. The most gruesome discovery was that he had lost both testicles.[47]

Forty-five years earlier in the London Zoo, Solly Zuckerman observed competition among one hundred hamadryas baboons placed in an enclosure less than thirty meters square.**[48] In the wild such conflict is almost always non-lethal, but in the zoo two-thirds of the animals died as a result of stress or injury as males fought over access to females.[49] The reverse happened however, when de Waal deliberately placed juvenile stumptail monkeys, which are large but socially rather laid back, with smaller, but more socially aggressive juvenile rhesus monkeys.[50] When the rhesus monkeys threatened the stumptails, instead of the expected return of hostility the stumptails simply ignored the aggressive posturing of the rhesus. After five months in the same cage the juveniles of both species slept together happily and the rhesus were so much less aggressive that de Waal dubbed them "new and improved" rhesus monkeys. The point again is

---

* Notably, all primate societies allow for some degree of social mobility, while most human cultures have developed restrictions based on lineage, gender, and status at birth. Modern democracies, in which individuals are at least in theory judged on their own abilities and achievements rather than on their parentage, say, could almost be seen as a return to a more ape-like state of affairs—though usually without the more violent and explicitly sexual nuances of chimpanzee competition.

** Not surprisingly, the study of other primates seems to have prepared a number of individuals to assume posts with considerable authority over people. Zuckerman became Chief Scientific Advisor to the U.K. Ministry of Defence in the 1980s and was elevated to the peerage. Jane Goodall's and my mentor, Robert Hinde, became master of St. John's College, Cambridge.

that even deeply embedded "genetic" behaviors can be influenced, and even reversed, by environmental conditions.

Some of our DNA is a template for making proteins but much is devoted to promoting or retarding the action of individual genes. The environment can modify gene expression, and differences in genetic information can modify a person's response to the environment: It is a two-way set of interactions. The way in which gene expression is altered by the environment has been beautifully illustrated by the work of Darlene Francis, a colleague at University California, Berkeley. In rats, some mothers lick and groom their offspring more than others. The pups that receive most grooming react more calmly to stress than those with less nurturing mothers, and they even have anatomical differences in their brains. This behavior is transmitted to their offspring, who transmit it to a third generation. Moreover, the effects are discernable in both the biological offspring of a "good" mother as well as pups born to another mouse but fostered by a "good" mother.[51]

Consilience

It's worth taking another moment to address Einstein's concern that the theory we have in mind when we investigate the natural world influences the questions we ask and the conclusions we reach. When it comes to warfare and terrorism, our theory—based on decades of work by hundreds of researchers in numerous disciplines—is that both nature and nurture play important, interlinked, and ever-changing roles. There is a temptation in science and in human affairs to dichotomize issues: Things are right or wrong and black or white. But in fact most social processes are complex and almost invariably involve the interplay of many factors, as the observations on monoamine oxidase and child abuse show so convincingly. This is a part of the consilience, or "jumping together of ideas," that E. O. Wilson has advocated—the idea that all the fields of human inquiry can inform each other, leading to a new and more complete understanding of ourselves and our world.

Unfortunately, the idea that our environment is the sole determinant of violence remains widespread in academic circles. In 1986, twenty international scholars drafted the *Seville Statement on Violence* at a UNESCO meeting asserting, "It is scientifically incorrect to say that war or any other violent behavior is genetically programmed into our human nature." Several scientific associations, including the American Associations of Psychology, of Anthropology, and of Sociology voted to endorse the Seville Statement. But true science proceeds by observation, experiment, and debate and not by endorsing written statements.[52] The observation that male chimpanzees—like human beings—band together to deliberately and systematically kill other members of their own species most strongly suggests that some violent behaviors are indeed inherited.

The billions of years of competition between rival DNA sequences, which has produced the genes coding our anatomy and the framework for many behaviors, is the result of processes that are as impersonal as a detergent removing grease in a dishwasher. We perceive such behaviors to be moral or immoral, and perhaps this is just a way our brains, as a key part of our genes' survival machine, reinforce certain important behaviors. As E. O. Wilson has written most eloquently,

> *Homo sapiens*, like the rest of life, was self-assembled. So here we are, no one having guided us to this condition, and no one looking over our shoulder, our future entirely up to us. Human autonomy having thus been recognized, we should feel more disposed to reflect on where we wish to go. In any such endeavor it is not enough to say that history unfolds by processes too complex for reductionist analysis. This is the white flag of the secular intellectual, the lazy modernist equivalent to The Will of God....It is enough to get *Homo sapiens* settled down and happy before we wreck the planet. A great deal of serious thinking is needed to navigate the decade immediately ahead. We are gaining in our ability to identify options in the political economy most likely to be ruinous. We have begun to probe the foundations of human nature, learning what people intrinsically most need and why.[53]

Genes can provide essential, quick-response frameworks for having an appropriate reaction to certain key situations but they do not write immutable rules. The Seville Statement also concludes "that biology does not condemn humanity to war"—and that is something with which we totally agree. It may be true in some senses that we are survival machines for our genes but we are not puppets whose strings are inevitably pulled by our DNA.

The mere fact that a set of impulses or behaviors has evolutionary roots neither implies that those impulses and behaviors are unavoidable nor that they are something we "should" or were meant to do. It simply and only means that those impulses gave our ancestors a reproductive boost over their competitors at some point in history. Neither we nor our progeny, however, have any guarantee of survival and subsequent repro-duction as a result. It is very much the rule in biology that times and en-vironments change—and those species that do not change with them are destined to extinction. Simply put, the successful survival equation our genes hit upon yesterday could bring us doom tomorrow.

The deeper fact is that as humans we have many competing impuls-es—for good and evil, aggression and compromise, anger and forgiveness, selfishness and altruism—and all of them are to one extent or another the legacy of evolution. Perhaps uniquely among the animals, we also have the faculties to decide which of them we will act upon under given cir-cumstances and to shape our world to encourage others to do the same. We have culture as well as biology and we have knowledge, science, and the ability to reflect on and moderate our own behavior. In this era of global poverty, environmental destruction, weapons of mass destruction, and fundamentalist terror, we had better take Rose Sayer's plea in the *African Queen* to rise above our nature very seriously indeed.

3

THE MISSING LINK

The idea of genes for behavior is no more strange than the idea of genes for development. Both are mind-boggling, but nature has never found human incomprehension a reason for changing her methods.

—MATT RIDLEY, *Genome*, 1999

O NE OF THE MANY PHYSIOLOGICAL SIMILARITIES between people and chimpanzees lies in the hormone changes associated with the sexual cycle. Menstruation in the chimpanzee averages every thirty-six rather than twenty-eight days, but the hormone changes are identical.* When I practiced medicine, I never took blood samples frequently enough to get good at it, and I certainly wouldn't like to try taking blood from a chimpanzee. But animal researchers including Anne Pusey, a primatologist studying the chimpanzee colony at the San Diego Zoo in California, are much less squeamish. Instead of taking blood samples to study the menstrual cycle in females, she collects chimp feces, stores them in plastic bags, and takes them back to a laboratory to measure the hormone levels (the hormones that circulate in the blood also pass into the gut).

---

* Chimpanzees are so like us that at least one female primatologist who worked closely with them in zoos noticed that her menstrual cycle became synchronized with the animals she studied, just as women who live in a dormitory may come to menstruate in unison.

Chimpanzee society is built around males, and as females become sexually mature they leave the troop of their birth and migrate to another group. Zoos also transfer animals, to simulate the natural dynamics, and one female Anne Pusey had known as a juvenile was flown to Frankfurt Zoo when she approached puberty. Some years later, Pusey visited Frankfurt. The now mature animal from San Diego recognized the researcher, bent down, picked up a handful of feces, and presented it to her. The female chimp had traveled halfway across the world, she saw hundreds and sometimes thousands of different people every day, and yet she recognized Anne immediately and wanted to express her solidarity by presenting an appropriate gift.

The more we discover about chimpanzees, the more compelling the similarities and the more intriguing the differences become. Because we shared a common ancestor five to seven million years ago, our similarities generally represent characteristics both species inherited long ago, while our differences reflect the different paths evolution has taken us down since. The human brain is twice as large as a chimp's, but under a microscope there is nothing particularly dramatic distinguishing it from that of an ape. The argument that will be developed in the rest of this book is that just as we have the same number and pattern of teeth as chimpanzees, just as we have the same bones in our hands and feet as chimpanzees and the same networks of neurons in our brains as chimpanzees, so we also share with chimpanzees a predisposition for adult males to team up, attack, and kill other groups of our own species. Understanding this predisposition is the missing link to understanding human war and terrorism. In order to do that, we'll need to get to know our biological cousins a little better.

The Troop

Chimpanzees are intensely social, highly intelligent animals. In the wild, no male has ever been seen to leave the community into which he was born and join another troop. As a result, the troop is built around a group of brothers, cousins, uncles, nephews, and other related males. Troops consist of thirty to sixty animals of both sexes and all ages. The size of the territory they control determines the amount of food available.

There are two species of chimpanzees—though Jared Diamond among other authors has suggested that we humans are so similar that we might be considered a third. The scientific name for the chimps we commonly see in zoos, and which Jane Goodall has studied in Gombe for almost half a century, is *Pan troglodytes,* [54] which for clarity we'll call simply "chimpanzees" or "chimps." Together with their smaller cousins the bonobo chimpanzees—*Pan paniscus* or "bonobos," again for clarity—the two species are almost like a personification of the Rousseau-Hobbes split on human nature that we saw in chapter 2. We'll visit the bonobos later, but for now, let's take a closer look at the decidedly Hobbesian *Pan troglodytes.*

Chimpanzees live on fruit, seeds, and plant pith, supplemented by insect grubs and occasionally meat from other animals they hunt and kill. They engage in what is called "scramble feeding," often with a good deal of noisy competition. The males commonly move about their territory in foraging parties of four or five, looking for fruiting trees, and groups break up and rejoin one another depending on the density of food sources. Female chimpanzees breastfeed their infants for thirty to fifty months—a long time, but not unusual for great apes. (Our hunter-gatherer, or preliterate, ancestors probably breastfed for at least fifteen months or longer.) Initially, chimpanzee mothers carry their babies everywhere, and later the maturing infant may ride on its mother's back. Burdened by pregnancy or childcare, females are more likely than males to forage alone. They are also more likely to confine their search for food to the core area of forest occupied by the troop. The males tolerate the young, but do not take part in feeding or nurturing them. When they can't see one another, chimpanzees hoot and call, sometimes over long distances. Every night each adult makes a fresh nest of bent branches in which to sleep, high in a tree. The young stay in their mother's nest, and family bonds can last a lifetime. Chimpanzees, like people, have local cultures. For example, troops living in the Tai Forest in the Ivory Coast in West Africa use stone hammers and anvils to crack nuts, while East African chimps have not yet discovered and passed on this skill.[55, 56]

When Jane Goodall began her study of chimpanzees in Gombe Stream National Park, four distinct troops occupied an area of thirty-two square kilometers (twelve square miles). Ultimately, the reproductive potential of a chimpanzee troop, like that of any other animal, is limited by its food supply. Jeanne Altman, studying baboons in Ambuseli National Park, Kenya, in the shadow of Mount Kilimanjaro, found that females spent fully 70 percent of their waking hours just searching for food to stay alive. Chimpanzees also spend more than half their time finding sustenance—

one very rarely sees an overweight ape in the wild. When the population gets too large for the size of its territory, or if a harsh season of shortage goes on for too long, some animals will be pushed into starvation. Like us, chimps can be fastidious—wiping blood from wounds or menstruation with wads of leaves, for example. But staying alive and passing on your genes is a tough business, and evolution does not permit behaviors that leave potential food sources untapped. Once, when Martha and I visited Gombe in the dry season, the chimps were so hungry that they were turning over their own excreta, looking for undigested fragments of food.

Chimpanzee society, like human society, is hierarchical. Both sexes remember perceived wrongs and expect favors to be returned. Males compete with one another continually, seeking the top, or alpha, position. They bluster, bully, tear off branches, hurl rocks, thump the drum-like buttress roots of the forest trees, and generally make a nuisance of themselves. Females also have a social hierarchy, which ensures that the most senior has access to the most reliable food sources, but unlike the male bluster, the female hierarchy depends on low grunts and subtle body language. Relationships between females are also longer lasting and less volatile than those between males. Male chimpanzees mate promiscuously, and often quite roughly, with any female who is ovulating. The alpha male gets plenty of sexual opportunities, but the other males also enjoy access to fertile females. For the most part, copulation is a public activity, but sometimes an ovulating female and a male leave the troop and wander as a pair, alone in the forest. The female may go willingly, or she may be coerced by threats and blows. Most of the time however, females exercise some choice in deciding with whom they will mate. Indeed, DNA studies show that females sometimes even slip away from their own troop and mate with males from another troop. In one paternity study, it was found that fully half the babies born in a chimpanzee troop in the Ivory Coast were sired by males from outside the troop.[57]

Chimpanzee social life is remarkably complex. Males in particular have a tendency to establish coalitions, and two adults can usually outmaneuver a single individual, however strong the latter. In all other primate societies, power follows the social hierarchy. But among male chimpanzees some lower-rank individuals may exercise considerable influence, increasing their access to food and sex. In rhesus monkeys or baboons, when a fight breaks out other individuals will try to join in on the winning side. In a chimpanzee troop, as often happens in human society as well, about half the time spectators side with the underdog—in fact we use the term "underdog" for the very reason that in a canine pack nobody ever helps the lowest-ranked animal. Chimpanzees have long memories and reward loyalty. When one male helps another achieve the alpha position, this "vice-president" may also come to exercise considerable power. One of the alpha male's roles is to break up fights, and a successful top-rank animal often intervenes to help the weaker individual in a scrap. Chimpanzees of both sexes also engage in a great deal of grooming and other rituals that help establish peace and cancel out the negative effects of competition.[58] It is common for the loser in a dispute, which may have included vigorous screaming and hitting, to reconcile by grooming the victor soon after the fight.

Team Aggression

It is in the chimpanzee equivalent of international relations, however, that we start to see the real seeds of human war and terrorism. Each chimpanzee troop recognizes its own territory, and the adult males become more alert and jumpy as they near its boundaries. Sometimes a stream or gully will represent a border, and then a party of males may advance confidently to their side of the divide but proceed with noticeable caution after crossing over.[59] Males spend from 10 to over 20 percent of their time "patrolling" the boundaries of their territory, based on direct observations. They feed on only about one third of these patrols, and forgoing valuable opportunities to eat implies that this behavior must be very important indeed.

There are two kinds of chimpanzee border patrol. The first is regular guard duty, in which the chimps make the rounds to ensure that no invaders are entering their territory. The second type is focused on a specific area, and seems designed to harass a neighboring troop. If a rich food source is near the edge of another troop's territory, then a feeding group usually pauses at some high point and scans the horizon. They may display by rushing about, swinging from branches, and uttering a series of screams called a "pant-hoot." They then seem to listen, presumably to discover if any chimps in the next territory respond. If all is quiet they go forward in the belief it is safe. If the invading team finds a nest made by a chimp in another territory, they may climb up and inspect it. [60]

When parties of foraging chimps from different troops do happen to meet, three things can happen. If the groups are of unequal size, the smaller group will usually withdraw quietly. If they are similar in size, then the males of both groups will engage in noisy displays, drumming the ground, dragging branches, hooting, and throwing rocks and sticks, but direct physical attacks are rare. After a while the frenzy calms down and each group watches the other carefully. As the competitors withdraw, each group makes a great deal of noise as if to say, "See—you didn't really frighten us!" But occasionally, when an invading force of half a dozen or more males meets a smaller group, it may attack for real.[61]

There is always a danger in ethology—the study of animal behavior—of anthropomorphizing the animals we are seeking to understand, which is to say, we have a tendency to see human-like motivations where there are none. But when the species share a direct common ancestor, and the behavior is so immediately recognizable—at least to anyone who has ever been a young boy playing in a neighborhood group, or indeed a patrolling soldier—it would be a mistake to disregard the obvious connection. Chimpanzee border patrols and raids follow the same template as human raids and battles, and both are built upon the key behavior that links chimp attacks to human war and terrorism. Wrangham and other behavioral scientists call these lethal chimpanzee raids "coalitional aggression," but that term strikes us as overly remote. When a small raiding party of male chimpanzees invades the territory of another troop, finds an isolated individual of that "foreign" clan, and sets about killing, it is showing us a version of our own most aggressive behaviors. These attacks are intentional, coordinated, lethal, and most of all, based on teamwork. In honor of this last, crucial element, we'll call these lethal raids "team aggression."

Chimpanzee team aggression involves extraordinarily vicious behavior in the attacking males. Their human observers may have spent hundreds of hours watching the individuals in more peaceful situations, and so witnessing such raids can be viscerally disturbing—perhaps a little like unexpectedly catching a favorite nephew in the act of torturing a stray cat. In the raiding situation, two or three males may hold down their adversary while the other team members use their limbs and teeth to attack. The males pound their victims and attempt to tear off strips of flesh. They may rip the testicles off of a male, or bite digits off his hands and feet. These killings take the form of brutal, drawn-out torture and the victim usually does not die immediately, but suffers loss of blood and infected wounds that lead to death only after several days. Notably, lethal attacks are most common in the dry season, when resources are most limited and potential victims are sometimes forced to forage alone in order to find sufficient food.

The precise logic of chimpanzee team aggression isn't fully understood, but the attacks do not appear to be spontaneous. Sometimes a raid begins with the alpha male displaying inside his own troop's territory and then leading a party of males on a raid into another territory.* According to Goodall, "To some extent all adult males show a keenness to participate in these exciting events [raids], although old males take part less often as it becomes harder for them to travel long distances. Even so among prime males, however, there are distinct individual differences." Some males are born warriors, apparently, always leading the groups in displays. Others may still be eager to join a patrolling party, but tend to travel toward the rear of the group and are often the first to turn back.

Lethal raids are generally brief events, and unanticipated by human observers. In an aggregate of 170 years of chimpanzee study by human beings across the whole of Africa, only seventeen episodes of team aggression have been observed directly. Scientists can't know where all members of a dispersed troop are at all times, but even including raids that are inferred to have taken place, team aggression seems to be a rare behavior. When the first observations of inter-troop violence occurred in Gombe Park, some critics argued that it was caused by the researchers themselves, who had been providing bananas to attract the chimpanzees to a central point for observation. However, the consensus among primatologists is that team aggression by male chimpanzees is a genuine and consistent behavior, and it has now been observed at five different sites in Africa, many hundreds of miles apart and most without artificial feeding.

One of the first observations of team aggression is also one of the most heart-rending. In 1974, Jane Goodall and her staff saw Madam Bee, an old female of the Kahama troop, attacked by a team of males from the rival Kasakela troop on at least three different occasions. (The assaults were part of a war-like series of raids observed at Gombe between 1974 and 1977.) With one arm paralyzed by polio, Madam Bee was unable to de-

---

* In order to study chimpanzees, observers follow the same animals patiently for months on end, until all the troop members become used to the human presence. It is not known whether having a primatologist in tow gives an attacking team of males an added advantage during team aggression.

Credit: Martin N. Muller

This male chimpanzee was brutally attacked and killed by a group of 10 males from a rival neighboring group in Kibale National Park in Uganda in 1998. Chimpanzee "wars" share striking similarities with human conflicts, including depopulated border zones between adjacent territories, attacks based on surprise and overwhelming force, and the fact that raids can recur intermittently for years on end. Perhaps most important, both human and chimpanzee conflicts are generally fought over territory and access to resources, and depend upon the ability to turn off all empathy for the enemy.

fend herself. During the first attack, she sustained a deep gash on one leg and her daughter, who had just become sexually mature, was led away by the assailants. The final fatal attack took place one year after the first and it is described in the observer's field notes, put together by Goodall:*

Jomeo…then turned and stamped on [Madam Bee] and slapped her. The [accompanying] females rushed out of the way into the trees…Madam Bee was unable to stand; she was shivering all over. At once Satan displayed up, threw her to the ground, stamped on her, and dragged her a few meters. Figan then attacked her with ferocity, hitting, stamping on her again and again. She was too much hurt (perhaps winded) to scream. When she

---

* "Display" in the following description is a term primatologists use to describe a behavior in males, and occasionally females, in which an individual erects its hair and runs forward (sometimes upright, perhaps dragging a broken branch) in an aggressive show of strength. Such displays may or may not be accompanied by hoots and screams.

stopped moving, he displayed away. Jomeo returned, pulled her inert body toward him, half picked her up and slammed her down, stamped on her, rolled her over and over along the slope. Finally he stopped and sat a few meters away. Madam Bee tried to stand, but fell and lay still. Again she tried to move; this time she started back up the slope, screaming and heading for a thicket. Satan displayed up, smashed her to the ground, pulled her towards him, pushed her away, he pounded her with hands and feet continuously for two minutes. The observers thought she was dead. Satan, his hair fully erect, stood by and swayed branches at her. Goblin also sat close, watching intently. Satan waved branches until she finally moved again. Goblin left at this point, joining the three other males who were displaying nearby. Satan remained for another minute, following and watching as she slowly moved into the dense undergrowth. She disappeared 15 minutes after the start of the assault, and Satan left. [It took the human observers three days to find Madam Bee. She had] deep wounds in the following places; left ankle, right knee, right wrist, right hand, back (several wounds), and big toe of left foot (hanging by a strip of skin). She died on the fifth day after the attack.*[62]

How to interpret such a brutal series of attacks? The original raid, resulting in the capture of a fertile female, seems logical enough in evolutionary terms—one more potential mate means more opportunities for the attacking males to reproduce. But why would the raiders return time and again, only to eliminate a post-reproductive female who could not have represented much of a threat to the attacking team? We may never know for certain, but it is worth keeping in mind that evolution is focused on overall outcomes, rather than the logic of individual events. In the sad case of Madam Bee, her death may not have provided any particular benefit to her assailants. But their unconscious evolutionary predispositions, to band together into a team, to recognize a vulnerable outsider with hostility, and to attack ferociously and without mercy, amount to an impulse to kill without logic. All it takes for evolution to favor such an impulse is for the attacking chimps to gain some advantage—such as captured females or access to new territory—frequently enough that their reproduction increases as a result.

Team aggression may be unconscious, but it still demands significant

---

* Goodall pioneered naming wild chimpanzees, instead of simply assigning numbers. She often chose names that she thought fit the animal's character, such as Goliath or Satan, and used names beginning with the same letter of the alphabet to link a family, such as Fagin, Fifi, Flame and Flint, or in the case of Madam Bee's daughters, Little Bee and Honey Bee.

brainpower and sophistication. Lethal raids require a long memory and relatively complex planning. Raiders must learn to move as a group, to advance stealthily, and to observe their victims prior to attack and make judgments about what other animals might be near and ready to help a potential victim if attacked. When male chimpanzees raid a neighboring troop, each will benefit if all act as a coordinated group. Four or five males in their physical prime, provided they can identify and isolate a single member of a rival troop, can destroy that individual with relatively little risk to themselves. It is notable that in none of the raids followed by human observers were any of the attacking males seriously injured.

Still, team aggression is no trivial undertaking, and the individuals in an attacking group are all in danger as the victim fights back. By acting as a team, however, they share the risk, turning a potentially terminal strategy into a relatively low-cost way to slowly but steadily wear down and eventually eliminate a rival troop. The shared risk also results in shared benefit. Anne Pusey observed that as the Kasakela troop expanded its territory it had more access to fruiting trees and the males grew stronger and heavier. The females also enlarged their foraging areas at the core of the group's territory and their babies were born closer together. Juvenile females from other troops seemed more likely to join the Kasakela troop when the males had been successful as warriors and enlarged their territory.[63] In short, controlling more resources means you are likely to have more progeny surviving to future generations: Team aggression is one way that both chimpanzees and humans have hit upon to reap that evolutionary reward.

Intriguingly, intergroup competition and team aggression are built upon a solid foundation of altruism—toward one's own team. Recognizing the difference between "us" and "them" (sociologists call these "ingroup" and "outgroup") is a key survival skill for social animals, as well as one of the primary drivers of team aggression, war and terrorism. Chimpanzees, like human beings, can be generous to their companions, as well as diabolically cruel to outsiders. Once, shortly before she was killed by the Kasakela males, Madam Bee was too tired to climb a fruiting tree which her daughters Little Bee and Honey Bee had found. As she lay on the ground, Little Bee climbed down from the tree holding one ripe fruit in her hand, and another by its stem in her mouth. Gently, she approached her mother and handed her a fruit. Mother and daughter lay side by side eating. This act of tenderness may be every bit the evolved impulse that team aggression is, but that does not make the one less fearsome or the other less sweet. As philosophers have observed of humanity, it is remarkable that such extremes can exist in a single creature.

Ultimately, team aggression relies upon altruism as much as it does upon ferocity, with each member of the attacking team risking its own safety for the benefit of all the others. In 1964, the English evolutionary biologist William Hamilton pointed out that the more closely related individuals are the more willing they are to take risks for each other, and the males in a chimpanzee troop are all blood relatives. This sort of altruism and reciprocity evolves because individuals that inherit the behaviors also benefit from them, of course.[64] The classic example of altruism and shared risk is the white tail on a rabbit or some species of deer. If one member of a group notices a predator, it signals its awareness of danger by displaying its white tail. By doing so, it increases its own risk of being seen and killed, but if all the animals of the species behave in the same way, then the benefits increase and the risks are shared. In chimpanzees and human beings, with their acute awareness of social interactions and long memories, much more complicated examples of altruism can arise.

Team aggression evolved because males who first manifested the behavior improved their chances of transmitting their genes to the next generation. Many social animals, from meerkats to rhesus monkeys, will work together to defend their territory against attacks by other members of the same species. But they lack the strategy of deliberately killing members of the opposing group by invading their territory and outnumbering them.[65] In fact, among the approximately 4,000 species of mammals, team aggression has been observed only among chimpanzees, wolves, spotted hyenas, lions, possibly the red colobus monkey—and most certainly in our own species. In Alaska one study found that 40 to 65 percent of adult deaths among wolves were the result of attacks by packs of animals near the border of their territory.[66] Social life among wolves, hyenas, and lions is built around groups of related females, however, so their version of team aggression lacks the sexual edge of chimpanzee raiding, which can be as much about securing new females as it is about capturing new territory. The harem social structure of gorillas means that pairs of males may fight, but genetically related adult males do not gather into groups, so team aggression has not evolved.[67] Whales and dolphins also have large brains and in some cases related males stay together as adults,[68] and some young male dolphins even join together to try to coerce females into copulating, but territory means a different thing in the three-dimensional vastness of the world's oceans and therefore, once again, team aggression has not evolved. But in chimpanzees, where forest living means defined territories and a troop means a clan of related males, the blind chance of evolution has set the stage for an especially successful form of team aggression against other members of the same species.

Outgroups

During one raid in the four-year-long chimp "war" in Gombe, a team of Kasekela males came across a female from a neighboring group and her infant. They had become separated from their companions, and were attacked viciously. The female adopted a sexual posture in front of one of her assailants, the same Satan who destroyed Madam Bee. During one attack the female actually reached out and touched him. "Satan actively rejected these contacts—and the second time he picked up a large handful of leaves and scrubbed his leg where her hand had rested," Goodall wrote. The lone female was attacked again and her baby fatally wounded.

We can judge what goes on in chimp brains only by watching them. But we have seen them wiping away bodily waste, or blood from wounds, with wads of leaves. Did Satan feel the strange female was in some way "dirty"? Certainly, he did not accept her sexual advances. Have chimpanzees evolved a behavior where they identify members of another troop as fundamentally different from themselves, and even defiled in some way? Do they "de-chimpanzeeize" other chimps as we so often "dehumanize" other human beings? The evolution of such an attitude would certainly make it easier to engage in planned attacks on members of your own species. Indeed, it might well be impossible for an intelligent, highly social animal to kill its own kind systematically *unless* it evolved some sort of neural machinery to de-identify those it is about to kill. The ability to dehumanize others may well be another necessary key to killing our own species.

Certainly, our tendency to form groups and be suspicious of outsiders has roots as deep as human history—if not much, much deeper than that. And it doesn't seem to matter how similar to ourselves these others may actually be. The terms "ethnocentrism," "ingroup," and "outgroup" were coined by the American sociologist William Sumner in 1906.[69] "Each

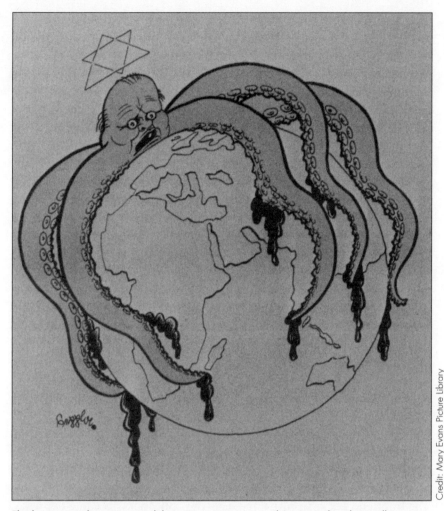

Credit: Mary Evans Picture Library

The human predisposition to dehumanize an enemy is deep-seated and virtually universal. In World War II, American propaganda portrayed the Japanese as apes while in Germany, the Nazis depicted Winston Churchill as a Jewish octopus with his tentacles around the globe. Similar efforts to make the enemy seem less than human continue to this day.

group," he wrote, "must regard every other as a possible enemy on account of the antagonism of interests, and so it views every other group with suspicion and distrust, although actual hostilities occur only on specific occasions." Sumner speculated that "the same conditions which made men warlike against outsiders made them yield to the control of chiefs, submit to discipline, obey laws, cultivate peace, and create institutions inside." Darwin made the same observation, writing that "the tribe including many members who from possessing a high degree of patriotism, fidelity, obedience, courage and sympathy were always ready to aid each other and to sacrifice themselves for the common good would be victorious over other tribes."[70] A strong loyalty to one's own group is another form of altruism, and it would have been a highly adaptive trait among our warrior hunter-gatherer ancestors; today, we call it patriotism.

As for hostility toward outgroups? That seems to be almost a corollary to ingroup loyalty, and virtually universal, at least among men.* Israelite and Philistine, Greek and Persian, Roman and Carthaginian, Crusader and Saracen, Aryan and Jew, Manchester United and Glasgow Rangers, Cowboy and Indian, Shiite and Sunni, Oxford and Cambridge, Berkeley and Stanford—the world falls into "them" and "us" always easily, and often lethally. As Rudyard Kipling wrote during World War I, "there are only two divisions in the world today—human beings and Germans."**[71]

Kipling's observation was no just-so story. Sociologists have conducted experiments to illuminate the predisposition to divide the world around us into ingroups and outgroups. It is a predisposition that manifests itself early in life. For example, groups of boys arrive at a holiday camp and become competitive merely because they arrive at different times.[72] In the 1970s, Philip Zimbardo at Stanford University advertised for student volunteers and on the toss of a coin they were assigned at random to the role of prisoner or prison guard.[73] A facsimile prison was laid out in a corridor of the psychology building. Those designated as prisoners were "arrested" at home, transferred to the "prison," had a chain affixed to one ankle, and were dressed in de-individualizing smocks with numbers on the front and back. In just three days the "prisoners" became depressed and ten out of twelve wanted to forfeit the money paid for joining the experiment in order to leave it. The "guards," however, all reported to work on time and some "went beyond their roles to engage in creative cruelty

---

* At an amusing but nevertheless revealing level, I find I identify with motorists and pedestrians according to the group I belong to at any one moment. When driving, I am surrounded by irresponsible jaywalkers who leap without warning into the road in front of my car; as soon as I lock my car door and proceed on foot, I am surrounded by sedans and SUVs traveling near to the speed of light and clearly intent on running me down.

** My own friendships with Germans, the universally loathed outgroup of my youth, are just one small example of a heartening trend—the expansion of the ingroup. This is most obvious in Europe, where more than a thousand years of often astoundingly brutal warfare have given way to economic cooperation and even a growing sense of shared European identity.

and harassment" of the "prisoners." The experiment had to be stopped prematurely because over the course of only a few days those playing the role of the prison guards became unacceptably arrogant and sadistic, while the prisoners identified themselves by their numbers and accepted their fate. Interestingly, it was Zimbardo's girlfriend who finally pushed him to abort the experiment. As Zimbardo and his colleagues wrote later, "healthy American college students...seemed to derive pleasure from insulting, threatening, humiliating and dehumanizing their peers."

The insights provided by the Stanford Prison Experiment have been confirmed in the real world many times. When the U.S. Air Force Academy tried to teach cadets how to survive enemy capture, the program rapidly degenerated into simulated rape of a female cadet by the male cadets, and poorly supervised guards at a New Jersey immigration center beat and ill-treated detainees. Exactly as happened with the Stanford students, "the midnight shift [was] particularly abusive."[74] The stress of war and insufficient training do not have to be invoked to explain the atrocities committed by American military prison guards at Abu Ghraib in Iraq—unfortunately, their deplorable behavior seems to be as much a default human behavior as is caring for our own children.

Interestingly, given the example of chimpanzee team aggression, being part of a crowd may also sharpen hostile behavior. We seem to have an inbuilt desire to go with the flow. In 1951, the pioneering social psychologist Solomon E. Asch set up a simple experiment in which volunteers had to match the length of lines drawn on paper. When a setting was contrived wherein actors choose the wrong line and were overheard by the experimental subject, about one third of the subjects went along with the false choice even though they "knew" they were making a mistake.[75]

In another famous experiment, in 1963, Stanley Milgram at Yale demonstrated the human predisposition to follow leaders.[76] He recruited individuals from different walks of life to take part in what they were told was a study of how people learn. The volunteers were given a set of switches labeled 15 to 450 volts and told they had to give a subject an electric shock every time he gave a wrong answer to a simple word-learning test. Unbeknownst to the volunteers, the machine was a dummy, and the "test subject" an actor who displayed every sign of increasing distress as the "shocks" supposedly grew stronger. The volunteers manifested internal conflict and argued that it was wrong to cause pain, but nearly all obeyed the experimenter's instructions to continue to turn up the voltage, even as they saw the victim writhing and screaming. Women were as obedient to authority as men, although they showed greater tension. In the 1970s, Albert Bandura linked the authority and "groupthink" effects with a similar fake shock experiment; he found that the decisions made by a group were more severe than those made by an individual.[77]

For over half a century, sociologists have conducted an ingenious and insightful array of such experiments. They have demonstrated time and again the human ability to dehumanize others, the power of authority, the importance of allies, and how shared responsibility helps us lose our individual identity, making for harsher, crueler actions. It seems possible to awaken these behavioral predispositions in almost anyone, although they do seem stronger in men. Such behaviors can be modulated, but we live under the shadow of an evil ability to dehumanize other members of our species. The standard social science paradigm does not explain why or how these behaviors arose, from an evolutionary perspective; however, it could not be more clear: These behaviors and impulses are exactly those needed to produce a form of team aggression every bit as effective as that of chimpanzees.

Hunting

Prior to Jane Goodall's patient and prolonged observations, chimpanzees were assumed to be exclusively vegetarian, subsisting primarily on fruits. The discovery that chimpanzees hunted and killed other animals predated the observation of team aggression, but at the time it was almost as astonishing. Hunting is also a male activity involving teamwork, foresight, and reciprocity, and it may have preceded the evolution of team aggression—biology has a way of building new structures and behaviors up out of existing parts. Regardless of which behavior came first, there is clearly a link. An experienced older male named Brutus in the Tai community in Ivory Coast was not only the best hunter in the group, he often led the younger males into episodes of team aggression.[78]

Goodall describes first witnessing hunting by chimpanzees:

One morning when Rodolf, Mr. McGregor, and Humphrey [three mature chimpanzees], and an adolescent male were sitting replete with bananas and the baboon troop was passing through the camp area. All at once Rodolf

got up and moved rapidly behind one of the buildings, followed by the other three. They all walked with the same silent, purposeful, almost stealthy
pace...I followed, but even so I was too late to observe the actual capture.
As I rounded the building I heard the sudden screaming of a baboon and a
few seconds afterward the roaring of male baboons and the screaming and
barking of chimpanzees. Running the last few yards, through some thick
bushes I glimpsed Rodolf standing upright as he swung the body of a juvenile baboon above him by one of its legs and slammed the head down
onto some rocks.[79]

Rodolf and the other chimps then ate the dead baboon. Wherever
chimpanzees have been studied, mature males have been seen to hunt,
kill, and eat colobus and other species of monkey along with bushbucks,
bushpigs, and baboons. Ovulating females often travel with males and
when they do they may point to prey and get excited when meat is shared,
though they rarely become part of the male team doing the actual hunting.* The successful hunter usually divides part of the kill with other
members of the troop. Chimpanzee hunts have been observed more than
one hundred times and the insights gained throw some light on the subtlety and complexity of cooperation between males bent on violence.[80] As
with other aspects of chimp life, there appear to be cultural differences
between various groups of chimpanzees, with the Gombe chimps in East
Africa focusing on snatching baby colobus monkeys from their mothers
while the Tai chimps in West Africa are adept at killing adults. Shared
evolution has provided each group with the impulses and abilities needed to become hunters, but individual experience has resulted in distinct
expressions—once again, nurture (culture, experience, the environment)
has a way of shaping the gifts of nature.

Chimpanzee hunts, like lethal raids, can be painful to watch. The
chimps may begin to eat the captured prey while it is still alive, perhaps
beginning by breaking open the skull to scoop out the brains or taking
mouthfuls of warm flesh from a twitching limb. All the chimpanzees in
the troop get excited once the prey is caught. Meat decays quickly in the
jungle, and the lead hunter can rarely eat the whole carcass at one sitting.
The hunting party has had to risk injury from falls and from prey fighting for their lives. Other members of the troop beg for a share of the meat

* With humans also, males are usually the hunters. Among 179 preliterate groups studied by anthropologists, men were the only hunters in 166; men and women hunted together in 13 and in
no case did women hunt alone.

Comparison of Hunting and Team Aggression Behavior in Chimpanzees	
Hunting	Team Aggression
Cooperation within a small group of related males	Cooperation within a small group of related males
Surprise potential prey	Surprise potential victim
Short duration (minutes not hours)	Short duration (minutes not hours)
Out-flank prey	Out-flank and outnumber victim
Risk of injury small as long as the group members support one another	Risk of injury small as long as the group members support one another
Immediate benefit: food	Long-term benefit: larger territory = food + more females for sex
Lack of empathy for prey	De-individualize and "de-ape" the victim
Females (if present) excited and males may share food with them	Females (if present) excited
Victim is other species; objective is food	Victim is same species; objective is causing severe injury and death

and the individual who makes the final kill often divides the corpse once he has eaten his fill. The males seem to be rewarded according to the effort and risk put into the chase.[81] Sharing meat with others of either sex may enhance a male's status, and sharing with females may lead to an exchange of sexual favors later.

Aggression Within the Group

Anger is an emotion, while aggression is a behavior. In social animals, there is an important difference between aggression outside the group, and aggression within. Ingroup aggression, while common, is rarely lethal, and there are almost always ways for the loser to cry uncle, or "tap out" like a defeated wrestler calling an end to the bout before serious injuries are sustained. Still, ingroup anger does lead to violence often enough. As with hunting and in parallel with most other mammals, men on the whole tend to be more violent than women. As pointed out in chapter 1, this makes evolutionary sense. If a mother dies, the lives of her offspring are also put in jeopardy.[82] For men, so long as they have reproduced early and often, it doesn't much matter from an evolutionary standpoint if they also happen to die young. Women have certainly also evolved to be competitive, but they are less likely to take potentially le-thal risks than men. Wherever criminal records have been kept, men are convicted of crimes five to twenty times as frequently as women, and cases of homicide in which a man kills another man are about ten times as common as cases where one woman murders another.*

Human murder is a special case, and one that illustrates how evolution may provide impulses, but it can't guarantee their outcome. Most homi-cides involve people who know each other, sometimes intimately—that is to say, they are an example of ingroup aggression. Murders begin quite frequently with a trivial altercation involving a perceived threat to a man's status. Men evolved to compete for wealth and status because those prox-ies tended to provide more opportunities for sex.[83] Remembering that evolution cares not a whit for morality, it has provided human males at the bottom of the social pile ample reason to risk everything, including vi-olent death, rather than live a passive, sexless life without passing on their genes—or at least this was the case historically. Social, economic, and technological changes have largely de-linked violence from success, suc-cess from sex, and sex from reproduction. In almost every society today, it is now the least wealthy, least powerful men who tend to father more children. But evolution's impulses remain the same, whether empowered women, financial planning, or condoms are involved or not.

Competition between human males is a form of sexual competition, with our exaggerated sense of pride, our sensitivity to perceived disre-spect, and our sexual jealousy taking the place of a stag's horns or a pea-cock's tail feathers. Perhaps that's why sexual control and sexual jealousy remain such common causes of human violence.[84] In Afghanistan women have been beaten to death for merely showing the flesh of an arm in pub-lic—a perceived threat to male control over female sexuality. Worldwide, between 5 and 20 percent of murders involve men killing their sexual partners, and such cases are more frequent when the man has more rea-son to worry about infidelity, such as in casual or common-law unions, when the age difference between the murderer and his wife is greatest, and when the woman has another sexual partner.[85] Age plays a particular-ly interesting role in matrimonial murder, which is most common when the wife is under twenty, and falls off the closer she comes to menopause.

---

* Whether such aggression ends in a fistfight or in murder is partly influenced by the available technology. In São Paulo, Brazil, murders average one an hour and 90 percent are with firearms. The United States has more murders with guns in one day than Japan has in a year.

And of course for every murder, there are many, many cases of non-lethal domestic violence of men against women.

The evolutionary logic behind male sexual jealousy certainly doesn't justify any of the depravity done in its name, but it does help to explain the dual standard of sexual fidelity so common throughout the world and across history. Many cases of domestic violence, spousal murder, and sexual control of women do have their roots in the same biology that underlies the male predisposition to team aggression, and understanding that is the best path to finding ways to bring these destructive behaviors under control. Infidelity by both sexes is relatively common in humans, but the reproductive implications differ. When a man commits adultery he may do his wife great emotional injury, but so long as he abandons the other woman and any children he may have produced with her, neither his wife nor their children together suffer any loss of resources. When a woman cuckolds her husband and becomes pregnant, however, he may wind up investing in that child for years or decades to come: Biologically, this is a "wasted" investment, as his own genes have not been passed on. A child that comes from a woman's womb is guaranteed to carry her genes; a man can never be so certain; and thus, men in many cultures have gone to great lengths to try to control women and their sexuality. Lord Kilbrandon's aphorism that "Maternity is a matter of fact, paternity one of mere inference" sums up the asymmetry between adultery and cuckoldry— and helps to explain the popularity of daytime television shows featuring paternity tests.

Honor and Chivalry

When animals live in herds or other large social groups there are more eyes and ears to detect a predator, and more bodies to work together to drive off an attacker. But competition is correspondingly more severe when animals live in close proximity, much as it is in confined spaces like a jail or zoo. One solution to the tradeoff between mutual support and cheek-to-cheek competition is to develop social hierarchies in which each animal "knows its place."[86] The social cues that let animals signal when they know they're beat are a useful safety valve in these hierarchies, letting individuals jostle for a new position without getting killed. In many species, aggressive behavior within the social group ends in a display of submission, as when a dog involved in a fight rolls on its back exposing its vulnerable underbelly, or men surrender by holding their hands above their heads. When physical attacks occur among monkeys or apes, they are often followed by episodes of grooming that help restore social bonds. Ingroup aggression does lead to death in rare cases, but unlike team aggression, it doesn't have deliberate killing as its goal. This actually represents quite a fine level of control over our evolutionary impulses—the anger and excitement we feel in a fight are no doubt the same, whether it is a battle to the death or merely a competitive boxing match. Social rules, including customs, laws, and morals, help us to modulate the outcome of that anger, and it's a good thing. The fact is, we would not have gotten very far as a social species without ways to limit our killing impulses.

If ingroup fighting can become lethal, so too can team aggression morph into ingroup behavior. And the reasons are often cultural. Medieval knights who found themselves fighting for opposing armies were still bound by the code of chivalry, which ensured that emotionally, all knights in truth belonged to the same troop—an ingroup courtesy that did not extend to foot soldiers, who could be killed with impunity. Chaucer's "true and parfit knight" was supposed to spare the life of a defeated opponent. Once captured, a medieval knight might live in comfort and even enjoy the pleasures of hunting deer with his captors, but he was honor-bound not to attempt an escape. Nor was this cultural control over killing impulses lost in days of old. In both World Wars, for example, fighter pilots sometimes behaved in unusually chivalrous ways. In June 1917, the twenty-one-year-old German pilot Ernst Udet, who had already shot down six Allied planes, met the French ace Georges Guynemer, who had thirty victories to his name. Flying in tight circles where each could see his opponent clearly, they fought until Udet's gun jammed. Guynemer saw the German hammering on his guns with his fists in frustration—and waved, spared Udet's life, and flew home.*[87] In World War II, Erwin Rommel, Hitler's "Desert Fox," was known for his scrupulous treatment of prisoners, based in part on the Teutonic ideal of medieval knighthood

---

* Martha's grandfather Douglas Campbell, mentioned earlier as the first U.S.-trained ace in WWI, engaged a two-seater German plane that had been photographing American artillery positions. The gunner on the German plane ran out of ammunition. "I'll be damned if he didn't hang his empty cartridge belt over the side, stand up with arms folded and glare at me," Campbell recalled. "Naturally, I didn't relish the thought of shooting this brave, unarmed chap. But I had no choice...if they returned to their base with those photos, it might mean the loss of many American lives." In this case, the ingroup ideal of chivalry broke through for a few seconds, but outgroup fears won out, and Campbell brought the German plane down.

(*Ritterlichkeit*).[88] In all of these cases of chivalry, the upper social level of the outgroup becomes implicitly part of a crosscutting ingroup.

Conversely, a member of an ingroup who not merely respects but actively helps an outgroup is universally detested. It is no accident that in Dante's *Inferno* the lowest rank of Hell is reserved for traitors, who are eaten repeatedly and eternally by Satan (the real one, not the devilish chimp). Chimpanzee raids and human warfare depend on being able to trust every other member of the team on any aggressive mission, so perhaps it's unsurprising that the sense of honor is especially strong among young men at the prime of their physical strength. Leading the Serbian forces against an overwhelming Turkish host at the Battle of Kosovo in 1389, Prince Lazar said, "It is better to die than live in shame." Whether it's an American street tough flashing gang symbols on a street corner, or a chimpanzee flashing his testicles on a bridge in Gombe, the message is the same. Honor, in the words of historian Ivan Perkins, "is the proud, aggressive, forceful, physical, anger-laced, testosterone-based, masculine demand for respect."[89]

Evolution and Choice

Of all the traits we share with chimpanzees, the predisposition to war may be one of the closest. The two basic tactics of both human and chimpanzee warfare are surprise and superior force. And of all the abilities and impulses that predisposition is built upon, from physical strength and aggression to planning, teamwork, and trust, perhaps no element is more crucial to the enterprise than the ability to dehumanize members of our own species. As we'll see in chapter 11, this ability to exclude fellow humans from our emotional ingroup may well be the trait underlying much of what we call evil, from slavery to the Holocaust, to the September 11, 2001, terrorist attacks.

At first consideration, the idea that warfare may be partly inherited makes the horror of human violence seem unavoidable, or worse still, in some way justified. But let us not forget consilience, the drawing together of information from many fields of inquiry. Evolution doesn't make morality obsolete, any more than being hungry excuses a violent mugging. Besides, evolution only provides the building blocks for our behaviors. It is the immensely complicated interaction between these evolved behavioral frameworks and an ever-changing environment—physical, cultural, and otherwise, right down to free will—that gives our behavior shape. Human beings have the gift of living in a remarkably wide range of environments, each requiring a different expression of our inherited predispositions. We are neither cockroaches nor robots, and we can choose to use our predispositions differently in different settings. Human warriors can cook their victims and eat them, or extend medical care to a wounded enemy as they would one of their own comrades—even though the same impulse of rage and hatred courses through their veins.

This is where we have the advantage over our chimpanzee cousins. For all the similarities we both inherited from our common ancestor, humans have gone on to evolve the unique ability to observe our own behavior and choose to modulate it. We can bomb our enemy's cities and enslave members of the next tribe, or we can abide by international treaties that regulate attacks on civilians; we can play team sports such as cricket according to complex rules of fairness, or we can be soccer hooligans fighting with broken beer bottles. And we can build a world in which the conditions that drive us to attack each other—poverty and overcrowding, perceived injustice and lack of opportunity, or even just firmly drawn lines between in- and outgroups—are kept to a minimum. Biology can never tell us how we ought to behave in a civilized world.[90] But it can help us to understand why we do certain things both constructive and harmful. Knowing where we come from is vital to understanding human behavior, and to finding a pathway toward a future that is safer and more secure for us all.

4

WE BAND OF BROTHERS

We few, we happy few, we band of brothers;
For he today who sheds his blood with me
Shall be my brother. Be he ne'er so vile,
This day shall gentle his condition;
And gentlemen in England now abed
Shall think themselves accursed they were not here.

—HENRY V, Act 4, Scene 3

A S A YOUNG BOY growing up in Cambridge during World War II, I was fascinated by the many Royal Air Force and American airbases built on the flat countryside around the city. I would lie on my back in the garden and watch the little yellow Tiger Moth biplanes that young pilots flew for their first solo flights, and also the four-engined bombers leaving on raids over Germany. My mother volunteered in a canteen for U.S. airmen, some of whom also came to services at our church. And so I met some of the American fliers, who were only a little more than ten years older than me. Half of them, statistically, would have been among the 47,000 U.S. Army Air Force (USAAF) aircrew killed, dismembered, or burnt to death bombing Germany. I remember them as intelligent, polite, almost gentle volunteers. They had a one in five chance of completing a full tour of thirty bombing raids, and yet less than one in 200 chickened out in the face of the appalling risks.

What kept the bomber crews fighting through so many long, cold, terrifying flights? What combination of forces drove pleasant, loving men to rain terror and destruction on people, including thousands of women and children, they had never seen? True, obedience to orders and fear of shame at home played their part—but these men were passionate and sometimes even eager warriors. Could it have been the primate predisposition for small groups of men to show enormous loyalty to each other, great courage in the face of death, and a lack of empathy in attacking an outgroup?

This book is in no way a repudiation of maleness, nor an attempt to do away with its more problematic aspects, were such a thing even possible. But as we continue with our diagnosis of war and terrorism, it becomes impossible to ignore the central role of masculine behavior. In this chapter, we'll take a closer look at the males of our own species, and how some of their most valued evolutionary impulses, including loyalty, courage, and perseverance, have been expressed and exploited in warfare down through the ages. Doing so will get us closer to working out how those same predispositions can be channeled into much more positive outcomes for men, women, and society in general.

Shakespeare's speech for King Henry V to his soldiers before the Battle of Agincourt captures the ingroup bond of men in battle[91]—and it probably also describes with some accuracy a band of chimpanzees on a raid into neighboring territory. Of all the impulses and behavioral frameworks that contribute to the male predisposition for war, the bond of brothers-in-arms may well be the most powerful. Camaraderie is celebrated in art, literature, and fable, and loyalty is rightly enough seen as a key human virtue. But let's take a step back for a moment and look at these powerful impulses through the lens of evolution. What we discover might be more than a little surprising.

The Joy of War

From a biological perspective, the willingness to risk limb, and sometimes life, for another makes most sense when that person is a close blood relative. Even if you die saving your brother in battle, for example, so long as he goes on to reproduce, then the many specific genes you share are still passed along. And as we saw with the bands of raiding chimpanzees in the last chapter, nature has provided a series of strong emotions, impulses, and behaviors that help reinforce the evolutionary logic of group loyalty and courage. Our thesis is that evolution has likewise bequeathed men, and especially young men, with a related assortment of impulses and emotions. Together, they make up the broad behavioral framework—loyalty, courage, aggression, and camaraderie—that underlies much of military life and honor. The small bands in which we usually now fight (squads and platoons, terror cells, football teams) are no longer made up of blood relatives for the most part, but the old evolutionary system of bonding and mutual support is still very much in place. Training and shared experience, as we will see, can shape and intensify the team aggression predisposition; they can also create a surpassingly strong sense of kinship among men who share no genes at all.

Our emotions often serve the purpose of encouraging or dissuading particular types of behavior, and pleasure plays an especially strong role in this regard. Good food, a warm safe domicile, and sex all give us pleasure, and of course they all contribute to our survival—though we still enjoy a rich meal when we're in no danger of starving, and humans are one of the very few species that indulge in sex primarily for pleasure rather than for reproduction. As for war, our own most brutal expression of the impulse for team aggression, it is remarkable that for all the wartime accounts of terror, pain, and boredom, there are also many genuine, surprising confessions of joy. Or perhaps not so surprising, for if the predisposition to raid and exterminate members of an enemy troop is, like eating and sex, partly a survival instinct and driven by our genes, then the experience of intense excitement and even overt pleasure on the battlefield is almost to be expected. Though it's an experience with which the authors of this book are happily unfamiliar, the evidence suggests that Ernest Hemingway had a point when he wrote in *For Whom the Bell Tolls,* "admit that you have liked to kill as all who are soldiers by choice have enjoyed it at some time whether they lie about it or not."

It would be a mistake to play down the fear, horror, and emotional trauma that often accompany service on the battlefield. But at least for some, combat can also be transforming and exhilarating. Nathaniel Flick received a BA in classics from Dartmouth College in New Hampshire before joining the U.S. Marines in 1999. He led an elite Recon unit in the

2003 invasion of Iraq, and his Humvee was ambushed on the road to Baghdad. Flick returns fire:

> Aside from insects and plants, I'd killed one living thing in my whole life. While mowing my parent's lawn as a teenager, I'd accidentally wounded a chipmunk with the mower blade. Gritting my teeth, I'd cut off its head with a shovel. Even this mercy killing has bothered me. I'd never been hunting and had no desire to go. Now, shooting grenades at strangers in an unnamed town, I was kind of enjoying myself.[92]

Two hundred and fifty years earlier, George Washington, writing to his brother about his first experience under fire, reported, "I heard the bullets whistle, and believe me, there is something charming in the sound."[93] Surely part of that charm is the fact that the bullets missed, for as Winston Churchill, who had been under fire in the Boer War, wrote, "Nothing in life is so exhilarating as to be shot at without result."[94] Still, there seems also to be a positive pull toward the battle, not just relief at surviving it. Theodore Roosevelt claimed that he "would have left my wife's death bed to go and fight," and even as president he claimed his 1898 charge at San Juan Hill "was the greatest day of my life."[95] In World War II, a British Spitfire pilot, Flight Lieutenant D. M. Cook, described the moments before aerial combat as "the most gloriously exciting moments" of his life.[96] And a night fighter pilot, Roderick Chisholm, described shooting down two German planes in one night as "sweet and intoxicating." As the Civil War General Robert E. Lee remarked, "It is well war is so terrible—or we would grow too fond of it."[97]

Joanna Bourke in her book *An Intimate History of Killing: Face-to-Face Killing in Twentieth Century Warfare* was not seeking a biological explanation of war. The quotations she has culled from firsthand accounts of fighting, however, tell a story that fits well with the theory that human beings share with chimpanzees a deep-seated predisposition to kill members of an outgroup. As just one example, Henry de Man, a Belgian socialist and intellectual, wrote of his experiences during World War I:

> I had thought of myself more or less immune from this intoxication until, as a trench mortar officer...I secured a direct hit on an enemy encampment, saw bodies or parts of bodies go up in the air, and heard the desperate yelling of the wounded or the runaways. I had to confess to myself that it was one of the happiest moments of my life....What were the satis-

factions of scientific research, of successful public activity, of authority, of love, compared to this ecstatic moment?*[98]

It is important to note that the great majority of military recruits are neither bloodthirsty nor motivated by a desire to kill—indeed, the modern professional military works hard to distance itself from the undisciplined killings of thugs and murderous sociopaths. And certainly not all men experience a thrill when they do kill in battle, and certainly not in all situations—although perhaps many could in certain circumstances, especially when very young. Still, women rarely, if ever, use words of elation to describe killing, and the "joy of war" that so many male combatants do report is one important clue in our search for the biological roots of warfare. Another line of evidence comes from a more universal experience of men at war, the particularly intense bond of brothers-in-arms.

Camaraderie

When Audie Murphy, the most decorated U.S. soldier in World War II, was asked why he was able to take on an entire German company single-handed he answered plainly, "They were killing my friends." Of all the reasons given for extraordinary valor on the battlefield, none is more common than simple, absolute dedication to one's comrades. Men have many reasons for joining the military, from idealistic patriotism to financial need to indeed, conscription. But once the shooting starts, Murphy said, "It's about the guy next to you."[99]

A group of chimpanzees scrambling through the forest in search of members of another troop to kill may consist of only four or five animals. Raiding parties among preliterate societies are often equally small, but modern armies can mobilize hundreds of thousands of men. At first blush, the difference is so large that the deep connections are not obvious. But on closer analysis, even the largest fighting force is built up from many smaller units, be it the platoon in the slit trench, the infantry squad

---

* Explicitly sexual metaphors also often emerge in accounts of warfare. A World War II British officer, Anthony S. Irwin, says of being dive-bombed by German planes and miraculously escaping death, "I felt like a man who had just had a perfect and shattering union with a woman. I sweated and wanted more." It's no wonder some authors suggest that warfare sometimes seems to be a surrogate for sex, but we suspect the connection is more basic: Both sex and war fulfill profound male evolutionary drives, so it's not surprising they might come to serve as metaphors for one another.

patrolling in an up-armored Humvee, the bomber crew in a cold moon-less sky, or submariners under an unfriendly ocean.

Human beings function best in small groups, and this fact has been known to military leaders for millennia. In the modern army, nine to fif-teen men comprise a squad, three to four squads make a platoon, and three or four platoons make up a company, totaling one hundred to three hundred men. In the second century B.C., Philip of Macedon's army was made up of platoons of sixty-four men, and the Roman army was split into groups of one hundred or less, commanded by a centurion. Genghis Khan may have become father to millions, but he kept his Mongol troops divided into units of ten, one hundred, and one thousand.[100] In the First World War, C. E. Montague wrote, "Our total host might be two millions strong, or ten millions, but whatever its size a man's world was his sec-tion—at most his platoon; all that mattered to him was one little boatload of castaways…"[101] The primary unit that provides physical and mental support, and inspires loyalty and courage, is a band of men not all that much larger than a chimpanzee raiding party.

There is a saying in Arabic, "Me against my brother; me and my broth-er against my cousin; me, my brother and my cousin against our common enemy." Just as it is with chimpanzees, we compete and bicker with those who are closest to us, but we also band together with our inner circle to compete and bicker together against other, similarly constituted groups. We have the Stone Age emotions of hunter-gatherers, which in turn may not be as far removed from the emotions of chimpanzees as we might like to think.

Friendship can take many forms in human life, but it is never stronger than among men who have been in battle together. Glenn Gray was draft-ed into the U.S. Army on the same day in 1941 that Columbia University awarded him a doctorate in philosophy. Analytical by training, Gray ex-plored the enduring appeal of combat at three levels: the spectacle of the battlefield, the lust for destruction, and most important, the camaraderie of war. In his 1959 book *The Warriors: Reflections on Men in Battle,* Gray describes how numberless soldiers die,

> …not for country or honor or religious faith or for any other abstract good, but because they realized that by fleeing their post and rescuing them-selves, they would expose their companions to greater danger. Such loyalty to the group is the essence of fighting morale.[102]

A few might run, and a few might choose not to run out of fear of punishment. But most don't run, even under the most horrific and immediately lethal of circumstances. Elsewhere, the philosopher-soldier writes of "this sense of comradeship [as] an ecstasy. Each is ready to give up his life for the other, without reflection and without thought of personal loss." Such behavior is often called "duty," but it is a duty men feel to their peers more than to any larger group—and it often lasts long after the immediate flush of battle. Men who have been wounded, earning the right to be sent home, regularly demand to return to the front to rejoin their squad, preferring to risk their life again rather than abandon their comrades. One person who jumped the safety of hospital to return to the firing lines of World War II's Pacific theater was the American author William Manchester. Looking back more than thirty-five years after the fact he wrote:

> It was an act of love. Those men on the line were my family, my home. They were closer to me than I can say, closer than any friends had been or ever would be. They had never let me down, and I couldn't do without them. I had to be with them. Rather than let them die and me live in the knowledge that I might have saved them. Men, I now know, do not fight for flag or country, for the Marine Corps or glory or any other abstraction. They fight for one another. Any man in combat, who lacks comrades who will die for him, or for whom he is willing to die, is not a man at all. He is truly damned.[103]

Such strong feelings, which can put us so explicitly in harm's way, demand an explanation—we hear the words of the survivors, but they stand in for the untold thousands of men who, similarly motivated by camaraderie, found only death. Evolutionary logic can seem convoluted at times, but remember that the survival instinct is not always focused on the individual who carries it. Sometimes its point of action is the group, and this ecstatic feeling of loyalty that maintains fighting morale is precisely the drive we would expect evolution to select in order to sustain a raiding party of apes, be they *Pan troglodytes* or *Homo sapiens*. The comradeship that makes young men risk their lives in a raid or a battle, in other words, is almost certainly a behavioral predisposition, shaped by evolution and released by a particular set of circumstances. It was—and is—unconscious and not related in any immediate intellectual way to its evolutionary goal of enabling a troop of related animals to extend their territory and create an environment capable of supporting more offspring. But once it evolved, the predisposition gave those males who had it an advantage in the acquisition of resources for reproduction. And it has persisted even though that advantage is now largely gone.

Killing

The emotional and psychological frameworks of the warrior are complex and varied, and the bonds of camaraderie are just one element— necessary to war perhaps, but certainly not sufficient on their own. To understand the success of team aggression as a survival strategy early in our evolutionary history, we'll have to unravel a few more aspects of the behavior. Among the most important, surely, is the ability to kill.[104]

In the previous chapter we saw how being part of a crowd increased obedience to authority and the willingness to be cruel. Again, this is the type of behavior we would expect evolution to encourage in a warrior primate. Team aggression depends on total reliability and mutually shared selflessness within the attacking group. But it also demands a willingness to kill members of our own species—and for a social animal especially, that is a very dangerous strategy indeed. Evolution, in its blind experimentation, seems to have found a degree of control over the killing urge by providing our minds with what might be called an empathy switch. We seem to have an innate mental ability to treat our fellow humans with either great compassion or cold disregard, depending on whether we've assigned them to ingroup, or out.

The fact is, killing another human is never easy—so long as we truly see that person as human. Colonel S. L. A. Marshall, who had been on the front lines in both World Wars, wrote that many a soldier about to kill "becomes a conscientious objector." [105] After the 1863 Battle of Gettysburg, in which 7,000 men died and 33,000 were wounded, 27,574 muskets were picked up from the battlefield. Ninety percent were loaded and one musket had twenty-three powder charges and pieces of shot.[106] How many men were killed before they could pull the trigger, and how many could not bring themselves to fire on an enemy that had been, until the start of war, part of their national ingroup? One commentator claims that 1 percent of the fighter pilots in the World War I U.S. Army Air Corps accounted for 30 to 40 percent of German planes shot down. Were they simply better pilots, who survived and became disproportionately successful, or were others reluctant to press home the attack? In World War II, Marshall found that only 15 to 20 percent of front line troops in the European theater fired their rifles. In Korea, the rate rose to 55 percent and it is thought that 90 to 95 percent of soldiers fired in combat in Vietnam. Could it be that the more different the enemy is perceived to be, the more readily the trigger is pulled?

Chimpanzees are also social animals, and they too seem to show ambivalence about killing their own species—and a similar diversity of individual responses. In the Kasakela troop raid described in chapter 3, Figan attacked the infant torn from its mother's arms. "Holding it by one leg, [he] leaped thorough the tree, smashing the infant against branches and trunk as he did so. He jumped to the ground and continued to flail his victim against the rocks as he ran." Clearly, in human terms, this was a chimpanzee eager to pull the trigger. But then a raiding companion picked the infant up and groomed it, applying in-troop ethics. Later, one of the immature Kasakela chimps "rescued" the injured baby and carried it for over an hour, though the baby ultimately died from its wounds.

Perhaps because it is ultimately an artificial distinction, the boundary between ingroups and outgroups of a single species can become unexpectedly porous, even in the heat of battle. On Christmas Day 1914, the Bavarian troops opposite the trenches of the Welsh Fusiliers sang the Welsh folk song *Ar hyd y nos* ("All Through the Night") in German. The Welsh responded by singing *Good King Wenceslas*, and soon the troops were mingling in no-man's land, exchanging British bully beef for German sausages and kicking a football around. There was not "an atom of hate on either side during these truces." Next day the soldiers went back to killing one another.[107] During the American Civil War, each army dug forward trenches, or picket lines, that were sometimes only a few tens of steps apart. Lonely men on picket duty began to fraternize and even agreed not to fire on each other. When the Confederate General Stonewall Jackson visited the picket lines in Virginia, a Union soldier remarked, "We could have shot him with a regular revolver but we have an agreement that neither side will fire as it does no good and is simply murder."[108] Living within spitting distance of one another, the men had adopted the ingroup rules of behavior.

The Reivers and the Hoplites

History provides many vivid examples of whole societies where the men were totally focused on warfare. In many cases, the cultural or genetic differences between even the most hated of enemies are slight enough to seem as artificial as those dividing chimpanzee outgroups. But as with chimpanzees, we seem to have inherited a genius for exaggerating our differences, and then using them as a basis for war. Certainly there is no shortage of raiding, plunder, rape, and territorial expansion in our past. But time and again, war and warring seem also to have taken on a momentum and logic of their own. We'll take a closer look at two groups that came to be identified almost entirely with wars and raiding, often at great cost to the societies in question. It's hard to think of an explanation for their behavior, other than as particularly powerful expressions of the evolutionary impulses and predispositions of team aggression.

I was born near Newcastle, in northern England. For centuries, the nearby border between England and Scotland had been the scene of constant fighting between small clans. Reivers were clan-based border raiders, and they took their warring seriously: during baptism, they took care to hold the right hand of a baby boy away from the font so that he could still strike "unhallowed blows" in battle as a man.* The Reivers were true raiders, and if not much territory was exchanged overall, a good many horses and sheep were stolen back and forth. Raiding, after all, is about acquiring resources. Their border raids were so consistent and lethal that we owe the word we use for the death of a loved one—*bereavement*—to their name. And as with chimpanzee team aggression, Reiver raiding parties were made up of genetically related men, who had grown up playing together in the windswept Cheviot Hills. In 1603, when England and Scotland were united under King James, many Reivers were hanged or imprisoned and a semblance of order was imposed from the outside. Some entered the British army and under the discipline of a formal military, became brave, respected soldiers. Others emigrated to the American colonies, where they continued their border traditions by warring amongst each other, and with the Native Ameri-

---

* The surname Armstrong has its origins among the Reivers, appropriately enough. Grahams, Nixons, Kerrs, Fenwicks, Johnstones, and others also can trace their roots to these warring groups. Fortunately, and as one more bit of evidence that a history of constant warring need not predict our future, there is no reason to suspect that today's descendents of Reivers are more prone to aggression than anyone else.

can tribes of the Appalachians. The posse of men made famous in cowboy movies is a direct descendant of the "hot trod" raiding bands that the Reivers introduced to America.[109]

The Reivers were classic raiders, but the hoplites of Classical Greece, 2,500 years ago, show how the same underlying predispositions can be reinforced and enhanced when they are formalized. The hoplites were heavy infantrymen, and they fought in larger, more disciplined groups than the Reivers, but their social structure, courage and warrior behaviors reveal the same underlying impulses and organization. Hoplites entered battle in long phalanxes, arrayed in ranks eight to twelve men deep. In direct manifestation of the mutual protection of team aggression, each man's shield protected the warrior on his right. Total commitment was ensured by placing the bravest men at the front, the most determined at the rear, and any waverers in between.[110] The hoplites sweltered and chafed in the heat, each man carrying up to half his weight in bronze helmet, breastplate, and leg armor, plus heavy shield and an eight- to ten-foot, iron-tipped pike. A hoplite muster might involve up to two-thirds of all males over age eighteen, making the force much more like a raiding party or militia than an elite, professional army. During the Greco-Persian wars of 499 to 448 B.C., hoplite soldiers from an assortment of Greek city-states with a combined population of perhaps only two million turned back the centralized power of the Persian Empire with an estimated population of seventy million. At the Battle of Marathon in 490 B.C., a force of hoplites killed over 6,000 Persians with a loss of just 192 of their own men—a stunning testament to the power of group cohesion, mutual sacrifice, and profound loyalty to one's comrades.

As is so often the case, the definition of "outgroup" can change depending on the circumstance, and the Greeks fought each other as well. The Persian historian Mardonios commented on the way in which the Greeks fought one another. "For as soon as they declare war on each other they seek out the fairest and most level ground, and then go down to do battle on it. Consequently, even winners leave with extreme losses; I need not mention the conquered, since they are annihilated."[111] Mardonios might also have added that these battles, by agreement, normally took place in the afternoon, after a late breakfast of cheese and dates laced with wine. A hoplite battle was somewhat like a British rugger match or an American football game, with the opposing phalanxes arranged along a line of scrimmage almost a mile long. If the form was more stylized than a bor-

Credit: Herve Lewandowski—Réunion des Musées Nationaux/Art Resource, NY

The training and group cohesion of hoplite warriors were central to classical Greek culture. In battle, thousands of these citizen-soldiers would face one another in long phalanxes several ranks deep. Each man carried a circular shield or hoplon, a spear, and sword; only wealthier landowners could afford the expensive bronze armor and helmet. Wars between city-states could occur every year or two, but Greek armies also faced powerful external foes. The courage and organization built up through hoplite training helped turn back superior forces during the Persian invasion of Greece at the Battle of Marathon in 490 B.C.

der raid, however, the risks were no less severe. Men were literally shit-scared before battles, which began with a screaming, ferocious charge, with spears shattered and shields crashed together. Once the armies collided, the fight developed into simple pushing matches, with the men in the rear ranks, beyond the reach of their opponent's spears, adding support as well as pressure by shoving their convex shields into the backs of their comrades to the front. The first rank, fighting "chest to chest" and "helmet to helmet," slashed with short swords and stabbed with broken spears. If a man fell he was as likely to be trampled by friend as by foe.

The phalanx formation wasn't just clever tactics, though certainly it was that. Fighting cheek by jowl, there could be no question of whom you were relying upon, and whom you would be letting down if you faltered or failed. Historian Victor Davis Hanson of the University of California at Fresno describes how the hoplites "went into battle for the man on the left and right, front and back, brother and cousin, father and son: Out of respect for, and in fear before, men of like circumstance, they forged some code of honor and salvaged a certain dignity (if not pleasure) from killing."

The Vikings also sometimes fought in a phalanx and would form defensive shield walls, as did the invading Norwegian king, Harold Hardrada, at the Battle of Stamford Bridge in England in 1066. The phalanx capitalized on the same deep loyalty as a result of growing up and working together as the Reivers had. It could lead to amazing acts of courage. Greek hoplites and sailors were free men, yet there were no conscientious objectors. The Greeks we still remember today, the poets, scientists, and philosophers, almost all fought in hoplite battles, and even the blind sometimes took their place in the phalanx. At Salamis, in 480 B.C., the Greeks who rowed their triremes against a much larger Persian force were all free men from the first democracy, motivated by the mutual bonds of choice and a shared fate. The 40,000 Persian sailors who drowned that day were literally whipped into battle.[112] When the Greek poet Aeschylus composed his own epitaph, he did not mention the writings that made him the greatest playwright of antiquity—one whose dramas are still performed two and half millennia later. Instead he wrote, "The grove of Marathon, with its glories, can speak of his valor in battle."[113] Aeschylus's brother was killed beside him at Marathon.*

As long as the phalanx held and the men covered one another with their shields, the close-packed formation was almost impossible to defeat—as the Persians found at Marathon. Sooner or later, however, one side would break ranks. The Greeks had a word for it—*panic*, from the unreasoning fear that the mythical shepherd god Pan might have observed in his flock of sheep. Socrates was well over forty when he fought

---

* In an extreme example of the dedication of warriors for one another, one hoplite group fighting at Chaeronea, called the Sacred Band, was composed exclusively of homosexual couples. The Roman military historian Plutarch wrote that the hoplites fought best because "brother is in rank with brother, friend with friend, lover beside lover." When Philip of Macedon saw after his victory at Chaeronea that every member of the Sacred Band had died fighting, he proclaimed, "May all perish who suspect that these men did or suffered anything disgraceful."

on the losing side in the battle of Chaeronea (338 B.C.), but he managed to rally his retreating friends, organize a stand, and escape annihilation, choosing, as he wrote, "to stay put there and face the danger without any regard for death or anything else rather than disgrace."[114]

If Greek hoplite battles had a stylized aspect, it would be a mistake to think that they lacked lethal consequences. Classicist Peter Krentz estimates that one in twenty of the victors and one in seven of the vanquished died in the average Greek battle. And many city-states were engaged in war every second or third year. Fighting the Persians at Thermopylae in 480 B.C., the Greek town of Thespiae saw one-third of its adults die; five decades later at Delion (424 B.C.) the same proportion was killed again, and the same yet again at Nemea in 396 B.C.[115] For all that today's battles are fought with fearsome weaponry, modern warfare has never seen carnage on such a scale relative to the population. Hanson makes a horrific but plausible calculation concerning the battle of Pydna, in 168 B.C. When the Roman legions finally defeated the Macedonian phalanx, he estimates that the battlefield would have been soaked with more than 10,000 gallons of human blood: 20,000 Macedonians were killed for 1,000 Roman dead.[116] Yet, however terrible the clash of arms in hoplite battles, it is sobering indeed to contemplate that the aggregate fatalities in relation to the total population may well have been higher yet in the repeated raids of pre-state societies, as in New Guinea fifty years ago or among the Reivers 500 years back.

Boot Camp

When the Reivers stole sheep or the hoplites fought pitched battles, each man was related to, or knew personally, those around him—often from childhood. To the extent that they were united by birth and trained by daily life, the behavior of hoplites parallels rather closely that of the chimpanzee raiding party. When men are intimately related by blood and daily social commerce they may be able to dispense with basic training. But societies became more complex and armies started to draw men from unrelated groups, and so more formal training developed to teach skills, certainly, but even more importantly, to build bonds of camaraderie. As early as the time of the Pharaoh of Amosis (1580–1557 B.C.), Egyptian soldiers were trained to fight in phalanxes and to scale siege ladders.[117] Military training can help shape and intensify the impulses behind team aggression, but its greatest power lies in forging true teams from groups of strangers. We mentioned earlier that military training can help build an artificial, but very strong, sense of kinship among unrelated males. All men may carry the predisposition for team aggression, but they must feel they are part of a tight-knit group of mutually dependent fellows before the real power of team aggression kicks in.

Military leaders through history have found ingenious ways to bond disparate men into cohesive fighting teams. Fighting the Crusaders, Saladin (1137–1193) used as his shock troops slave soldiers called Mamluks. They captured or purchased Kipchak Turks, Circassians from the Caucasus, and Christians from various areas. In order to unite this diverse group of men shanghaied from the very enemy they were then trained to fight, they were subjected to an intensive twelfth-century version of boot camp. Young Mamluk archers were taught to ride bareback, and to drop their reins and control their mounts with their knees as they shot behind them. They had to slice lumps of clay with their swords up to 1,000 times in a day to build their arm muscles, and some died from the rigors of training.[118] It is exactly this sort of elite training and shared hardship that builds strong bonds of mutual support and respect between warriors, simulating the bonds of blood relationship. Modern traditions of military training and discipline date in part from the reign of King Gustavus Adolphus, who came to the Swedish throne in 1611 when he was only seventeen years old. He combined flintlock musketeers with pikemen and created a permanent army which trained frequently, under strict discipline. Gustavus also developed much of the modern hierarchy of command.

Perhaps nowhere, and at no other time, was the art of military bonding taken to more extreme lengths than in ancient Sparta. Common experience and family bonds were reinforced by extraordinary training, spawning a fighting machine the courage of which is still unparalleled. Sparta survived as a city-state from 1000 to 142 B.C. At age seven, Spartan boys were plucked from their families and inducted into what has been called "the Scout Troop from Hell." Older boys, who were called "the whip carriers," beat their juniors. The kids went barefoot and were allowed only one thin blanket (it can get very cold in winter in Greece), the food was famously atrocious. At nineteen, men joined what amounted to a Spartan secret police, the *krypteia*, which murdered serfs without redress and, often enough, for minimal cause. The Greek historian Plutarch wrote that "Sparta is held together by fear,"[119] but this was a fear that enhanced the loyalty and dedication of the warriors to each other. Authoritarian control can build resentment, but it can also build upon instinctive drives for cohesive aggression and make them even stronger. Only at thirty was a Spartan man allowed to marry, but even then the hoplite regiment remained his home, and he only visited his wife after dark to copulate. [120, 121]

Modern boot camp is not as openly sadistic as Spartan training, but it too is designed to disorientate, physically exhaust, and demean, and to break the links between the recruit and the civilian world. There are skills to be learned, of course, and physical conditioning to be achieved. But the main point is to awaken and intensify the instinctive predispositions of team aggression, producing warriors imbued with intense loyalty to those who have gone through the same training.* William Manchester provides a vivid account of U.S. Marine boot camp on Parris Island, South Carolina, at the beginning of World War II. "It was hell...tolerable even at its worst because you were all in it together, and you knew you would all make it together."[122] Today, Marine recruits are still scheduled to arrive in the middle of the night. Their heads are shaved, shearing them of past appearance and starting to build group identity. Every minute of the day is filled with exhausting marches and exercises. During training, if one person messes up a task or fails to qualify on a physical test then the whole squad is punished. This group punishment reinforces the key lesson of team aggression—every member of the team depends on every other member, and the failure of one can mean disaster for all. In the words of one drill instructor, "If you don't break them down, stir their souls, and maybe clobber the recalcitrant, you can't build them back up again."[123] The end result of boot camp can be something like a religious conversion. Recalling his drill instructor, one Marine said, "He trashed me three times a day. I'll always remember him. I think he's a great guy. 'Cause I started giving up, and he never gave up on me. I'll remember him to the end of my life." Ultimately, the rigors and shared misery of boot camp are shortcuts to

---

* This type of bonding is not limited to the military. As a medical student, I went through a strong socializing process. Beyond the long hours, make-or-break examinations, and ingroup language of medical terminology, medical students are further set apart by the fact that they dissect cadavers—something you do in a team—and dress up in masks, gowns, and gloves before entering the surgical theater. There are pragmatic reasons for all this, of course, but it also builds lasting bonds not just with one's fellows, but with the broader medical fraternity as well. I can go to any country in the world and my medical qualification immediately opens the door to trust and friendship with other doctors.

creating a feeling of kinship among the men in the squad. Once that perception is firmly instilled, the predispositions that drove our primate ancestors to risk their lives for their relatives on a raid come to the surface for these surrogate brothers.*

The ability to recognize kin, the ultimate ingroup, is critical for social animals. It is a prerequisite for avoiding incest and, as noted earlier, it underpins altruistic behaviors. In some animals, kin recognition is automatic; tadpoles, for example, use chemical messages in the water to identify their own siblings and then shoal with them for mutual protection.** Primates use voice and physical appearance to identify one another. Chimpanzees have very different faces and even the human observer can soon learn to tell individuals apart. When humans lived in small hunter-gatherer bands, everyone we met or saw about us would have been either a kinsperson or the sexual partner of a blood relative. We needed no special sense, other than the ability to identify individuals, to establish who were our kin and, by implication, who was a potential enemy or sexual partner and who was worth taking a risk for in war or some other enterprise.

An important insight into the way in which we treat those we grow up with as kin comes from the work of Joseph Shepher.[124] He followed 2,769 marriages of couples who had grown up as children in an Israeli kibbutz. The children spent many hours in communal kindergartens while their parents worked in the field or on other tasks. Not a single case of marriage between individuals who had been children together in the same kibbutz was found. It seems that the virtually universal human taboo against having sex with close relatives is based on whom we knew as children. Familiarity would have been an effective proxy for kinship during our hunter-gatherer past, because shared genes and acquaintanceship were largely synonymous. In the same way, recognizing and working closely with a small band of men of roughly the same age seems to spark deep impressions of kinship, even when the men around you are not in fact your blood relatives. In the case of U.S. Marine boot camp, the physical exhaus-

---

* Deliberate efforts to build an aggressive force of young men can also exacerbate other aspects of violence. Among U.S. servicemen in Fort Bragg, North Carolina, the rate of domestic violence is twice the national average. In a single month in 2002, four soldiers shot or stabbed their wives in fits of temper shortly after returning from Afghanistan.

** Perhaps most remarkable of all, Susan Dudley, a botanist at McMaster University in Canada, observed that a plant called a sea rocket can recognize its kin. In the presence of kin the roots of the sea rocket grow slowly but beside an unrelated plant they grow aggressively, taking as much nutrition from the ground as they can. Hostility to strangers and loyalty to kin are the hallmarks of evolution.

tion, the ever-present brutality, the petty rules, and ultimately the pride of having survived the ordeal act as a high-speed way of molding a group of unrelated young men so that they feel they are indeed a band of real brothers. Marine sergeants claim their squads leave boot camp with "two families"—the ones they grew up with and the Corps they will join on the appropriately named "Family Day," which marks the end of boot camp.

There is no question that the resulting comradeship nourishes courage. Loyalty to kin, or perceived kin in the battle line, feeds the supreme sacrifice of dying for others. In an analysis of the 207 Medals of Honor awarded in Vietnam, Joseph Blake of Virginia State University found that sixty-three were cases where men had thrown themselves on an exploding hand grenade in order to save the lives of their comrades. Miraculously, four of these men survived, but all must have known when they made their move that they were about to die.[125] Such acts of selflessness are most common in those parts of the armed forces that are subject to the most intensive basic training routines. Among thirty-seven Medals of Honor given to U.S. Marines, twenty-two were for smothering grenade explosions, compared with six of seventy-five medals given for the same act of courage to non–elite forces.

Courage

Given the hoplites' extraordinary bravery and dedication to one another in battle, it's not surprising that "hero" comes from a Greek word (meaning demigod or protector). In classical literature, heroes are generally egotistical, destructive men who gain material possessions and poetic glory by placing themselves at the center of any battle they can find. It's not an entirely charming image, perhaps, but the message is clear: Put your life on the line for the team, and you'll gain the admiration of your own people and if you survive, the worldly goods of your neighbors. From the Greeks to the ancient Chinese, and the Vikings to the 2,351 Japanese soldiers who died in an attack on U.S. troops on the Aleutian Island of Attu in 1943, heroes are defined by intense loyalty, a fighting spirit, and above all, immense personal courage. They're the ones who risk their own lives, often without hope of survival, for the good of the group. It's the ethos of team aggression in overdrive, and so long as the enemy is vanquished, whether the hero dies in the process hardly seems to matter.

Glenn Gray, the soldier-philosopher, thought deeply about courage on the battlefield. In part, he felt, men pressed on under brutal conditions because the fatigue, lack of sleep, inadequate food, and cold—or excessive heat—diluted their self-awareness, to the point where questions about the proximity of death were pushed aside. Just struggling to the next day or the next night was enough to think about. Some men, he says, dealt with the carnage by choosing to believe that whoever else might die, they would be safe—and he pointed out that when a man believes it is his particular destiny to survive, he may do things that others would consider heroic. Others felt, in Gray's words, "that all bullets are intended for them and every shell likely to land on the particular spot they have selected as temporary shelter. Insatiable death lurks everywhere ready to pounce on them." Such men tended to be loners, in Gray's experience, and were likely to be classified as cowards, or at best a target for their fellows' jokes. Some men became deeply religious under fire, but for others war was a game—and win or lose, an exciting one. For these, Gray wrote, war "offers something peace cannot: the opportunity to telescope much experience into a short compass. If death be the issue, they are normally fatalists by instinct and can accept it more calmly than the prospect of a boring, empty period in their lives."[126] It is common enough to hear that you never feel so alive as when your life is in danger, and one suspects that this vitality, too, is one of nature's incentives to not just enter into battle, but to give it your all.

Honor is a close cousin of courage, at least in the sense of valuing one's ideals above one's personal safety or comfort, and is valued just as highly in military life. One of the longest histories of exceptional loyalty, courage, and honor was written by the sepoys, or Indian privates who fought for the British Raj in India in the eighteenth and nineteenth centuries. After his victory at the 1803 Battle of Assaye in India, Arthur Wellesley (later the Duke of Wellington) said of the sepoys, "I cannot write in too strong terms of the conduct of the troops; they advanced in the best order and with the greatest steadfastness, under the most destructive fire against a body of infantry far superior in number."[127] Such courage is all the more revealing because these troops earned a pittance, wore an inappropriate woolen uniform in temperatures of 100 degrees Fahrenheit, and were subject to harsh discipline—to say nothing of the fact that they were fighting for an occupying power, which worshipped a foreign god and for the most part did not even speak their language.

Between 1825 and 1833, 60,000 sepoys served in the Indian army and there was not one case of desertion in front of the enemy, and only thirty-five courts martial for other offenses, such as sleeping on duty. What drove this military sense of honor? The sepoys came from a system of kinship and group identity that went back over 3,500 years. When the European explorers reached India they used the Portuguese word *casta,* or lineage, to describe the strict social subdivisions they saw. Hindu society was—and is—divided into several main, and thousands of minor, castes. Priests and scholars, or *Brahmans,* can consort with other Brahmans; warriors, or *Kshatriyas* (or *Rajputs*) with other *kshatriyas*; and *pariahs* or "untouchables" only with other pariahs. Marriages can take place only within caste lines. The caste system uses social pressures to reinforce a basic biological drive to support kin, and to be wary of and show hostility to outgroups. The British in India, probably through a mixture of intuition and accident, tapped into the existing caste system and used it to their own advantage. They recruited sepoys from the same towns and villages together, and let each caste follow its own rituals. Natural inclinations, reinforced by caste identity and intensified by rigorous military training, produced intense loyalty within the group and remarkable courage in the face of privation and death.

Surrogate Wars

There is probably no human activity more intense than full-on warfare. But the same impulses and emotions that drive it can be expressed, with varying degrees of intensity, in situations ranging from competitive sports (by both players and fans), to firefighting squads and emergency crews, to proxy competitions between nations, such as the nuclear arms race and the space race. The bonds that tie a football or soccer team together share their origin with the camaraderie of a Marine platoon or a hoplite phalanx, and are sometimes quite powerful. Men who have done both say the sense of brotherhood they share with fellow firefighters can be just as strong as that shared by comrades in a military firefight.

Adventure sports, such as mountain climbing or wilderness exploration, provide an interesting example of how the emotions and impulses of war can come into play under quite different circumstances. George Mallory, who died trying to climb Mount Everest in 1921, had written his father from France during World War I saying, "my instinct is to want more fighting."[128] The transition from warfare to non-violent but still dangerous activities is something young men can make very easily.

Mountaineers, like soldiers, do not wish to die, yet without its lethal danger the sport would lose much of its appeal and for some, become meaningless. Each time climbing equipment improves, climbers have themselves new and necessarily more dangerous challenges. Once the extremely difficult North Face of the Eiger in Switzerland was finally conquered in the summer of 1938, enthusiasts set about repeating the climb in winter. Safety gear is much more advanced now, but climbing is still dangerous. In the first six months of 2001, for example, sixty climbers died in the Alps alone. Prior to World War II, seventeen German Alpine climbers had died trying to scale the Eiger. During the German invasion of Crete, mountaineers from that same group displayed outstanding courage in capturing the key airfield of Maleme from the British. The mortality rate on the Eiger was 24 percent; in full battle on Crete it was only 18 percent.[129]

Like men in battle, climbers focus on a single goal—in this case, conquest of a difficult mountain peak. An icy, rocky, gale-torn, dangerous terrain becomes the enemy to be defeated. Falling stones, a constant danger on exposed peaks, are called "mountain artillery" by German Alpine climbers. Like war, a serious mountain climbing expedition can be hell. Plodding through deep snow carrying forty pounds of provisions to the next camp at 4:30 A.M. in a howling gale with the temperature 20 degrees below freezing is a long way from a pleasant Alpine stroll. Yet some men (as well as some women) are eager to climb. Even though the ultimate goal is never a violent one, the shared effort, mutual trust and sacrifice, and lethal danger of mountain climbing can produce friendships and shared experiences every bit as profound and lasting as those of the battlefield.

Climbing a 26,000-foot peak in the Karakoram Range between India, Pakistan, and China in 1958, a team of American mountaineers came across a small cairn marking the grave of climbers who had died on another attempt some years earlier. They added a few more rocks to the cairn and a note in memory of the dead. "It was not very much," wrote the recorder of the climb, "but it was tangible proof they had not been forgotten. The history of the mountain is told in terms of men, and at this moment we felt a small but integral part of the continuing sweep of human history in this remote area, a link between the past and the future."[130]

The same words could have been written by men in battle, or by old warriors paying homage to fallen comrades at a war memorial.

Revenge

Honor, courage, and loyalty may count among the most treasured attributes of the ideal soldier or mountain climber, but war, ultimately, is about killing. And unlike chimpanzees, most humans, especially the honorable ones, blanch at the thought of killing for mere land or booty. But from self-defense to divine right and manifest destiny, there has always been a ready supply of urgent reasons for war available to the king, general, or political theorist who would start one. Amongst the most popular is revenge, and this aspect of war too seems to have roots in our deep, evolutionary past.

Revenge is actually quite a sophisticated concept, and it requires a considerable amount of brainpower to pull off. A long-term memory is essential, as is a sense of justice and the ability to conceive of plans and project them into the future. It just so happens that all three are specialties of social animals, and none more than ourselves and our chimpanzee cousins. Chimpanzees have an acute sense of fair play and a sharp sense of outrage when it is offended. And just as they might remember a long-ago researcher with a gift of excrement, they also have the memory needed to harbor a grudge and the ability and inclination to retaliate, no matter how long it takes. In *Our Inner Ape*, Frans de Waal relates a story about zoo chimps that were not fed until the last animal came into the cage from an outdoor area at night.[131] On one occasion, two young chimps returned late, delaying the evening meal for the entire troop. The chimps ate that night, but the next morning the two dawdlers were roughed up. On the second night, they were the first back into the cage.*

Revenge is also a significant factor in human raids and warfare, and quite often the only reason—or at least the only reason anyone can remember—for family feuds and the multigenerational wars between preliterate peoples. The flame of revenge still burns strongly in the modern world as well. The U.S. did not invade Iraq until a year-and-a-half after 9/11. Set aside for the moment the fact that Iraq had nothing to do with

---

* There is also a clear connection here to the group punishment ethos of boot camp. In this case, ingroup revenge is a means of discouraging behavior that hurts the group, though in chimps at least, it may also help to build loyalty and enforce team cohesion. One suspects that in humans, other than those in boot camp, resentment would be the more likely response.

that terrorist attack, and also the supposedly urgent official reasons given for the war. For the men and women actually heading into battle, revenge was a significant motivating factor. Nathaniel Flick for one recalled that

> a feeling of profound gratitude that I was in a position to get revenge for 9/11 surged through me. Its intensity was startling. It wasn't just a professional interest in finally doing what I'd been trained so long to do. It was personal. I wanted to find the people who had planned the attack on America and put their heads on stakes.[132]

We suggest that this urge for revenge is part of the bundle of inherited impulses which facilitate team aggression. Whether it is carried out in a hot rage, or as part of a cold, calculating plan, revenge not only provides motivation to press the attack—it also helps justify the absolute destruction of the enemy, rather than mere domination. Revenge is one of the keys to breaking through the taboo against killing members of our own species. The wartime examples, even from highly disciplined, professional soldiers, are as numerous as they are chilling.

In 1935 a Gurkha regiment was fighting for the British near the Khyber Pass in what is now Pakistan. A British officer was wounded and captured by the enemy, members of an important tribal group called the Pathans. The following morning his corpse was found, flogged mercilessly and desecrated, with his testicles stuffed into his mouth and his skin pegged out on a rock. When a Gurkha captured a wounded tribesman a few days later, the British commanding officer went into a rage, had the prisoner staked out on the ground and ordered his men to boot him in the testicles until he died. His mutilated body was then placed on the rock where the British officer's corpse had been deposited for his fellow tribesmen to find.[133] During the World War II London Blitz, the commanding officer of a Royal Air Force bomber airfield pinned a photo of the London bombing on the wall and with a Biblical turn of phrase exhorted his crews to "Go and do likewise to the Hun." When William "Bull" Halsey saw the destruction at Pearl Harbor in December 1941, the Admiral muttered, "By the time we're through with them the Japanese language will be spoken only in hell."[134] Halsey captained the carrier group that launched B-25 bombers, which took part in the Doolittle raid on Tokyo in April 1942. Fifty Japanese civilians were killed, and to avenge their deaths the Japanese army in China slaughtered tens of thousands of Chinese civilians—revenge gone twice wrong, resulting in the senseless deaths of people who had nothing to do with either of the attacks. The day before D-Day, Sergeant "Wild" Bill Guarnere learned that his brother had been killed in the Italian campaign. "I swore when I got to Normandy there ain't no German going to be alive," he recalled. "I was like a maniac. When they sent me to France, they turned a killer loose, a wild man." Once in France, Guarnere captured some Germans, but while they were being held, a German machine gun opened fire and the prisoners tried to run. Guarnere shot them: "No pity, it was as easy as squashing a bug."[135]

Child Soldiers

Once, while crossing from Israel into Gaza in 1990, the Palestinian doctors accompanying me were turned back by the Israeli guards. My colleagues were embarrassed, but there was no violence. However, the crossing, when not barricaded, has been the scene of almost daily conflicts for years and many tens of teenagers have been killed at this one border crossing.[136] The Palestinians call those who belong to groups such as Islamic Jihad, Hamas (the Islamic Resistance Movement), and the Al-Aqsa Martyrs' Brigade "The Boys." While it may not be literally true in all cases, the nickname sums up neatly the age and sex of those who across the ages have been willing to risk death during aggressive raids.

For boys, the rise in sex hormones at puberty replaces the innocence of childhood with the sexual drive, competitiveness, and risk-taking characteristic of early adult life. Their experience is limited, the drives are strong, and the rewards can seem great and death remote. Young men, in short, seem almost designed for dangerous raids. This phase can pass innocently enough, but given the right—or really, wrong—conditions, the volatility, aggression, and ego of young male adulthood can be shaped and manipulated into an extremely deadly force. In recent years, this vulnerability has been exploited time and again in the most horrifying ways imaginable.

I spent a little time in Liberia, in West Africa, during the 1990s. While there, I met boys as young as twelve who were veterans of that country's recently ended First Civil War. "Child soldiers" is the term, but of course they were all male,* strong-armed or coaxed into taking up weapons they

---

* This should hardly be taken to mean that girls and young women were unaffected by the violence. Liberia is yet another instance where violence against civilians, rape, and sexual slavery have been part and parcel of warfare.

sometimes could barely hold and turned against the population in a frenzy of violence and senseless destruction. One humanitarian organization was struggling to get these same boys back into school and, I suppose in some way, back to being boys. It was uphill work—for many boys, staring at a blackboard doesn't come naturally under the best of conditions. How much less appealing it must be when compared to the thrill and camaraderie of working with a group of friends to blow things up—to say nothing of wielding absolute power over people's lives and deaths compared to obeying a teacher. The psychological trauma inflicted on these underage warriors is dramatically revealed in *A Long Way Gone*, Ishmael Beah's first-person account of life as a boy soldier in Sierra Leone during the early 1990s. It is chilling to think of the 300,000 in Africa alone who have been pressed into service as soldiers, but who, unlike Beah, do not manage to rebuild their life.[137] Perhaps evolution can explain why these boys make such effective warriors and terrorists, but it's hard to think of a word other than evil to describe the adults who exploit them.

The military indoctrination of young boys does not always involve actual fighting, but the dynamics are the same. The Nazis, much like the Spartans, involved young boys in military-style training from an early age. By 1936, 90 percent of ten-year-old German children were in the Hitler Youth and a few years later 900,000 took part in a single Nuremberg Rally. It was a seductive program: Boys who had never had a holiday went camping, and there were games and singing. "We are the Future; In front of us the Sun is shining" was one marching song. The natural tendency of youth to rebel against their parents was actively encouraged. "What mattered," said one Hitler Youth member sixty years later, "was the company of other boys; we were together and that's what counted."[138] The Hitler Youth was not a new invention, it merely copied and adopted what the boy's brigade, religious societies, and Scouts had done before, but on a vastly larger scale and with increasing brutality. Obedience was drummed into the children, cowards were ostracized, there were tests of strength and rough games, and older boys were given authority—predictably abused, often enough—over younger boys.

If at first the Hitler Youth were not marched onto the battlefield, their contributions to Hitler's war effort increased as the Führer became more desperate. First, the boys were drafted as firefighters, then as antiaircraft gunners. In 1944, the High Command began to recruit seventeen-year-old boys for the defense of Normandy. They fought a furious battle, and of the

20,000 adolescents who went to Hitler's Western Wall, only 5,000 survived. In 1945, as Germany was collapsing, the official recruiting age was dropped to sixteen, and on the Eastern front, boys as young as twelve entered the lines. Endowed with emotional and psychological predispositions for war and shaped by years of drilling and propaganda, the Hitler Youth fought a stubborn and effective battle against a Russian army of veteran soldiers. The boys believed their superiors' promise of "final victory," even when more mature minds recognized the inevitable defeat. A testament as much to the courage and loyalty of youth as to its painful naïveté, the life expectancy of the young recruits was one month. While grown soldiers of the Wehrmacht wept to have children next to them in the trenches, some boys remained enthusiastic for the fight and dreamed of dying as heroes. A few even committed suicide rather than give up the Führer's last battle.

Whether we look at Sparta two and half millennia ago or the Hitler Youth that sprang from an (initially) democratic, pluralistic, industrial society, we see that practically all young men have in them the predisposition to form bonds with their peers that are so strong that they are willing to risk their lives for one another. And the evidence shows time and again that it often doesn't take much more than permission, or a little encouragement, for them to kill and maim those they believe to be their enemies. It is an ugly picture of male adolescence, perhaps especially for those of us who have been boys, or have raised them. And it is difficult to think of an adequate explanation other than that we are looking at an inherited tendency that once upon a time served the reproductive advantage of those that manifested it. That it so clearly no longer does so is unfortunate, but evolution is not only remorselessly amoral; it is also not nearly as efficient as we might like in pruning branches that come to bear toxic, destructive fruit.

The Importance of Population Size and Structure

Young men are consistently more volatile and potentially more aggressive than older men. It is no wonder then that so many societies have traditionally kept a tight rein on their young males, both to feed their military machines and to keep potential chaos in check. The energy, drive, and sense of invincibility that so often accompany a young man's transition into maturity hit at just the time that young chimpanzees are struggling to establish themselves in the troop's hierarchy, and the human experience was probably not much different for much of our history. But if evolution's competitive turbo boost gave individual young men a comparative leg up early in human history, in today's more complex social systems the volatility of young manhood can have significant negative impacts not just for the individual, but for society as a whole. So far, we've focused mainly on how the particular nature of young men plays a role in their behaviors and predispositions. But it's important to look also at how those young men affect the broader community, and how society can in turn moderate the behavior and influence of youth.

In 1979, twenty-four-year-old Muhammad Ibrahim Asgharzadeh was a bearded radical student* who climbed over the railings of the U.S. Embassy in Tehran. He helped precipitate the Iran hostage crisis that consumed the U.S. and ensured that President Jimmy Carter lost his bid for reelection. By the time Asgharzadeh was thirty-seven, however, the fundamentalist student had become an advocate for religious liberty, and was sentenced to solitary confinement in prison by the Guardian Council of Iran, which his youthful fervor had helped bring to power. Within another decade, Asgharzadeh was leading a reformist political party, advocating closer relations with the U.S., and favoring Western-style suits. Reflecting on his youth in 2002, he said, "I made a mistake climbing over that wall once."

One man's tale doesn't necessarily tell a universal story, but Asgharzadeh's arc, while extreme in its particulars, will likely resonate with many men his age or greater. As a young student myself in 1956, I was all for the British attack on Egypt after President Nasser nationalized the Suez Canal to finance the building of the High Aswan Dam. Half a century older, I see this same invasion as extremely ill advised and counterproductive, as well as involving needless deaths on both sides. The transition that we make with age reflects not only our growing experience and shifting philosophies, but also a changing willingness to engage in or condone violence. Young men are the revolutionaries, the superstar computer programmers, the best athletes, the most courageous soldiers, the bravest mountaineers, and the most creative musicians, but they are also the most vicious gang members and nearly all the suicidal terrorists.

The impact of youth is not just a story of individual life changes, how-

---

* From the widespread riots and demonstrations of the 1960s to the Algerian War (1954–1962), the massive 1980s student protests in South Korea, and certainly the Iranian hostage crisis, it is interesting to note how frequently youthful students have been at the heart of the most chaotic events of the last fifty years—even the word "taliban" means student. While it is possible that exposure to new ideas may contribute to this phenomenon, we suspect that the combination of evolutionary predispositions and a permissive, highly social community of youths is all that it takes.

ever. Relatively minor shifts in demographic trends can actually precipitate major political events and help to shape the course of history for
entire communities and nations. When a society has a high birth rate
and a somewhat lower death rate, as occurred in the nineteenth century
in the West and in twentieth-century developing countries, the population structure becomes skewed toward youthfulness. Imagine dividing
a population into groups by age, as demographers do. If the population
is growing, there will be more newborns than five-year-olds, more five-
year-olds than teenagers, and so on. If each age group were represented
by size in a diagram it would take on the appearance of a pyramid, with
relatively few old people at the top and successively more individuals as
the age groups become younger and younger. In Europe, North America,
Japan, Korea, and other modern societies, where infant deaths have fall-
en to low levels and women have access to effective contraception and
safe abortion, birth rates are low and the population structure looks more
like a lean skyscraper, with more balance between youth and age. Importantly, in quickly growing populations with pyramidal population structure, there are generally more young men aged fifteen to twenty-nine than
there are men thirty years old and older. In more stable populations, the
number of men over thirty exceeds the number of younger males.

It turns out that population age structure can tell us a lot about a country's political stability. At a 1993 conference organized by the U.S. Central
Intelligence Agency, Gary Fuller of the University of Hawaii presented
a paper entitled *The Demographic Backdrop of Ethnic Conflict*. In it, he
showed that episodes of civil strife in Sri Lanka occurred at times when
20 percent or more of an ethnic group's population was aged fifteen to
twenty-four.[139] At York University in Toronto, Christian Mesquida and
Neil Wiener used an index of civil violence around the world in the late
twentieth century and found a sharp rise when the ratio of younger men
aged fifteen to twenty-nine years old equaled or exceeded the number
of older men in the population.[140] They also looked at the historical pattern of warfare in North American tribes. The Mohave Indian population
grew from 3,000 in 1770 to 4,000 in 1872, during precisely the time that
raids on their neighbors became most frequent. In the twenty years before World War II, the population of Germany expanded by 35 percent,
while those of Britain and France grew by only 2 percent. Youths formed
the killing squads during the 1994 genocide in Rwanda, at a time when
the country had one of the highest population growth rates in the world.

In Liberia, which has been torn apart by civil war, half the population is under seventeen. In the U.S., a country with a younger population than most Western nations, the median age is thirty-five. In countries such as France, Finland, and Britain, with stable populations and generally non-violent politics, there are more than twice as many men over age thirty as in the fifteen to twenty-nine age group. With the exception of Bosnia in 1994–1995, every time the U.S. military has intervened on the ground since 1990, it has been in a country where the average woman has four or more children—in other words, political instability and violence often follow hard on the heels of high birthrates.[141]

In 2008 Michael Hayden, the director of the U.S. Central Intelligence Agency, identified rapid population growth as the number one problem in national security.[142] Until recently, population growth and age structure have been major driving forces behind violence, insurrection, and war.[143] They are not the only cause, of course. As Richard Cincotta and Robert Engelman point out in their 2003 monograph, *The Security Demographic: Population and Civil Conflict after the Cold War*, "Demographic factors do not act alone, [but interact] with non-demographic factors, such as historic ethnic tensions, unresponsive governance and ineffective institutions." A high ratio of young males can be conceived of as a national "risk factor" for violence, just as smoking is a risk factor for lung cancer—not all people who smoke die of cancer, but many do, and not all nations with a high ratio of younger to older men start wars or spawn terrorists, but many do. Or, to use a different metaphor, a large number of potentially warring young men in a society might be likened to the uranium in a nuclear reactor, while women and older men are like the graphite rods that are pulled in and out of an atomic pile to slow the rate of the reaction. If the rods are too few, then political and social meltdown can occur. If there are sufficient rods, then the reactions are stabilized and the pile will produce the energy we need without exploding. Without women and older men, a critical mass of youthful male energy can erupt into conflagration. In early twentieth-century China, population growth was extremely rapid, and the ratio of young to older men peaked at exactly the time Mao Zedong launched his Cultural Revolution (1966–1976). Using statistical tests, Mesquite and Weiner concluded that the age structure of a population accounts for about one third of the variables associated with the likelihood of warfare. In social science research terms, this is a strong correlation to be taken seriously.

There is another important factor. The ratio of young men in a population varies not only with the birth rate and the population structure, but also with changes in the sex ratio. The sex ratio at birth is tilted natural-ly toward boys (about 105, to every one hundred girls) but because little boys have a slightly higher mortality rate than little girls, by puberty the natural ratio is almost one to one. Humanity has found ways to tweak that balance, however. In nineteenth-century China, female infanticide was a widespread response to abject poverty. A poor family could not always af-ford to invest in a girl, who would marry into another family soon after puberty and not return any benefits to her own family. In Huai-pei prov-ince, in Northern China, so many girl babies were killed that the sex ratio rose to 129 men for every one hundred women. Up to a quarter of all men never married, and these "bare branches," as the Chinese called them, "had nothing to lose except their reputations for violence." They formed teams of bandits and thieves, and by the 1880s the largest group—the Nien Jun—controlled an area containing six million people and posed a serious threat to the Qing dynasty.[144] Valerie Hudson of Brigham Young University in Utah and Andrea Boer of the University of Kent argue that such distorted sex ratios produce authoritarian forms of governments, as elites respond to the threat of angry, violence-prone young men.[145, 146]

It is a sound generalization, although as usual there are exceptions. In nineteenth-century America, migration altered the population structure as men moved westward in search of new lands or to join the gold rush. By 1880, in Dodge City, Kansas, there were 124 men for every one hundred women and there was a great deal of violence. In contrast, in some other frontier societies with equally skewed sex ratios, such as New Zealand in the nineteenth century, women worked beside men with some degree of equality and a good deal of peace. In 1893, New Zealand became the first democracy to grant women the vote. In nineteenth-century Britain, with large numbers of men emigrating to the colonies, most women stayed at home, pushing the sex ratio in the opposite direction. The 1851 census found, in the words of Victorian commentators, that there were 500,000 "superfluous women." The statistic captured public attention, and for the first time in British history, hundreds of thousands of women were no lon-ger swept up in continuous childbearing and childrearing. Adult women without children helped influence legislation relating to women's prop-erty, the search for companionable marriage based on love, the suffragette movement, and the admission of women to universities. The word "femi-nist" was not coined until 1895, but the search for female autonomy and rights began decades earlier, and it was partly driven by the differential migration of young men in search of opportunities overseas.[147]

In parts of contemporary Asia, the ratio of men to women has been changed by infanticide, neglect of the girl child, and sex-selection abortion.* In India in 1993–95, the sex ratio at birth reached 113.8 males to one hundred females, and it has been calculated that there may have been over 16 million more men aged fifteen to thirty-five than women in that age group by 2006. The sex ratio in Hubei province, China, is reported to be as high 130 males to one hundred females (there may also be some underreporting of female births) and the 2000 census there found about twenty million more men aged fifteen to thirty-four than women.[148] Sex selection and infanticide are illegal in China and India, but the laws are difficult to enforce and many young men are already searching in vain for wives. Some older men will marry younger women, but that can actually exacerbate a sex ratio imbalance.** Perhaps others will seek brides from other countries. The skewed sex ratios are already giving rise to social problems. Thankfully, it is difficult to think that a contemporary Chinese leader, however ruthless he might be, would be able to incite the violence, havoc, and chaos of the Cultural Revolution. China's One Child policy was designed to bring population growth under control. But it has also had the consequence of shifting China's age structure from a youth-dominated pyramid to an age-balanced pillar. Even with sex selection, the ratio of young men to old has dropped, and though China is going through momentous economic and social changes, the country remains remarkably stable given its twentieth-century history.*** Given China's cur-

---

* A male fetus can be identified using relatively low cost ultrasound machines, which became available in the 1990s. A more expensive sex test, based on cultured cells from the amniotic cavity surrounding the fetus, can be performed much earlier. Unscrupulous abortion providers sometimes charge a woman for a test, but never complete it. They then abort the fetuses of all the women who ask for selection, regardless of the sex of the child. They keep a female fetus from a previous operation to show women who were in fact carrying a male child.

** South Korea also has a skewed sex ratio, but the major factor leading to fewer women is not sex-selection abortion (although it has occurred) but the fact that the birth rate has fallen rapidly, and men traditionally marry women about five years their junior. The decreasing size of each birth cohort leads to an increasing deficiency of women at the time of marriage.

*** It can't be stated too strongly that population structure is just one factor in social stability. An authoritarian government also obviously has an impact, and economic opportunity is another crucial factor, in China and elsewhere. In the 1970s, Ireland and Palestine seemed to be on parallel paths, in terms of terrorism. Today, the violence in Israel and the occupied Palestinian territories continues to escalate, while terrorism has all but disappeared from Northern Ireland. A political agreement was reached between warring Republicans and Loyalists, but the role of the "Celtic Tiger" boom economy can't be discounted. Simply put, people who feel they have something to live for are much less likely to risk their lives in terrorism or war.

rent population structure, with a much lower ratio of young people than existed at the time of the Cultural Revolution, a case could be made that there simply are not enough angry young men to generate the type of un-thinking political adherence that made Mao's long dominance possible.

Whatever we think of the morality of abortion, it can have two quite distinct societal impacts when it is widely available. When used as a method of sex selection, abortion may increase the possibility of conflict by increasing the ratio of young men to women in the society. But when abortion offers a woman the option of not having a child of either sex, the opposite may occur. In 1999, John Donohue of Stanford Law School and Stephen D. Levitt from the University of Chicago startled the world of criminology by pointing to a dramatic decline in criminal offenses in the U.S. at precisely the time the first generation of children born to moth-ers who enjoyed the choice of safe abortions reached eighteen to twenty-four—the age associated with maximum criminal activity as well as the likelihood of warfare. They suggested that without access to safe abor-tion following the Supreme Court ruling in *Roe v. Wade* in January 1973, the crime rate would have been 10 to 20 percent higher.[149] A weakness of this argument is that in some locations, such as New York,[150] there were many abortions before *Roe v. Wade*—only those were done in dangerous, exploitive ways. Still, there may well be a kernel of truth in Donohue and Levitt's observation.

In January 1965, at the height of the Cold War, I visited Eastern Europe on a mission to find out all I could about family planning there. In Po-land, Hungary, and Czechoslovakia, I suspect that I was the only Western visitor my hosts had seen in many years, and the welcome I received was as warm as the air outside was cold. In Prague, I came across a study of 220 children born to women who had been refused an abortion, and an-other 220 carefully matched children from the same communities whose mothers had not sought an abortion.* An American colleague, Henry Da-vid, followed up on this contact and over the next thirty-five years it grew not only into a fruitful example of East/West cooperation, but into one of the most remarkable studies of human development. Over the decades, research showed that children born to women who were refused an abor-

---

* The Czechs were virulently anticommunist, and their hatred of Moscow would boil over a couple of years later during the Prague Spring. Bright young people at the time I visited had few options— they could not become entrepreneurs, and did not want to join the centrally controlled media, so many became doctors. The medicine and science I found there were of the highest quality, and not contaminated by politics.

tion felt less positive about themselves, were rejected by friends more often, and were less likely to perceive themselves as happy. By their early
twenties, the "unwanted" children had more problems with alcohol and
were twice as likely to wind up in prison.[151] Such studies are about averages, and some individuals whose mothers had sought an abortion did
better than their peers. But overall, being "unwanted" had deleterious,
seemingly life-long effects, including a greater chance of being convicted
of a crime. Perhaps Donohue's and Levitt's controversial findings are indeed telling us something important.

Animal Lessons

One more line of evidence supporting the importance of population
and age structure in shaping the stability or volatility of human societies comes from animal studies. Older male elephants experience periodic surges of testosterone called musth, when they dominate all other
males and aggressively seek mating opportunities. In South Africa, orphaned male elephants from the Kruger game park were transported to
another, distant game park. Lacking older, dominant males in the herd,
they entered into musth at an unnaturally early age. They went on destructive rampages and gored to death forty rare white rhinos. Proof of
the stabilizing role of older males came when six mature bull elephants
were introduced to the herd of seventeen younger males, and the destructive attacks stopped.[152] The Monterey Aquarium in California has
a popular program for rehabilitating orphaned sea otters. When one
young male was returned to the wild after being hand-reared in captiv-
ity, he was seen to attack seal pups and force them under the water until
they drowned. Sea otters had never been observed to kill other mammals before these well-intentioned attempts to nurture orphaned pups,
nor had elephants been observed to kill other species until their natural
population structure was altered by human culling, poaching, and the
creation of national parks. It appears that in other social species as well
as in human beings, older individuals provide an important counterbalance to the aggressive instincts of young males.[153]

Even more striking are the observations of Robert Sapolsky, who has
studied savanna baboons for over twenty-five years. Baboons live in
troops as large as fifty to one hundred animals. Social life is noisy and often violent: The males regularly fight each other and they frequently ha-

rass the females. In one troop of olive baboons that Sapolsky was studying in Kenya in 1983, the risk-taking males of one troop learned to forage from the garbage of a tourist hotel, and became infected with bovine tuberculosis as a result. Half the adult males died, leaving the adult females in the majority—and the social life of the troop was transformed to one of considerable harmony.[154] Baboon society, unlike chimpanzee society, is built around related females, with the young males migrating and fighting their way into a new troop and dominance hierarchy. Interestingly, once the more peaceful culture the females established was in place, it persisted into the next generation, even as new young males (who in normal baboon society would have been much more aggressive) entered the troop.

A Universal Emotion

Combat, particularly prolonged fighting, belongs to a world of experience that is literally unknowable, except for those who have survived it. Yet the evidence demonstrates that at its core is a set of deep emotions, identical on both sides of the firing line and consistent across the pages of history. A shadow of that behavior seeps through to the rest of society, and even as a child during World War II I sensed a little of it. The scientist C. P. Snow captured the mood in 1940s England from an adult perspective: "Oddly enough, most of us were very happy in these days. There was a kind of collective euphoria over the whole country."

The loyalty, courage, camaraderie, fear, and elation of men in battle have been described a thousand times in books, plays, and private letters, and told to friends, colleagues, wives, and girlfriends with varying degrees of eloquence and poignancy. Given the necessary set of external circumstances, it seems that young men will almost always respond in the same way. It is a behavior that is so universal, and in one form or another so familiar, that we rarely step back and ask how it might have arisen. Yet the proverbial man from Mars—or woman from Venus—would surely find it odd that the members of one sex should behave in this way, hazarding their own lives and inflicting so much pain on others. (Women are increasingly involved in formal warfare, of course, but as we shall see, their motivations and behavior are often quite different from those of their male comrades.) But if, like *Pan troglodytes*, young male *Homo sapiens* also have an inherited predisposition to team up with kin—or perceived kin—and try to kill their neighbors, then the behavior of the Greek hoplites at Marathon in 490 B.C., a Russian or German tank crew at the Battle of Kursk in July 1943, the men on Omaha Beach on June 6, 1944, and Vietnamese troops in the Tet Offensive of January 1968 becomes explicable.

If men, and male behaviors, are at the core of the problem of war and terrorism, they are also an essential part of any solution. We can't emphasize too strongly that while many of our behaviors have evolutionary roots, biology is not destiny. As humans, we have many ways—as individuals, as societies, and as a global community—to shape our actions and change the outcomes. Biology merely provides the starting point and the raw materials for human society. Culture, community, and free will are what determine its ultimate shape. In the chapters that follow, we'll start to see more clearly how understanding the evolutionary underpinnings and biological reality of war and terrorism can help us control and limit the impact they have on modern life.

5

TERRORISTS

*Terrorism: the systematic employment of violence and intimidation to co-
erce a government or community.*

War: the employment of armed force against a foreign power.

—Oxford English Dictionary, 1993

T HE TERRORIST ATTACKS OF 9/11 killed more than
2,900 people. They were innocent office workers,
airline passengers and crew, brave police officers and
fire-fighters, and military and civilian workers in the
Pentagon. There is something especially heinous about
unannounced attacks on innocent bystanders, and though we have
to assume that a marauding chimpanzee would disagree, no cause
and no perspective, evolutionary or otherwise, can excuse or justify
murderous attacks on civilians. But if we are to confront terrorism
in the long term, then we need to step back and ask—have terrorists
always been with us, or are they an unusual product of recent history? Are terrorists unique, or are they part of the spectrum of team
aggression in humans? And indeed, is a "war on terror" the most appropriate response, or is there some other more fruitful way of making terrorist attacks less likely?

What Is a Terrorist?

Just because war and raiding are older than civilization, this does not mean the basic behaviors and the impulses that drive them cannot take on new expressions as new conditions emerge. In today's world, a particular kind of team aggression—built very much upon ingroup identity, the desire for revenge, and a lack of empathy for the outgroup—has taken on a profile out of all proportion to the number of warriors involved. And that outsized impact, of course, is the very point of terrorism. Raiding chimpanzees—and invading superpowers—only attack when they outnumber their rivals, overwhelming them with mighty force in order to limit the attackers' own risk of death or injury. Terrorists stand this standard doctrine of team aggression on its head.

With the time-tested approach of overwhelming force denied them, terrorists have discovered new opportunities for attack in a modern, interdependent, technologically vulnerable society. Today's terrorist is more effective because he has plastic explosives and low-cost, cell phone–controlled devices to detonate them remotely, along with an enticing array of easy-to-attack targets in the vulnerable underbelly of modern life, from computer codes to shopping malls. Sometimes, as on September 11, 2001, the modern world even provides ready-made weapons in the form of large, jet-propelled civil airliners filled with explosive fuel. All that is needed is the ability to shut off empathy for the attacked, and as we have seen, that appears to be an ability we all share.

Ever since Classical times, team aggression has been organized, supplied, and controlled by the state. Terrorists are different. We should think of terrorists as independent squads, or a few independent squads loosely joined together, rather than being organized into a formal army. They are small, self-motivated bands—sometimes no more than a few tens or hundreds—and often not aligned with any state organization. When the attackers and the victims speak a different language, or worship a different god, then the hatred and fear are ratcheted up to a higher level.

Some aspects of "terrorism," broadly defined, have deep roots. The Assassins* were a radical sect of Islam active from the eighth to the fourteenth centuries. Formed from small bands of intensely religious young men, bound together with initiation ceremonies and elaborate symbols,

---

* The Assassins are said to have derived their name from *hasishin*—hashish, or marijuana—which seems to have been used to drug young recruits.

For female chimpanzees, the most practical strategy for survival is usually to join the troop of males that is most successful at killing its neighbors and expanding its territory. Human females through history have been confronted with similar choices. This sixteen-year-old Colombian girl has willingly joined a cell of the rebel group FARC (Armed Revolutionary Forces of Colombia) and sits outside the barracks with her pet parakeet, waiting for her soldier boyfriend.

they murdered political and religious leaders. As one poet put it, "a single warrior on foot" can leave a king "stricken with terror, though he own a hundred thousand horsemen."[155] At the beginning of the twentieth century, terrorists (called anarchists in those days) caused 2,691 deaths in Tsarist Russia.[156] At the end of the twentieth century, the Basque separatist group ETA (*Euskadi Ta Askatasuna*) in Spain killed 822 people in forty years of fighting, while the IRA (Irish Republican Army) killed over 1,800 people over the same interval. And just like their assassin forebears, terrorists have often targeted individual leaders. The assassination of Austrian Archduke Franz Ferdinand in Sarajevo in June 1914 sparked the worldwide catastrophe of World War I. In 1984, Indian Prime Minister Indira Gandhi was killed and in 1991 her son, former Indian Prime Minister Rajiv Gandhi, was also assassinated. Sri Lankan President Ranasinghe Premadasa was killed in 1993 and Benazir Bhutto in 2008.

Assassins and terrorists are different from raiders and soldiers, in that they care little or not at all for their own survival, and different again from highwaymen, brigands, and sociopaths, because terrorists perceive themselves serving a larger cause. Terrorists, after all, destroy themselves and the innocent not out of the desire for tangible goods, but because they think they are selflessly serving their people, their society, or their god. But we will understand terrorists best if we begin by recognizing that their basic emotional drive is that of any warrior; the suicide bomber taps into the same ancient predisposition that kept the hoplite fighting against impossible odds.

Modern terror campaigns are most usefully seen as yet another aspect of team aggression, only enacted on a much smaller scale than organized warfare. Iraqi insurgents, Tamil Tigers, Chechen rebels, Kashmiri fighters, the IRA in Ireland, ETA in Spain, Hezbollah in Lebanon, Hamas in the Palestinian Territories, and Al Qaeda in various parts of the world— no matter what *we* might think of these groups, *they* perceive themselves as warriors waging honorable campaigns of defense or liberation for a particular parcel of land or set of ideas. This is hardly a recent development in the world, but what is new about modern terrorism is that the fighters—usually relatively few in number—find themselves opposed by advanced armies, equipped with machine guns, tanks, and helicopters, which they know they can never defeat head on.

Evolutionary attributes run deep, and human ingenuity is another pervasive gift of our biological past. As a result terrorist acts are developing in new ways. The early twentieth-century terrorists lacked ways of sinking ships and there were no airplanes to highjack; today, new vulnerabilities and more powerful explosives have lead to new forms of attack. The 1980 bombing of Pan Am Flight 101 over Lockerbie, Scotland, killed 170 people; the 1996 Khobar Towers attack in Saudi Arabia killed nineteen American troops; the 1998 bombing of U.S. embassies in Kenya and Tanzania killed 224 people, and the 2000 attack in Yemen on the warship *U.S.S. Cole*, killed seventeen sailors. Then, on 9/11, nineteen terrorists brought down the World Trade Center in New York and rattled America to her foundations, triggering wars which according to the Congressional Budget Office will cost over one thousand million dollars.[157] Nineteen men with swords or bows and arrows could never have brought the Roman Empire to a halt, even for a moment, because there were neither the vulnerabilities intrinsic to modern living, nor easy access to large-scale explosive power.

A tiny cell of zealous enemies can find innumerable ways to attack the modern, interrelated world of tall buildings, transport terminals, water reservoirs, fuel storage tanks, ground and satellite communications, cruise ships and oil tankers, sports events and crowded arenas, highways, deep tunnels and long bridges. When terrorists found it more difficult to highjack airplanes they switched to attacking commuter trains, as in Madrid in 2004, when 191 died and 1,500 were injured, or to blowing up buses, as in London in July 2005, when 55 died and 700 were injured. Revealingly, the 2005 London attacks were not carried out by Al Qaeda but by British-born Muslims who had become radicalized partly because they heard of other terrorist attacks. The more complex modern society becomes, the more inescapably vulnerable it is to attacks by tiny, disaffected minorities. Tens of millions of foreign nationals cross sovereign frontiers every year, hundreds of millions of tons of cargo pass through the world's great ports every month, and billions of overseas phone calls are made every week. Europe's long Mediterranean shores facing Africa and the Middle East cannot be made impregnable, and any sufficiently persistent person can walk into the United States from Canada or Mexico.

As we will see in chapter 9, much of the world is awash with easy-to-obtain weapons from previous wars. In addition, contemporary science and technology continues to spin off explosives, lethal poisons, and complex devices capable of disrupting modern society. Who would have guessed when the Internet was invented that controlling unwanted "spam" would become a huge industry, or that clever young men (it always seems to be young males) would delight in designing software viruses to block computer networks?

Terror Emerges

It is a profound reflection of the power of ingroup identification that the use of the word "terrorist" usually depends on whose side you take. One person's "terrorist" is another's "freedom fighter," and the difference does not always come down to whether or not civilians and non-combatants are attacked. During World War II, members of the French Resistance were heroes to the Allies, but terrorists to the Germans and their Vichy collaborators. One of the first uses of the noun "terrorist" was to describe the Jewish attack on the King David Hotel in Jerusalem in 1946, when the British were still administering Palestine. Like numerous Palestinian attacks on Jewish targets in more recent decades, that raid by Jewish terrorists—or freedom fighters if you prefer—was made by young men who felt justice was on their side.* The Palestinian suicide bomber who straps himself with explosives and detonates them in a crowded café is a terrorist in Israel, but a hero in the Gaza Strip. Israelis describe a Palestinian terrorist as a "genocidal bomber" from the "disputed territories," while to a Palestinian he is a "warrior" from "the occupied territories" undertaking "a heroic martyrdom operation against the Zionist entity." The Israelis call their military actions in Gaza "pre-emptive strikes" or "pinpoint preventive operations," while the Palestinians call them "assassinations." Ingroups and outgroups think they are so different, but their vocabularies are actually mirror images of one another.

Terrorists may act alone, but they always emerge from a tightly knit group with whom they have been in close contact for long intervals. (The lone attacker, like the Unabomber Ted Kaczynski, borrows terrorism's techniques, but his attacks likely have more to do with personal pathology than with shared evolutionary predispositions.) Most terrorists have the full support of their ingroup—of fellow terrorists certainly, but also quite often of their families, their communities, and even their ethnic, national, or religious groups as a whole. In the case of September 11, two of the nineteen hijackers were brothers, exactly as individuals might be in a chimpanzee raid; many of the others lived with each other or trained in military-style camps in Afghanistan. The adolescents who throw stones at Israeli troops in Gaza, or who until recently harassed British troops in Belfast, grew up together playing in those same streets.

Those implementing 9/11 belonged to a network of several hundred, or perhaps a few thousand, like-minded men spread across several continents. In Palestine, Northern Ireland, or the Basque country of Spain, terrorist organizations have often made "declarations of war," but terror campaigns are closer in nature to tribal raids than to conventional war. One could almost argue that the various communiqués warning of a resumption of armed attacks, while the attacks themselves are made without warning, quite closely parallel male chimpanzees "displaying" their hostility on a regular basis, and then attacking only when the advantage is very clearly on their side. It's just that, given modern police forces and well-armed soldiers, the ancient raiding tactic of a stealthy attack by an overwhelming force no longer works.

---

* A radical Jewish group led by Menachem Begin (later prime minister of Israel) that was seeking to end the British administration blew up the administration's headquarters, killing 91 people.

Suicidal Killings

Robert Greenblatt was an obstetrician-gynecologist who often helped me when I began to work internationally in the 1970s. He served in the U.S. Navy in World War II and was one of the first physicians into Hiroshima after the Japanese surrender. But the most vivid story I remember him telling was from a time when his ship was under attack from *kamikaze* suicide planes.* An excellent raconteur, Bob described how he was transfixed watching the Japanese planes dive on the U.S. ships through a sky that was filled with defensive anti-aircraft fire.

In recent decades, some terrorist groups have become increasingly dependent on suicide bombers. Still, however much we may hate what terrorists and suicide bombers do, we will never understand their behavior or be able to defend ourselves against it so long as we label them "cowardly murderers." In a sense all war is murder—in that it involves killing members of our own species—and the reality is that it takes courage and a cool head to strap explosives around your waist, conceal a machine gun, drive a truck loaded with explosives, or learn to fly knowing you will die as a result. Suicide was an essential part of the September 11 attack, even though Osama bin Laden claimed most of the nineteen men involved did not know the suicidal nature of their work until they boarded the planes.

Japan's first *kamikaze* attacks were organized by Vice-Admiral Takijirō Ōnishi, who was tasked with defending the Philippines against American invasion in October 1944. The Japanese leadership understood the hopelessness of their fight and Ōnishi set up what he called a "special attack unit," or *Tokubetsukōgeki*, usually shortened to *tokkō*, to sink American aircraft carriers and supply ships by crashing their planes loaded with bombs and fuel into the decks. Before war's end almost 4,000 *tokkō* pilots died, killing almost 5,000 Americans. Each pilot who "volunteered" was invited to step down in front of their comrades if they had second thoughts about dying. None did, and to some extent their story also shows how easy it is to manipulate the loyalties of young men in a time of war. "I don't even remember telling my feet to move," recalled one pilot. "It was like a strong gust of wind whooshed up from behind the ranks and blew everyone forward a step, almost in

---

* *Kamikaze* means "divine wind," referring to the hurricane which destroyed the fleet of the Mongol Emperor Kublai Khan as he attempted to invade Japan in 1274 A.D.

perfect unison."[158] All of Japanese boot camp was brutal and the training of *tokkō* pilots was sometimes especially so, yet a culture built around a divine emperor and the uniqueness of the Japanese race triggered an intense sense of comradeship. The *tokkō* were young, bewildered, lost, and vulnerable—one *kamikaze* pilot wrote his mother shortly before dying, "Mother, I still want to be loved and spoiled by you...I want to be held in your arms and sleep."[159] But they were also supremely brave, resolute, and dedicated to their fellows. When one *tokkō* unit thought they would run short of fuel and only some pilots would be able to fly to their deaths, Corporal Bandō said to his commander, "If my plane can't fly on the day of our mission, I want you to let me ride in the fuselage of your plane so you can take me along."[160]

Most of today's suicide bombers, like the *tokkō* pilots of the 1940s, are young males, although, as always, there are exceptions. Daoud Abu Sway, the father of eight children, was forty-seven years old when he detonated a suicide bomb in Jerusalem in December 2001. Raed Abdel-Hamed Masq, who killed twenty Israelis in a suicide attack in August 2003, was thirty and had two children. (As a child Masq had memorized the Koran. When he died his wife, who was five months pregnant, said, "I am so proud God allowed him to be a martyr.") Female suicide bombers have also been recruited recently, in part because women are less likely to arouse suspicions. In Chechnya, female suicide bombers took part in fifteen attacks between 1999 and 2004, and in 2002 women were among the forty terrorists who took 850 hostages in a Moscow theater, demanding the withdrawal of Russian forces from Chechnya. The high ratio of female terrorists from such a patriarchal society may reflect the brutality and hopelessness of life in Chechnya, but overall female suicide bombers remain a minority.

It is difficult at first to see how suicide attacks could represent a successful evolutionary strategy, because those carrying any mutation facilitating this type of behavior would likely kill themselves before reproducing. Unless an individual with genes coding for an increased ability to overcome the very powerful urge to survive lives long enough to pass them on to the next generation, those genes would not be transmitted—although the strong sexual allure of *tokkō* pilots suggests that in some cases at least, embracing death in this way might actually create new reproductive opportunities. Suicide can also be a partially altruistic behavior aiding the survival of the killer's kin. For example, the rapid population growth and poorly managed economies in many parts of the Islamic world prevent a large number of young men from acquiring enough wealth to marry and support a family. As a suicide bomber's relatives are often rewarded with material gifts, by governments including when he was in power that of Saddam Hussein, or by less official bodies, then perhaps in such a setting the fittest thing to do is to volunteer as a suicide bomber. You yourself die, but your family—and thus some version of your genetic endowment—has a greater chance of survival. But this is speculation, and frankly questionable—evolution is much more likely to provide the deep, unconscious impulses that underlie our behavior rather than the more pragmatic, real-world reasons. What suicide bombers clearly demonstrate, however, is the strength of the bond between individuals—almost always men—who belong to the same group, troop, platoon, raiding party, or terrorist cell.

Those who recruit suicide bombers intuitively know how to emphasize the basic biological underpinnings of kinship. In Palestine, Hamas and Islamic Jihad identify possible suicide bombers and then draw them into intense spiritual and political indoctrination. Recruits are often isolated from their own families, given uniforms or distinctive hairstyles or tattoos to encourage ingroup identification, and immersed in the emotions and linguistics of kinship. Boys and girls chosen as suicide martyrs in the Iran-Iraq war (1980–1988) were called the "Children of Imam" and Al Qaeda recruits called Osama bin Laden "Elder Brother." And separation from one's biological family need not mean their disapproval. The father of the bomber who killed twenty Israelis at a disco in Tel Aviv in 2002 said, "My son has fulfilled the Prophet's wishes. He has become a hero. Tell me. What more could a father ask?" Belief in the supernatural is another powerful route to overcoming the very natural desire to stay alive. Al-Tamimi, an Iraqi insurgent who claims to have handled over thirty suicide bombers in Iraq, says that his volunteers "want martyrdom immediately." He describes one Saudi suicide recruit in his early twenties who couldn't "stop smiling and laughing, even singing. He is sure he is going to paradise and he can't wait."*

---

* An eagerness to embrace suicidal death can be found in any society. Paul Hill, an American and one-time Presbyterian minister who perceived himself to be fighting a war against abortion, was executed in Florida in 2003 for shooting an abortion doctor. He told the public, "The sooner I am executed…the sooner I am going to heaven. I expect a great reward in heaven. I am looking forward to glory."

Gangs

Many Western citizens are unaware of the street gangs in their midst. Other than seeing their graffiti on buildings and bridges they may as well not exist. I was taught about gangs by my University of California students who had gone to school in downtown Los Angeles, where an adolescent boy could hardly avoid joining a local gang. Because street gangs are among us, at least in most large urban areas in the U.S., academics know more about the details of factors driving their violence than we do about Al Qaeda or Palestinian terrorists, but the two share many common features. Observing the one may help us better understand the other.

In the 1990s there were an estimated 80,000 gang members in the Los Angeles area, and over 500 gang killings a year, facilitated by easy access to cheap "Saturday night special" handguns and automatic weapons. Years of exposure to gang life can lead to the same acts of unspeakable viciousness terrorist organizations display. Sanyika Shakur, a.k.a. Monster Kody Scott, was a member of an L.A. gang and when a sister of his "homies" (fellow gang members) was raped and stabbed by rival gang members, Kody Scott's gang seized a member of the rival gang and chopped off both his arms at the elbow with machetes. One arm was taken, and one arm discarded down the street. "Later that night," wrote Kody, "we partied and had a good time. The arm was taken as proof of completion."[161]

Numerically, the L.A. gangs are more numerous than Al Qaeda. Indeed, the 18th Street Gang alone is thought to have 8,000 to 20,000 members.[162] Yet there are no U.S. troops in Humvees patrolling 18th Street in Los Angeles. Why is this?

On the spectrum of team aggression, street gangs are one step nearer to a preliterate raiding party than to terrorists and therefore they are perceived as less threatening—they only destroy themselves. Some aspects of gang behavior do parallel chimpanzee raids, but in an urban jungle. "Gang warfare" is driven by the two basic emotions that also fuel chimpanzee and preliterate warfare—territory and revenge, or just being from another neighborhood. Merely belonging to another gang is enough reason to justify an attack, as also happens with the chimpanzees of Gombe. Unlike terrorists who attack a perceived outgroup for ideological reasons, in the case of Al Qaeda by mounting attacks halfway across the world, gangs fight their neighbors for what they perceive as territory and resources.

Some gangs develop out of the loyalties built up in team games. The genocide that occurred in Rwanda in 1994 took shape when the Hutu president Juvenal Haryarimana and other leaders established gangs called *interahamwe*—literally, "those who attack together." They were a direct outgrowth of clubs of young football supporters; their leaders, like other young men, wore dark glasses, flamboyant clothes, and distinctive hairstyles, and rode motorcycles. In South Africa, Winnie Mandela, the wife of the then-imprisoned Nelson Mandela, had a vigilante gang called United Football Club, which killed at least one black person accused of being a traitor. Other gangs display more explicitly military attributes. The Hell's Angels, founded in 1948, sprang directly from war itself—the first members were pilots and gunners who had fought together in World War II. Some motorcycle gangs may deal in drugs, but many bikers can afford the most expensive Harley Davidson motorcycles through more legitimate means, and they too enjoy the camaraderie that is at the core of gang life. At regular intervals, tens of thousands of bikers gather from across the continent in cities such as Daytona Beach, Florida, and Sturgis, South Dakota, to parade, drink, joke, and show off their female partners.

Like trained troops graduating from boot camp, street gangs generally have initiation ceremonies. Initiation is often violent and may involve a severe beating administered by members of the gang the new recruit is joining. It is perceived as an expression of love and "the more you get beat the better." Unprovoked attacks on a rival gang, or even random civilians, are another common feature of initiation. Much of the time, as in warfare, street gangs may simply "hang out," waiting for something to happen. Like regiments, gangs take pride in their names, and they have "colors" and other identifying marks that act as uniforms. Tattooed symbols are almost universal, and brandings common. Some gangs use a secret language of hand signals, and identify their territory with graffiti. Members display healed gunshot wounds with pride. "I've been shot. I could walk around like Superman," said one youth.

Young people, perhaps especially young men in gangs, see killings but "don't worry about death." Attacks can be exciting, and even trying to escape the police can add to the "rush" the gang member enjoys. Success, according to one L.A. gang member, is "As many soldiers you can drop on the other side." Death, as always, is implausible and fascinates, and needs to be ritualized by the living. When a gang member is killed the gang contributes to an expensive funeral and turn out in force. In the autobiography Monster Kody Scott wrote while in prison, he describes how he was eleven-and-a-half at his initiation and how he shot a rival gang member the same night.[163] Gang life, his mentor tells him, is no casual commitment. "It's gettin' caught and not tellin'. Killin' and not caring, and dyin' without fear. It's love for your set and hate for the enemy." Recalling how he felt when he was seventeen, Scott wrote, "I felt nothing but a sense of duty....I had been to five funerals in the previous two years and been steeled by seeing people whom I had laughed and joked with and eaten with, dead in a casket. Revenge was in my every thought. Only when I had put work in could I feel good that day: Otherwise I couldn't sleep. Work does not always constitute shooting someone, though that is the ultimate."

Whether we are thinking about terrorist cells or street gangs, it is important to understand that the basic group to which they belong becomes a surrogate family. Gang members often come from single parent families, or families where one or both parents are in prison, and the gang provides unconditional acceptance. "Love" is a word used with unexpected frequency by gang members, just as it is by soldiers. Honor is central to the young male image, and it is intensified in the gang setting. "If you ain't got honor, you got nothing man. No amount of money can buy honor," remarked one Los Angeles gang member.

Kinship is important among terrorists and among gang members, whether based on ethnic origin, geographical location, or actual blood ties. American street gangs are almost always built around ethnic groupings and neighborhoods. A member of the IRA in Belfast, or ETA in Spain, would never recruit someone from another ethnic group, just as a Hispanic gang in L.A. or Chicago never recruits black members, and vice versa. Respect and loyalty go together and gang members, like the formal military, hoplites, Yanomamö warriors, and chimpanzees on patrol, will risk and give their lives for one another. Scott described another gang member, who had been shot thirteen times and lost one eye, as "loved by a few, hated by many, but respected by all." Scott even quotes the philosopher-soldier Glenn Gray, and writes of the allure of "total lawlessness" and "the sense of self importance, self worth, and raw power...exciting, stimulating, and intoxicating beyond any other high on this planet...*nothing* outside my set [the urban equivalent of a platoon] mattered."

Young urban men without economic opportunity may be the most likely to form actual street gangs, but it is not just the environmental influence—young men from even the most privileged backgrounds can also display aspects of gang behavior. Thirteen out of nineteen U.S. presidents in the twentieth century belonged to fraternities in college, as did the first president elected in the twenty-first century. And many belonged to secretive, supposedly elite societies, such as the Skull and Bones club at Yale, which claims both Presidents Bush as alumni, among many other prominent leaders. The very idea seems laughably quaint today, but even the late 1950s institution of "panty raids" by fraternity brothers represents a clear expression of male team aggression, albeit one of the least violent imaginable, with rape replaced by the symbolic capture of women's underwear. Innocent as they were, panty raids still displayed many aspects of team aggression: A close-knit group of male friends, often without long-term planning but only some sort of short build up of mutual dares and commonly lubricated with alcohol, proceed in a stealthy manner outside their normal territory and conduct a ritual capture of "resources." Such behavior can shade over into violence with disarming ease. Fraternities in Filipino universities are based on U.S. models, and at the prestigious University of the Philippines (UP) in Manila in the 1990s, vicious hazing ceremonies and "frat rumbles"—violent clashes between rival fraternities—killed eleven students and wounded 150.

There are no absolutes in biology and some women also crave the camaraderie of gang life, but technology probably plays a role as well: The ready availability of guns in the contemporary U.S. helps women overcome any physical limitations in relation to men of the same age. Even so, in an investigation of 200 consecutive gang homicides one Los Angeles detective did not find a single murder committed by a female gang member.

Street gangs in the America of the twenty-first century, with its wealth and technical sophistication, emphasis on individual liberty, and equality of opportunity, are more than just a stubborn social issue. They offer compelling testimony to the power of the inherited predisposition of young men to display intense loyalty and love for one another, and to dehumanize and attack their neighbors for the simple reason that they *are* neighbors, and so occupy an adjacent territory. A chimpanzee, were he able to do social research, would recognize his own behavior manifesting itself in a cousin species with whom he has not shared a common ancestor for millions of years.

Can There Be a "War on Terror"?

Terrorists wage an asymmetrical battle. They remain hidden until the moment of attack and cause damage far greater than their numbers—and fear far greater than that. Often we do not know who they are, where they live, or how large the group is. There is no battle front, except for the latest explosion or shooting. Terror groups like Al Qaeda know they can never win an outright victory over the West, but they strive to demonstrate that they themselves cannot be defeated either. In a "war on terror" it is impossible to define what victory might look like, other than a cessation of violence—which might or might not be temporary— let alone how it would be achieved. If we think of terrorists, or insurgents in Iraq, as a platoon of warriors outside the formal structure of an army and not directly linked to any state, then they can never be defeated in the sense the Saxons were defeated at Hastings in 1066, or the Confederates were defeated at Gettysburg in 1863. In some ways, a handful of angry men who are willing to die for a cause, drawing on inherited predispositions going back millions of years, are more difficult to defeat than whole nations. Four years after Pearl Harbor, Japan was defeated; eight years after 9/11, Al Qaeda was still perceived as a major threat.

Victory in a conventional war means killing or capturing so many of the enemy, or destroying their supply lines, to the extent that the remainder eventually surrender. In a "war on terror" it is difficult to find and kill all the terrorists, and as the leaders are killed new ones arise, hydra-headed, even angrier and even more radical than before. The illusion of a "war on terror" which the U.S. has pursued since 9/11 has increased rather than reduced the number of warrior teams willing to die fighting the West. The reprisals characteristic of terrorism-based conflicts often end up generating yet more terrorist volunteers. In an opinion poll of Palestinians conducted in June 2002, more Palestinians (78 percent) approved of suicide bombings than supported peace talks (60 percent). The invasion of Iraq in 2003 ended up creating new opportunities and supplying new recruits to Al Qaeda, which prior to the U.S.-led invasion had nothing to do with this particular brand of street-level terrorism—nor, of course, with Iraq. The Israeli bombing of Lebanon in 2006 increased public support for Hezbollah.

Raids and terrorist attacks will be repeated for as long as there are small bands of men sufficiently motivated to carry them out. Terrorists who want to attack the West know they cannot win against the firepower of a modern army But as long as there are angry people among the nearly seven billion population of the world, small groups will team up to exploit the vulnerabilities of modern living in order to hurt those with whom they disagree, of whom they are envious, or about whom they have woven some improbable conspiracy theory.

Responding to Terror

We dignify terrorists beyond their due when we declare a "war on terror," and we miss the point by calling them "cowardly murderers." An evolutionary perspective suggests a better way of categorizing and confronting twenty-first-century terrorism: It helps get inside the heads of those who hate us, which is an essential step if we are to confront and overcome Al Qaeda and similar groups.

All of science relies on the assumption that we can be dispassionate observers of the world around us. This is difficult enough for chemists and physicists. It becomes considerably more challenging for biologists, and ever more so as we work our way up the chain of being to its human peak. When it comes to same-species killing in our *own* species, well, it takes a phlegmatic observer indeed to keep emotions at bay. And yet, if we are to understand same-species killing, then we must be able to analyze any conflict from the perspective of both, or all, antagonists. There is every reason to assume that the emotions driving a modern terrorist attack are similar to those felt by other teams of warriors. Their tactics, timing, and horrific impact may have been uniquely dramatic, but the men behind the September 11, 2001, terror attacks on New York City and Washington, D.C., were a band of young male raiders like any other, and their behavior was part of the universal male predisposition to band together and try to kill perceived enemies. And no matter how heinous we may feel that their attacks were, it would be foolish in the extreme to deny or ignore the fact that they remain heroes to many millions of entirely sane, moral, and otherwise compassionate people.

The paradigm of evolutionary psychology suggests that acts of terrorism tap into our ancient emotions and therefore demand a particular response. We should begin by recognizing that an *unemotional* response is impossible. Anyone running for elected office in America in the aftermath of 9/11 was unlikely to have gotten very far unless they recognized the deep shock and desire for action among the overwhelming majority of people. But while understanding the anger we also need to build the level of objectivity required to establish clear, appropriate, and achievable military and political goals.

The first step is to put terrorist attacks into perspective. The Madrid bombing in 2004 killed 191 people and injured 1,500, and the London attack in July 2005 killed fifty-five and injured 700. It is revealing to compare these numbers with deaths during World War II. Continuous attacks during the London Blitz in early September 1940 killed over 8,000; German air raids on Britain in World War II killed 43,000 people and injured 139,000.[164] The Allied bombing of Hamburg in July 1943 probably killed 50,000 in one attack on one city. Overall, in World War II, one in every 300 Americans was killed; in Germany it was one in ten, and in the former Soviet Union an unimaginable one in seven.[165] The United States is a vast, populous, wealthy country. Horrible as it was, 9/11 killed just one person per 100,000 of the total U.S. population. Globally, in the 1990s, the risk of being killed by a terrorist was one in ten *million*.

If, as the standard social science model asserts, we lack inbuilt drives and responses to raids and warfare, then Americans could have regarded the destruction of the World Trade Center rather like an earthquake— a frightening, unpredictable event, unlikely to be repeated for decades, where we grieve for the dead and sustain those who have suffered physical loss. In such a scenario, there might have been minimal damage to the economy, and no national drive for revenge. The paradigm of evolutionary psychology, however, recognizes that war and terrorism are not earthquakes but visible, deliberate acts of human violence which resonate with evolved behaviors. They demand a response. The challenge is to devise a nuanced, proportional reaction. It is here that Western democracies have failed.

I do my writing in a house in Berkeley, California, about one mile from the Hayward Fault. This fault is probably more dangerous than the better known San Andreas Fault running past San Francisco, where my co-author does his. In 1868 the Hayward Fault jumped six feet, causing a magnitude 7.0 earthquake which killed thirty people. A few tens of thousands of people lived along the Hayward Fault 140 years ago, but today Martha and I are joined by some two-and-a-half million neighbors along the Hayward Fault. The fault slips on average every one hundred years, and it is estimated the next major temblor could cause damage amounting to $150 billion.[166] But earthquakes are not deliberate acts of human violence, so we underestimate the risk they present. I have some bottles of water and tinned food in the garage, and in the bedroom a mitten for handling broken glass and a crowbar to pry open jammed doors—but come to think of it, I am not sure I could find the mitten and the crowbar without a light. Acts of terrorism tap into evolved predispositions and, unlike earthquakes, we tend to overreact to such threats.

Politicians get elected not because they analyze the world objectively, but because they read accurately how the Stone Age emotions we all harbor will lead us to vote at the next election. Any analysis in the weeks immediately after 9/11 would have included uncertainties—for example, about the possibility of similar attacks in the future. President George W. Bush and Vice President Dick Cheney, however, had a Churchillian view of the world in which they genuinely believed they were leading a global war against terrorism and that any consideration of alternative strategies was dismissed as weakness. After millions of years of social living, human beings are programmed to detect sincerity in speech and body language and many American citizens and a majority of politicians were carried forward on a rising tide of fear and a desire for revenge. Five days after the attack, the White House sought and received Congressional approval for the President to use "all necessary and appropriate force against those nations, organizations, or persons he determines planned, authorized, committed or aided the ferocious attack that occurred on September 11, 2001." Only one person—a woman—voted "nay." Others who voted for the resolution would later claim they had been deliberately misinformed, but taking an evolutionary perspective would teach people to *expect* to be misinformed in such a situation. However it is understood, this resolution was not the result of sober analysis of how and why the attacks occurred, or of the best way to prevent similar attacks in the future. Rather the resolution empowering President Bush was chillingly similar to the Gulf of Tonkin resolution, which Congress passed in August 1964. Following a minor engagement when Vietnamese gunboats fired on a U.S. destroyer (which was feeding intelligence information to South Vietnam), the U.S. Congress voted to give President Lyndon Johnson the power to use "all necessary steps, including the use of armed forces" to fight North Vietnam. The Gulf of Tonkin resolution led to a major escalation of the Vietnam War. On that occasion, only two senators voted against the resolution.[167]

Inappropriate overreactions to military threats are remarkably common: In 1739 Britain declared war on Spain because Robert Jenkins, a sea captain, had his ear cut off when he was captured by a Spanish ship near what is now Florida.[168] Following 9/11, the U.S. government voted for a $40 billion anti-terrorist package—and the public purchased 116,000 American flags from Wal-Mart. Forty-three nations, from Australia to Zimbabwe, also passed anti-terrorism legislation, each item of which in one way or another restricted previously hard-won freedoms. The U.S.A. Patriot Act permits federal agents to search a person's home without the owner's knowledge, and to detain even American citizens labeled "enemy combatants" in a military prison without access to a lawyer. In many respects, the U.S. government response to 9/11 has been disproportional, inappropriate, and irrational. Some of the laws and regulations developed since 9/11 are exercises in futility. The cargo that goes into an airliner's hold is not screened for explosives, while the shoes of the passengers sitting over it are x-rayed.

We need to look at the big picture. The only plausible long-term strategy to defeating terrorism is to forestall the emergence of angry young men eager to take part in team aggression. Efforts to reduce gang activities have been disappointing. A few charismatic and dedicated church and social leaders have provided gang members with the sense of belonging they seek, but police efforts to suppress gangs in big American cities, such as Chicago or Los Angeles, have usually adopted the rules of the street, trying to suppress gangs by making life on the street as unpleasant as possible. In Scott's words, "The notion of a 'war on gangs' being successful is as realistic as the People's Republic of China telling Americans to stop being Americans."[169] Ninety percent of American street gang members are recruited from minorities in areas of high crime and persistent high unemployment, and gangs will persist as long as these problems remain intractable. Similarly, terrorism will remain a danger as long as schools cannot keep pace with growing numbers of students and young men cannot find jobs other than shining shoes.

An achievable set of policies to reduce the risk of terrorism, which we will explore later in this book in more depth, is to focus on those upstream forces that drive terrorism rather than its downstream results. As it has been for millions of years, a prominent goal of team aggression by young males is to increase resources and to gain access to sexual partners. Consider the Gaza Strip: It is a tiny rectangle of land, squeezed between the Mediterranean on one side, Egypt on another, and the state of Israel proper on two more. It is home to more than one million Palestinians, some of whom have been settled there since the Six Day War in 1967. Politically, the Palestinians are classified as refugees, although most people living in Gaza today were born there and know no other home. Refugee status has ensured good preventive medicine services, universal vaccination, and tolerable hygiene, thanks to United Nations help, but family planning has been treated as a political hot potato. Lack of family planning and an explicit political understanding that "the Palestinian womb is the one weapon the Palestinians have" has led to very rapid population growth. In 1950, there were one million Palestinians; today there are 3.1 million, and the average woman has six children. According to UN projections, the Palestinian population could reach between 9.45 and 13 million by 2050. Half the population is under the age of fifteen. The unemployment rate is 40 percent and rising. These demographics make any long-term solution to political problems exceedingly difficult if not impossible. Arab society is sexually chaste and few young men have sexual outlets prior to marriage, but they lack the financial independence to marry. Given the large pool of young men, the lack of opportunities, the sexual frustration, and the inevitable resentment toward a dominant outgroup, it is hard to imagine a more perfectly conceived breeding ground for team aggression, in this case taking the form of terrorism.

6

WOMEN AND WAR

WITH MARTHA CAMPBELL

This led to no-holds-barred evolution of any trait that helped males in this competition, even if it ultimately hurt the female. Male efforts to exclude other males, to herd and sequester females, were all outcomes of sexual selection. Often attempts by males to control their mates came at the expense of the viability of mothers and offspring.

—SARAH BLAFFER HRDY, 1999[170]

I N LATE 1970, the Palestinian Liberation Organization launched a group called Black September, named after the traumatic expulsion of Palestinian refugees from Jordan during that month. Black September grew rapidly into a dedicated and ruthless group of young men who gained world attention by seizing and killing Israeli athletes at the Munich Olympic Games in 1972. However, within a few years, when the PLO was offered observer status at the UN, Yasser Arafat needed to rein in the group. After much debate about what to do, the senior PLO leadership recruited approximately one hundred attractive young, unmarried Palestinian women and asked them to join "a critical mission of the utmost importance to the Palestinian people." But rather than embarking on terror attacks of their own, the women were brought in to test the impact of marriage, family, and economic opportunity on potentially violent men. The volunteers were introduced to key Black September militants, who were told that if they married they would be given $3,000, an apartment, a gas stove, a TV, and long-term employment, along with another $5,000 if they fathered a child.[171]

This unique but true story from a patriarchal society has an interesting ending. Like Stanley Milgram at Yale testing the willingness of his research subjects to inflict "pain" on another person, or Philip Zimbardo observing the "prisoners" and "guards" at Stanford, the authors of this real-life experiment also tested their subjects. Every so often they would offer the now-married terrorists a passport and ask them to visit a European city on PLO business. None accepted because they feared arrest and losing their new family. Marriage and having children had put them in a situation where violent, high-risk actions no longer made emotional, economic, or evolutionary sense, and Black September collapsed.

If men and their deep-seated behavioral predispositions are an important cause of war and terrorism, could women and their sometimes-opposing predispositions be part of the cure? In medicine, many poisons have an antidote, but such a direct cause-cure relationship between the sexes in relation to war seems very unlikely. And yet, just as we can't overlook how central men and their inherited impulses are in establishing conditions conducive to war, our diagnosis would not be complete if we were to ignore evidence of factors that may decrease war's probability.

Women relate to wars and terrorism in several interlocking ways. Both men and women compete for resources, although as pointed out earlier, men are more likely to resort to violence than women. Team aggression and raiding may have given a reproductive advantage to the successfully violent males in our human and pre-human lineage, and those females who closely allied with them also benefited from a larger foraging area, more food, and healthier offspring. But it's important to recognize that team aggression in all its forms is a zero sum game—that is to say, all the wars, raids, and rapes in the world have never yet resulted in a single additional ear of corn, scrap of cloth, or healthy baby* that could not have been gained through peaceful means. Wars do not generate new resources but take resources from other groups. For our male ancestors who were

---

* The result of war is usually quite the opposite, of course—depletion of resources, loss of life, and missed opportunities to grow food or produce goods. Counterintuitively, however, warlike pursuits have made positive contributions to society. From canned food to atomic energy, the needs and particular urgency of war have spurred numerous technological innovations. And as we'll see in more detail in chapter 8, the creative demands of outwitting and vanquishing our neighbors may even have stimulated humanity's unique and characteristic level of intelligence.

successful in attacking their neighbors, more territory meant more access to food, a larger troop with more females, and more progeny surviving in the next generation. But the violence did nothing to produce more food, and females would have survived and reproduced just as well if some other troop of males had controlled the same part of the forest.

Women on average are less aggressive than men—inherently so, we would argue—and their reproductive agenda is less directly competitive. But this hardly means that women are bystanders in human evolution, or in current affairs. It's just that sometimes their deep influence plays out in ways that are more subtle than bunker busters and machine guns.

Whatever intrinsic, evolutionary gap that existed between male and female strategies relating to violence and reproduction has been widened further by at least 5,000 years of social and technological change. As we will see later, technology has made male-driven conflicts increasingly destructive over time. As settled agriculture and domestication of animals became increasingly important, wealth was passed from one generation to the next, and marriage systems and sexual customs began to emphasize male control of female childbearing. Cultivated grains and domesticated animals increased the overall food supply, sparking more rapid growth in population and increased competition for land, as well as increasing the predominance of volatile, aggressive young men in the population. When ways were discovered to artificially control childbearing, these were consistently obstructed by men.

It is overly simplistic to say, as was popular a generation or two ago, that there would be no war if women ran the world. But women do, on average, make different political choices than men, who tend to lean further to the right, support more defense spending, and prefer a top-down "decision maker" style of leadership rather than the consensus-building approach more commonly favored by women. Intriguingly, even the sex of their children can influence parents' political views. Ebonya Washington, a Yale economist, has shown that members of the U.S. Congress who have a daughter are more likely to vote in favor of safe abortion, for example. Moreover, the more daughters a Congressional father has, the more liberal a Congressman's voting record tends to be. The trend carries over to the voting public. In the U.S., 37 percent of parents who have only girls register as Democrats, compared to 31 percent of parents with only boys[172]—a small but statistically significant difference. In Germany, two out of three parents who had a son, if they switched parties, moved to the right, while two-thirds of those having a daughter moved to the left.

These observations may reflect the differing biological agendas of men and women; women are more likely to network with other women and support social welfare, while men are more likely to compete with one another and vote for leaders who are perceived as strong and decisive. Either way, there is strong empirical evidence that when women have political influence and independence, the impact on violence at home and aggression abroad can be profound. In general, those countries where women enjoy individual freedom and play a political role in society are less likely to attack their neighbors or sponsor terrorism than are more male-dominated societies.* Patriarchal societies, where women are constrained by tradition and religion and most of their adult life is taken up by childbearing, are often violent.[173] When women are given the ability to have fewer children, competition for resources is reduced, the ratio of younger to older men in the population falls, and women, who then have more time and energy, can play a more equal role in civil society. However, in the man's world of politics, the first women to take leadership positions, such as Margaret Thatcher in Britain or Benazir Bhutto of Pakistan, are often unusually combative. But overall, legislatures with many women tend to be more peaceful. The manipulation of the Palestinian young women and the Black September terrorists shows vividly how raising a family can dampen team aggression in men, but in order to fully untangle all the ways in which war and terror are affected by the interactions of women and men, we need to begin by understanding the evolution of competition between the sexes.

The Battle Between the Sexes

Darwin first described how evolution can pit the two sexes against each another in his 1871 book *The Descent of Man, and Selection in Relation to Sex*. Based on thousands of observations, evolutionary biologists make the generalization that the sex with the greatest investment in reproduction will be more cautious and choosy during mating, while individuals of the sex that makes a lower investment are likely to compete among themselves for access to the opposite sex. In mammals such as us, fe-

---

* As the 2003 invasion of Iraq shows, democracies fed misinformation about weapons of mass destruction and driven by short-term patriotism can still make serious misjudgments. But it is perhaps revealing that as the American public became increasingly disillusioned with the war in 2006, 40 percent of men still supported the war compared to only 34 percent of women.

males make a profound investment of energy and time to carry, feed, and nurture their offspring. In contrast, the investment of a male mammal can be as limited as a single ejaculation. One male, if he defeats or outwits his rivals, can father the offspring of many females, while the defeated or outwitted males may father none at all.*

The asymmetry in male and female investment in reproduction is especially large in human beings, where rearing each child involves a long pregnancy, a dangerous delivery,** the demands of breastfeeding, and many years of childcare. In other words (and as any parent knows), human children are extremely expensive in terms of parental time, energy, and lost opportunities—and fertile females are the limiting factor for any male seeking to increase the number of offspring he leaves. Just as men who evolved the predisposition to band together and fight their neighbors increased their likelihood of successful reproduction, so too did those who evolved the impulses or devised new strategies to control women and their reproduction end up fathering more children. Any individual with a genetic tendency for non-lethal sexual aggression toward women, whether belonging to the ingroup or an outgroup, was likely to leave more offspring who would inherit that tendency. Sarah Blaffer Hrdy of the University of California, Davis, describes the "no-holds-barred evolution of any trait that helped males [compete], even if ultimately it hurt the female."[174]

The battle between the sexes may be unequal, but it is not totally one-sided. It took female biologists and psychologists such as Jane Goodall, Patricia Gowaty, Sarah Hrdy, Anne Pusey, and Anne Campbell to understand the full extent of female competition for resources and choice of mates.[175] In apes, as in our species, male competition is easy to see. Young male orangutans sometimes force unwilling females to mate with them. Watching wild chimps in the forest, Martha and I didn't have to wait very long before seeing males hooting and showing off in front of one another, and roughing up females to demonstrate their dominance. But careful observation reveals females' more subtle tactics. Mature male chimpanzees mate with most of the females of their troop, but those females of-

---

* Nature has provided an interesting role reversal, which helps illustrate the generalization that the sex making the smaller reproductive investment is likely to be the most aggressively competitive. In seahorses, males have evolved a brood pouch, into which females deposit their eggs with a penis-like ovipositor. The male fertilizes the eggs internally, and incubates them for two to four weeks. In this only known case of paternal pregnancy, males rather than females are the limiting reproductive resource, and females are the more colorful and aggressive sex.

** Without modern obstetric care, up to one in one hundred births result in maternal death.

ten escape male control by also mating with males from a neighboring troop. Chimpanzees make promiscuity an art form, with the average female mating over 130 times with a dozen or more different males for each conception. One female was seen mating fifty times in a single day.[176]

As with so many evolutionary drives, sexual competition between men is not always conscious, and it often has ancillary consequences, including domestic violence, repression of female sexual enjoyment, and the double standard of sexual morality that runs through so many societies.[177] Chimpanzee fathers tolerate the play of youngsters generally, but they do not nurture infants in any way, and don't appear to show any particular interest in their own offspring compared to others of the troop. In contrast, men in all human societies try to identify their own children and, in varying degrees and sometimes-uncertain ways, they contribute to their care. Identification of offspring is the key, however. From an evolutionary perspective, any investment a woman makes in her child is biologically worthwhile because she is certain that the child in hers. But if a man puts resources into rearing a child that is not biologically his own, then he is squandering an opportunity to support his own offspring and pass on his own genes to the next generation. From strict moral codes and sequestered harems, to chastity belts and female genital mutilation* men across virtually all human cultures have developed ways to control women's sexual choices and ensure their paternity.

As an evolved predisposition, the male drive to control women and their sexuality runs deep, and it can be extraordinarily difficult to reverse. Men all too often use their greater physical strength to coerce and beat women. In studies in Nicaragua, for example, 44 percent of all men self-reported assaulting their wives or girlfriends, and in a sample barrio in Quito, Ecuador, eight out of ten women said their husbands had beaten them.[178] But as we have seen with other forms of aggression, the drives that biology has implanted in us do not always need to be expressed, and indeed can often be channeled in more positive directions. In the case of domestic violence, it is certainly clear that biological explanations do not excuse human actions, and that evolutionary impulses would make a very poor basis for our morality or our legal systems. As we have suggested earlier, civilization at its best often overcomes inherited, hardwired

---

* During this procedure, which is common in parts of Africa and the Middle East, the clitoris is removed and in some regions, parts of the adjacent labia as well. Sexual pleasure and desire are greatly diminished or totally extinguished, and the woman is left with life-long memories of extreme pain associated with the genitals.

predispositions, replacing them with other, often complementary or contradictory behavioral patterns. At the same time, the better we understand the origins of the predisposition for men to control women or for team aggression by men, the more capable we may be of mitigating it. In chapter 3, we mentioned that nature has provided an interesting contrast to the Hobbesian behavior of *Pan troglodytes*, the chimpanzees that Jane Goodall first studied and that we have spent our time with so far. Rousseau couldn't have known it, but bonobos, the second species of chimpanzee, offer an impressive measure of support for his more benign view of human nature.

Bonobos

The species *Pan paniscus* was recognized as distinct from *Pan troglodytes* as recently as 1929. It used to be thought that these "pygmy chimps," or bonobos, were separated from their larger cousins by only a few minor anatomical and behavioral differences, such as the fact that bonobos are more slender, walk upright more readily, and unlike the *Pan troglodytes*, will paddle and play in water.* But the behavioral differences especially turn out to be quite profound. Bonobos of both sexes have a playful, exuberant sex life, and females are not as universally subservient to males as is the case with *Pan troglodytes*. Like *Homo sapiens* but unlike *Pan troglodytes*, adult male and female bonobos sometimes share the same nest at night. Most crucially, and despite thousands of hours of observation, bonobos have never been seen to engage in team aggression.

As we saw earlier, the males of most social mammals leave their birth-troops at puberty. As a result, herds of elephants, packs of wolves and hyenas, and prides of lions all live in matriarchal groups consisting of related sisters, aunts, and female cousins. Both *Pan troglodytes* chimpanzees and bonobos reverse this pattern, and the primary social unit is made up of males, with females leaving the birth-troop at puberty. In the case of *Pan troglodytes*, when the young, inexperienced female moves to a new troop she has no friends. She is smaller than the males, she does not know the territory well, and she is under considerable stress. We remember watching a young female in Gombe whose sexual swelling advertised what may have been her first ovulation. She was chased through the branches by an

---

* The origin of the name "bonobo" is uncertain, but it is increasingly used as the vernacular for this fascinating species.

aggressive male until she fell about fifteen feet to the ground with a great thump. Fortunately, she picked herself up and made an escape, but it was a clear lesson in the difficulties facing a young female when she joins a strange troop.

After puberty, human beings are interested in sex throughout the year, and unlike practically every other species, most of our sexual activity is not directly associated with procreation. But chimpanzees and most especially bonobos take florid sexual activity one step further and, unlike us, do it all in broad daylight and in front of their peers. When a new female bonobo enters the troop, she first establishes a bond based on sexual pleasure with a female already at home in the troop. Female bonobos engage is what primatologists call a *hoka-hoka*, or mutual genital rubbing. They intertwine their arms and legs, and one hugs the other and sometimes lifts her completely off the ground while they move their hips rapidly and press their clitorae together. They may have *hoka-hokas* on the ground, or swinging from branches. Male bonobos also engage in same-sex behaviors such as mutual masturbation using their hands and feet (chimps have an opposable toe as well as an opposable thumb), oral sex, and what has been called "penis fencing," in which they swing from branches and knock their long thin penises together like sabers. Bonobos have no clear dividing line in sexual orientation—all adult bonobos also engage in heterosexual intercourse.[179]

Bonded by sexual pleasure, female bonobos come to one another's assistance if either is threatened or attacked by a male. While females are smaller than males, two females who trust each other and act together can outwit a single male and, if the need arises, physically defeat him. This bonding changes the balance of power between the sexes. In troops of ordinary chimpanzees, the males consistently dominate and often physically beat the females. This behavior is much rarer in bonobo society, where there is greater sexual equality and the males do not simply get their way. If I am a zookeeper at San Francisco Zoo, where there is a *Pan troglodytes* troop, and I throw a bunch of bananas in the enclosure, a male is likely to push in and pick the food up. If I do the same in San Diego, where there happen to be captive bonobos, the females may well collect the bananas. Perhaps because of the dramatically different state of competition between the sexes, bonobos seem to be altogether less aggressive than their more rough-and-tumble cousins. In an odd but vivid example of their differing natures, bonobos may even be more sensitive to high-stress sit-

Credit: © Frans Lanting/www.lanting.com

While the chimpanzees Jane Goodall studies (*Pan troglodytes*) and the rarer bonobos (*Pan paniscus*) are similar in anatomy, they differ in social behavior. Bonobo females develop strong friendships as shown here, where two females embrace one another. Such pairs also sexually pleasure one another. Among *Pan troglodytes* the males bully the females, but in *Pan paniscus* society a pair of females can resist male domination even though the males are larger. We can see reflections of our own behavior in each species.

uations. The first bonobos to be kept in a zoo arrived in Munich in the 1930s. It is said that during an Allied bombing raid during World War II, three bonobos died of heart attacks,[180] while all the *Pan troglodytes* chimpanzees in the same zoo survived.

It is tempting to think that the less patriarchal nature of bonobo society compared to other chimpanzees could explain why bonobos don't practice team aggression. That is how many evolutionary "just so" stories get told, however, and it is well to remember Einstein's caution about our preconceived notions influencing the questions we ask and the conclusions we reach. In the case of the bonobos, a careful look at ecology as well as behavior suggests that in-troop patriarchy and out-troop aggression are probably unrelated. Indeed, the lack of team aggression in bonobos can be explained by an important difference in how and where the two species live.

Remembering Darwin's finches and the beak variations that allowed each to tap a unique food source, similar species living side by side often evolve specialized diets. This permits them to occupy the same area without overt competition for food. In much of Africa, chimpanzees and gorillas originally lived in the same forest habitat, without competition. Ordinary chimpanzees eat mainly fruit, and a little meat. Gorillas are strict vegetarians, and largely subsist on fleshy leaves, which they digest slowly in their large stomachs and voluminous intestines.

Some years ago, one of us (Malcolm) crossed the Congo River 500 miles inland from the sea. Even so far upstream, the mighty river was a formidable barrier several miles wide, and it was not surprising that gorillas were never able to cross it—they live only north of the Congo. Somehow, however, some chimpanzee ancestors made the crossing. Lacking competition from gorillas, those ancient chimps began to eat leaves as well as fruit, and they evolved eventually into the species we now call bonobos. With a better, more diverse, and more stable food supply, bonobos can afford to forage in groups, whereas ordinary chimpanzees are often forced by scarcity of food to feed individually, exposing them to attack by male groups from neighboring troops. In bonobos, where individuals are less likely to be found alone, team aggression either never evolved or has been lost, not so much because it wasn't necessary as because it wouldn't provide any benefit—the strategy depends entirely on finding lone enemies to attack.

It does raise the question, however—did the lethal team aggression evolve twice, once in *Pan troglodytes* and once in *Homo sapiens*, or was it passed down from a common ancestor to all three species, only to disappear from the repertoire of bonobo behavior more recently? Because the evolution of new structures or behaviors is quite improbable, any distinct anatomical feature or unusual behavior is more likely to have evolved once rather than twice. Statistically speaking, the most plausible explanation is that the common ancestor of human beings and both chimpanzee species had a predisposition for team aggression and bonobos lost this behavior because they no longer needed it. Differences in food sources may well explain this switch away from same-species killing.* Bonobos that forage in a group eating leaves cannot be ambushed and killed one at a time. It follows that bonobo males still practicing team aggression would

---

* Interestingly, male bonobos also seem to be ambivalent hunters at best. On the few occasions when male bonobos have been seen hunting, the females seem to end up with most of the meat. In one intriguing observation, bonobos were seen capturing guenon monkeys but not eating them. Instead, the bonobos carried the smaller primates around for hours as playthings until they died.

have met with fierce group resistance and many defeats, would not have increased their opportunities to breed with females, and indeed, may not have survived their attempts. Aggressive males would not have passed the tendency for team aggression on to future generations. The conclusion has to be that team aggression died out in bonobos, while persisting in *Pan troglodytes* and *Homo sapiens*.*

Women on the Firing Line

There is a fanciful literature that interprets early human society as matriarchal and intrinsically peaceful, or alternately, that there were societies in which women were the primary fighters. Intriguing as it might be to think such a world existed, there is no hard evidence to support either idea. Sarah Hrdy writes, "I know of no evidence that any matriarchal society ever existed."[181] In sixty-two out of seventy different societies studied by anthropologist Martin Whyte, wars and raids were an exclusively male business,[182] and in all the Norse sagas, for example, there is not a single female warrior.[183] This is not to say that women have played no role at all in warfare and fighting through the years. In many cultures, including the Central American Aztecs, the Asante of nineteenth-century West Africa, and indeed, throughout the West during World War I, women taunted hesitant warriors, goading them into action. Among some tribes in New Guinea, women could wander unmolested through the middle of a pitched battle and pick up arrows to resupply the men, "as if they were harvesting potatoes or cucumbers." During the American Revolution, Mary Hays McCauly became famous as "Molly Pitcher," carrying water to the soldiers. When her husband was wounded next to his cannon, she loaded and fired the artillery— that is, if she actually existed. But it doesn't really matter whether Molly was a legend, a composite, or an exaggeration. The fact that a woman's presence on the battlefield inspired such awe is proof enough that war was an undeniably male domain.

The role of women in warfare has increased dramatically over the last

* It is also possible that bonobos retain the capacity for team aggression, and that the behavior is silenced by social and environmental conditions—though remember that bonobos have *never* been seen to display it. If bonobos have lost the genetic underpinnings of team aggression, humans too could one day follow that evolutionary path. Given that evolution generally works on timescales of hundreds of thousands to millions of years, we don't recommend the wait. Intriguingly, though, humans and bonobos do share a gene associated with socializing which other chimpanzees lack.

century. In the Great War, with its unprecedented demand for recruits, women were brought in to replace men in clerical positions in the army and navy, and they entered manufacturing, service, and farming in significant numbers, freeing men to fight. During World War II, the Allied forces engaged women even more extensively, not only as secretaries but also in servicing increasingly complex fighting machines and in communications and transport. Women in Britain's Auxiliary Transport Unit (ATS) became pilots ferrying aircraft across the Atlantic. In 1941, the U.K. Parliament gave women full military status, and a year later conscription was introduced for women as it had been for men. By the middle of 1943, half of the female population of Britain was employed in some way or another in the war effort. But women were kept out of combat and the King's Regulations in Britain forbade the ATS pilots from carrying even small arms, even though they did fire anti-aircraft guns.

Hitler strongly opposed women having anything to do with the armed forces. The female *Helferinnen*, or "helpers," played an important role as telephone operators and clerical assistants but, unlike the women in Britain, they were denied military status and were registered as non-combatants under the Hague Convention. By 1943, however, even Hitler was forced to change his policy, and German women began to play the same role in air defenses that British women had commenced playing two years earlier. Still, only at the end of 1944, as the defeat of Germany was imminent, was conscription for women introduced.[184]

Russia established a very different tradition. Communism emphasized female equality, even though, following the pattern of so many Communist ideals in practice, equality usually meant that Russian women worked hard outside *and* inside the home. As early as World War I, the Russian army had some female battalions whose formation was saluted by the English suffragette leader Emily Pankhurst as "the greatest event in history." In World War II, the specter of defeat carried Soviet women into battle in large numbers. In 1941, fighter squadrons were formed in which the commanders, the pilots, and the mechanics were all female. Major Raskova told her new recruits, "If you're chosen, you may not be killed—you may be burned so your own mother would not recognize you. Do you really want to go through with this?" No volunteer backed out. Like men, they developed a sense of camaraderie, although that camaraderie involved subtle and important differences from the male pattern. One pilot said, "There was so much mutual respect that people just tended to get on with the job without having to be ordered. Of course, you need people in command, and we had that. But a group of really motivated females does not, I think, need quite the same sort of rigid discipline that men do." The women fighters perceived themselves as different from the men and took pride in reminding themselves "all the time that we were girls."[185] They altered their uniforms to fit their figures, defied regulations and kept their hair long, and even put on makeup before battle. Like any warrior, they could be so scared before a sortie that they vomited, but once in the air they got on with the task of flying and fighting with immense courage.

In the United States during World War II, some 150,000 women served in the non-combat Women's Army Corps, with strong support from General George C. Marshall. After the war in 1948, the Women's Armed Services Act formalized the role of women in the armed forces, although it did not specify their role in combat, or even if one existed. Notably, the uniforms that women wore—skirts and heels—hardly suggested a fighting role. Women have served in the same units as men since the late 1970s, and although they are still not allowed in direct combat units, they have served in combat zones since the early 1990s.

Over the past twenty years, the proportion of women in the U.S. Army has risen, and has reached about one soldier in seven. Over 160,000 women have been deployed in the U.S. armed forces in Iraq and Afghanistan. Unlike men, women tend not to form tight bonds with each other during deployment, perhaps because there are fewer women than men to choose from, but partly perhaps also because they lack some of the predispositions that bond men during battle. Certainly, men do not usually develop with women the same non-sexual trust and comradeship so characteristic of a unit of men. In fact, the military can be a hostile environment for female troops, who are forced constantly to prove themselves. Officers tell women soldiers to go to the latrine or showers in pairs because of the danger of harassment or rape, and some female soldiers confess to carrying a knife to protect themselves, not from *jihadists* but from fellow American soldiers.[186] In one sample of military women deployed in Iraq, 16 percent met the diagnostic criteria for Post Traumatic Stress Disorder, compared with just 8 percent of men. Decades of studies suggest that women on average handle stressful and crisis situations at least as well as men, but exposure to both the trauma of war and the threat—and all too often the reality—of rape, harassment, or neglect by fellow soldiers provides a toxic double whammy.

When warfare involves entering coordinates on a computer instead of carrying a rifle across a lethal fire zone, there is no reason to think that women won't perform as well as men. But much of the pressure to include women in every facet of warfare has not come from military leaders, but from the civil courts, women's groups, and politicians trying to read public opinion. Sometimes the outcome is confusing. For example, under the British Sex Discrimination Act, a civilian employer is required to recognize "risks specifically affecting women," and when mixed-sex training units were introduced into the army in 1994, they were based on a "gender-fair" policy permitting lower physical standards for women. Four years later, however, a "gender-free" policy with identical physical standards for both sexes was substituted. However, women have a broader pelvis and less upper body strength, and in trying to achieve the same standards as men during basic training they end up with four times as many musculo-skeletal injuries.

Interestingly, male resistance to mixed-sex military units is based only partly on concerns about women's abilities to perform heavy physical tasks. Women warriors argue that they fight as well as men and also point out that combat experience which they are excluded from obtaining, is the key to both respect and career advancement in the armed forces. In 2000, the European Court of Justice ruled that a clause in the German constitution limiting the role of women in the military violated European Union laws against sexual discrimination. But when women began to be integrated into the German army, many male soldiers felt their identity as warriors was somehow diminished. Whether their female comrades could perform under pressure or not wasn't the issue—these men perceived a loss of camaraderie, which as we saw in chapter 4 is one of the traditional pleasures of war and military service for men.

Given the direction of both legal and social changes, the role of women in armed services in many parts of the world is likely to continue to increase. The most significant factor may actually be technological, and so basic that it is often overlooked: access to contraception and safe abortion. Physically, a pregnant woman cannot not be part of a fighting unit.* The impact of women in the trenches is just now being truly tested, as is

* During and after World War II, a woman who fell pregnant in the Allied armed forces was automatically dismissed—although the man who made her pregnant suffered no career consequences—and in the U.S. even marrying a man who already had children could end a woman's military career. At the beginning of the twenty-first century, female military personnel who became pregnant were not permitted to obtain abortions from military hospitals, they had to return home and pay for their own operation; and as usual, the male involved suffered no consequences.

the effect of female comrades and commanders on the prevalence of wartime rape and looting. It is possible that a more fully integrated military, with more women in positions of power and influence, would develop its own ethos and culture distinct from the male-dominated paradigm that has persisted for thousands of years, if not much, much longer. The early evidence, however, including the participation of female as well as male guards in the (often-sexual) abuse of prisoners at the U.S. military prison in Abu Ghraib, Iraq—which was under the command of a female general at the time—suggests that such cultural changes remain a long way off. In the meantime, issues of sex discrimination persist because promotion and economic rewards are often tied to military leadership on the battlefield.

There are other models for female warriors, and they too suggest that even if women are less likely to initiate a battle, they can certainly be as fearless and as fearsome as men once the fighting starts. In ethnic conflicts and especially where minority groups are, by definition, outnumbered, women have played significant military roles. Women fought in the Algerian War against colonial France, and with the Tamil Tigers in Sri Lanka and the Kurdish Worker's Party (PKK) in Turkey. In the late 1990s, Ethiopia and Eritrea fought a brutal, costly war over their shared border—a culmination of decades of fighting back and forth. In Eritrea, about one third of those drafted into the army were women. The sexes were separated during basic training, but women had to undertake the same grueling exercises as the men, and at the battlefront, men and women fought in mixed platoons. Women commanded squads of fifteen soldiers and platoons of forty-five, but never the 400-soldier battalions.

The presence of women on the front line in Eritrea evidently did change the behavior of male soldiers, who, for example, would never withdraw as long as a woman was still firing at the enemy. One interesting difference is that the women are reported to have taken few prisoners, preferring to kill captured enemies instead. An Ethiopian soldier told a *New York Times* reporter, "Women are very bad. They don't capture at all. Most Ethiopians know that."[187] During the Liberian civil war in 2001 and 2002, a young woman who called herself "Colonel Black Diamond" controlled an all-woman troop of fighters. She carried a pistol and a cell phone and wore tight-fitting jeans, an embroidered strapless top, pink nail polish, and a red bandana. Black Diamond and most of the other women in her troop had been raped during the course of the war. The men agreed, "These women have no pity, no sympathy. You cannot beg them like you can with men."

Perhaps not surprisingly, the women had formed an ingroup of their own.

It seems possible that women as fighters are more focused on defense and survival than are men. This is what we would expect if they do indeed lack the male genetic predisposition to band together and, as we saw in the previous chapter, occasionally empathize and fraternize with the enemy—so long as there is an implicit ingroup connection. However, there is another way to see this pattern. Perhaps for Black Diamond and her troop, and the Ethiopian women as well, men on the whole are actually an outgroup. Accordingly they would have less empathy for the enemy males than their own male fighting partners would. In Eritrea, female soldiers did in fact take steps to protect women on the opposing side. This division in sympathies could be particularly significant in ferociously patriarchal societies such as the Eritrean, where most of the female soldiers had undergone female genital mutilation, a painful experience that might have helped cement a bond similar to that shared by Black Diamond and her cohorts.

For many years in Eritrea, sex between soldiers was forbidden, and affairs were punished by assigning the culprits to hard labor or latrine duty. After 1978, marriage was permitted but spouses were always assigned to different units and were not allowed to have children. Dispensing oral contraceptives was military policy. Toward the end of the war, premarital sex became acceptable, providing that the couple could be considered to be engaged. Knowledgeable Eritreans say that rape of women soldiers or of civilians was unknown in the army—while extramarital affairs and sex between unmarried soldiers was said to be especially common after a serious battle.[188] There is an impression that when peace came, divorce was more common among marriages where both spouses had been soldiers, and the grounds cited were often sexual incompatibility—possibly because their fighting experience makes women soldiers are less compliant than the average village wife. Many female fighters remain unmarried, because having had sex during war, they were considered by men to be no longer "pure," even by their former warrior colleagues.

Sex and War

Genghis Khan, as we have seen, combined sex and war with brutal efficiency. It's not saying much to suggest that all soldiers think about sex a great deal—there are few young men who don't, after all. But there does seem to be something about wartime military service that heightens the expected obsession with sex still further, whether it's masculine competition, the general absence of women, or indeed the proximity of death and a desire, conscious or otherwise, to leave progeny before it arrives. "The most common word in the mouths of American soldiers," writes Glenn Gray about World War II, "has been the vulgar expression for sexual intercourse. This word does duty as an adjective, adverb, verb, noun, and in any form it can possibly be used, however inappropriate or ridiculous in application."[189] That sort of word choice is hardly restricted to the armed forces these days, but at the time it was a significant deviation from social norms. Perhaps more tellingly, it is still common enough for even professional soldiers on leave from combat to frequent female sex workers and sometimes, when discipline is lax, the twenty-first-century soldier like his predecessors over the millennia engages in rape. Modern armies condemn and for the most part control the propensity of soldiers to rape. But there's plenty of evidence that this has not always been the case. Yanomamö men of the Amazon Basin consider it a given that successful raids lead to the acquisition of women, for example. If a raiding party ambushes a couple some distance from the protection of their home village, the man will be killed and the woman abducted. First, all the men in the raiding party rape her; then when their victim is brought back to the attackers' village, those men who were not present on the raid have their turn. Finally, she is allotted to one man as a wife. This particularly bald example of connection between war and rape is hardly unique. To the Maori of New Zealand, for example, the proverb "men die for women and land"[190] expressed a self-evident truth. After one battle, Maori warriors immobilized the women by running "sharp sticks through their feet to prevent their escape. A man then copulated with any one of them and the moment he rose from the act he killed her.[191] As in the case of poor Madam Bee's daughter, chimpanzees sometimes bring back a new female after raids on a neighboring troop—and we would have to be willfully blind to think that our own troops, however heroic or disciplined they may be, do not sometimes mimic this behavior.

There has probably never been an army of any size that did not use prostitutes. The slang word "hooker" reputedly comes from the Union General Joseph Hooker, who corralled sex workers in one area of Washington during the Civil War. Some camp followers who provided sex for soldiers went of their own free will, but more often they were driven by poverty and lack of choice. In some cases they were forcibly conscripted. The attitude of those in command has varied from naïve denial to active encouragement. In World War I, the British, French, and Germans supplied their soldiers with condoms. U.S. politicians at home, in contrast, pretended that American doughboys in France were chaste. An epidemic of sexually transmitted diseases proved they were not.

In the 1930s, the Japanese Imperial Army, fearful that the spread of syphilis would incapacitate their troops, instituted the notorious system of "Comfort Women." These women were usually the youngest daughters of the poorest Japanese and Korean farmers and fishermen. Conscripted as sex slaves, they were forced to service twenty-five to fifty men at a time, each for an allotted three minutes. The women received a small stipend, and were expected to send money home to their families. After the defeat of Japan in 1945, the Japanese cabinet was fearful that "sex-starved" American soldiers would arrive and rape Japanese women, and within two weeks of the surrender the government had decided to build a "sexual dike" in order to preserve the chastity of the rest of the population. A Recreation and Entertainment Association recruited 55,000 Japanese Comfort Women to serve the occupying Americans. Many had been prostitutes before, but an appeal was also made for young women "to sacrifice themselves."[192] The American administration closed the Recreation and Entertainment Association after only six months, not for moral reasons but because of rising rates of sexually transmitted diseases. In 1945 when Berlin fell to the invading Russian Armies, the troops were told, "Soldiers of the Red Army, the German Women are yours." "As the Russians came," wrote one woman, "they just raped women. They lined them up against the wall, dragged my mother and grandmother out...they raped her too...just like they did me, dear God." Sometimes a soldier might give a child chocolate and rape her mother.[193] In the Vietnam War, U.S. soldiers on "rest and recuperation" in Saigon or Bangkok exploited a sex industry that had existed before their arrival, but which became many times larger and more flamboyant under their patronage.

Did Warrior Amazons Exist?

If team aggression is simply an extreme form of behavior found in both sexes, then somewhere in the world, or at some time in history, we would expect to find a group of female warriors with all the characteristics of male coalitionary aggression—a "band of sisters" structured like the seemingly universal band of brothers, who come together spontaneously to attack and kill some outgroup. The trope is certainly common enough in legend and fantasy. Real-life examples, however, are starkly absent. Certainly all female mammals will defend themselves against attack with the same vigor as males, and women are no different. But there never has been—and it seems never will be—a female equivalent of the Greek hoplites, where a team of warriors come together voluntarily to fight, in the absence of immediate threat. In fact, the only all-woman group of warriors that is reasonably well documented is the female bodyguard of the kings of Dahomey in nineteenth-century West Africa, in what is now Benin.[194]

The Dahomean king had absolute power over life and death. Subjects could be executed on a whim, and beheaded corpses were salted and hung in the marketplace—a grim reminder of the royal prerogative. He could lay claim to any woman he fancied in the kingdom, and he and his courtiers did so with such frequency that the sex ratio in the rest of the country became seriously skewed. At one point, the king actually had to subsidize prostitution to stop the commoners from fighting. One group of women, "who for some reason or other came under the control of the king, but whom he did not desire because of lack of personal attractiveness," were recruited as the royal bodyguard. They were initiated in a ceremony where older guards made a cut in each recruit's left arm, collected her blood in a polished human skull, mixed it with alcohol, and fed it back to her. In drinking this she would be "reborn," swearing loyalty to the other women. The older guards carried whips, and enforced a strict training regimen. The punishment for breaking the required vow of chastity was death. The women wore short tunics, carried a club and a razor-like machete, and used muskets during battle. The all-female bodyguard was also eventually deployed in warfare, and any woman not returning with a prisoner or the head of an enemy would be punished. They sang songs with blood-curdling lyrics such as "We will return with the intestines of our enemies." They were said to soak their fingernails in brine and sharpen them as weapons—not a terribly practical measure perhaps, but an effective addition to the iconography of all things womanly turned terrifying.

It is difficult for twenty-first-century readers to comprehend the cruelties of ancient despotic kingdoms, such as the Assyrians, the Mongols, the Aztecs, or the Dahomeans. The Dahomeans, like the ancient Assyrians, had to prostrate themselves on the ground and throw dust over their heads when the king passed. One of the celebrations at the annual Dahomean Great Council was to decapitate one hundred slaves and prisoners of war. In 1889, when the new king Behanzin had several hundred people killed during his father's funeral, his female bodyguard acted as executioners. Only women and eunuchs were allowed in the royal compound, so in a way it is not surprising that the royal guard was comprised of excess concubines. Despite many similarities to male warriors, these women did not come together spontaneously in order to engage in team aggression as men do. Ultimately, the Dahomean king's bodyguard was an invention of a male despot with a voracious sexual appetite and a thirst for cruelty—and the women were forced to exist under unimaginably harsh conditions.

How Many Children?

If women have not usually been direct combatants in battle, their roles in war and peace have nonetheless been significant throughout history. To see how women have a key role in one of the most important factors determining the likelihood of war and terrorism, we first have to take a deeper look at how population and demographics have changed over the course of time.

Beginning 5,000 years ago or more, settled agriculture and the more consistent food supplies it provided led to a decline in the age of puberty. The availability of cereal grains and cows' milk provided new ways to feed infants, resulting in a decrease in the average duration and intensity of breastfeeding, which acts as a natural contraceptive. [195] There are modern examples of the effects of agriculture on reproductive potential, as well. !Kung hunter-gatherers in southern Africa took a century to double their population from the late 1800s to the late 1900s.* Meanwhile in Kenya starting in the 1960s, the population of agricultural groups grew at over 3 percent per year, doubling every twenty years. Girls became pregnant at an earlier age, and babies were born closer together. Rapid population growth

---

* Population growth of hunter-gather clans is typically slow. Puberty arrives late—among some tribes in the New Guinea Highlands, women do not begin to menstruate until they are eighteen to twenty years old—and pregnancies are spaced almost four years apart by breastfeeding. On average, a hunter-gatherer woman might have four to six children in a fertile lifetime, half of whom could be expected to die before reproducing. Even slow-growing populations exceed the carrying capacity of the environment eventually, however, which means either appropriating the resources of neighbors, or dying of famine and the diseases that stalk malnourished populations.

has caused the ratio of younger to older men to grow, and this has contributed to the kind of near-constant warfare in many countries that was characteristic of early city-states throughout the ancient world, including in the areas we now call Egypt, Iraq, China, and Central America.

Until recently, human population size was limited by scarce resources, as the population size of other species has always been. It follows then that the great events of history that made new resources available have touched off spurts in population growth. Settled agriculture and domestication of animals made larger populations possible. The Industrial Revolution touched off a particularly dramatic acceleration of global population growth: the human population did not reach one billion people until 1830, but quickly passed 2.5 billion by 1950. This explosive growth was not because people had more babies, but because many more people survived due to the huge triumphs of more successful crops, cleaner water, better hygiene and medical treatment, and widespread vaccination. Today, the world population is fast approaching seven billion people. China and India are each home to almost as many people as lived on the whole planet just 200 years ago. Largely as a result of increased access to effective family planning methods, the rate of population growth fell considerably in the second half of the twentieth century, but the population explosion is by no means over. Globally, there are one million more births than deaths every 110 hours. Unless India makes family planning widely available to people who really need and want it, its population (now approximately 1.1 billion) could almost double by the end of this century. China still grows by seven million people every year, and we will see an increase of a billion people on this planet within the next fifteen years.

Such large numbers boggle the mind, possibly for good biological reasons. We evolved in a world where life was lived out in a clan of a few hundred people at most; the need to conceptualize or deal with 10,000 people, let alone a million or a billion, never arose. Perhaps if we had evolved from bees such numbers would be conceivable, but for many people it is difficult to worry about an extra million here, or hundred million there—the numbers are simply too large. And yet humanity's dramatic population increase is not just *a* problem; with the possible exception of nuclear annihilation or another asteroid strike like the one that killed the dinosaurs, it is the single biggest problem we have now or have ever had. Much could be done to lift the bottom 2 billion of today's global population out of abject poverty, but a finite world would run out of oil, water, timber, and other resources (such as grain, for a diet with a lot of animal protein) before everyone became as rich as the person reading this book. The burgeoning population's needs for globally limited food, space, and raw materials, compounded by richer societies' ravenous appetites for consumer spending, are critical factors in everything from species extinction and global warming to global poverty, epidemics of disease, and as we're seeing, violence, war, and terrorism.

Starting in the 1960s, some economists and demographers began to warn of future catastrophes if population growth was not curbed. A number of highly successful, large-scale efforts were made in countries such as South Korea, China, Thailand, Sri Lanka, and Colombia to make family planning choices available, and in the case of China, and India in the 1970s, compulsory. Birth rates fell rapidly as a result. Elsewhere, as in Pakistan, Nigeria, and Palestine, family planning did not reach the majority of people and birth rates remained high.* Countries with smaller families, such as South Korea, China, Singapore, Malaysia, and Brazil began to prosper economically, while those where women were least empowered continued to have many children and consequently some of them have turned into today's most politically unstable and violence-prone nations.

When I first visited Bangladesh in 1969, the Pakistani authorities had imposed a twenty-four-hour curfew because, among other things, rioters had burned the government family planning clinic to protest the heavy hand of the West Pakistan government. My colleagues and I were driven from the airport to the hotel in an armed convoy, and the sessions at the scientific meeting we were attending were unusually full, because if you wandered outside the hotel you risked getting shot. The common practice of skipping meetings to take in the local sights just wasn't worth it.

At that time the average mother in Bangladesh had more than six children. The women were largely illiterate and infant mortality, at 150 per 1,000 births, was sky high. If I had said that in 50 years large swaths of the country would have replacement-level fertility, or just above two children per family,** I would have been laughed out of the room. But that is precisely what happened. Once modern family planning became

---

* In the single case of Karachi, Pakistan, there were one million people in 1950. Today there are 11 million, and by 2025 there will be 20 million—or three-quarters as many people in this one violent, disorganized city as lived in the whole of Pakistan when it was founded in 1947

** "Replacement level fertility" takes into account that some children will die (sadly, even in 2008 almost seven out of every one hundred Bangladeshi babies die in the first year of life) and some individuals may be infertile or never marry.

widely available in the villages, the birth rate plummeted. Then in the 1990s, when the focus was taken off family planning, the decline in fertility stalled for a decade and the average number of children rose to 3.3. By 2008, the average began to fall again, reaching 2.7 across the country and only slightly above two in two large districts.[196] Unfortunately, again because of lack of attention to family planning, the use of contraception has not risen in the past five years, so we must assume women are having more abortions. This is not the best way to lower fertility, but even poor, illiterate women in a conservative society can be so desperate that they will do anything to have fewer children.

When the children of a society with a high birth rate themselves have many children, and start at a young age, a great deal of demographic momentum is generated. Like the effects of compound interest on an investment account, it doesn't take much more than a few percent annual population increase to spur a runaway population explosion, which is exactly what happened in Asia, Africa, South America, and the Middle East in the twentieth century. Like a supertanker under way, once population growth gains momentum, it becomes more and more difficult to stop as well as more potentially damaging. Spiking populations quickly outstrip the budgets needed to keep pace with educational needs, and undereducated children soon become underemployed adults, struggling in stagnant economies or trying to make a go of farming on parcels of land that get smaller as they are passed to each new generation. There are 150 million unemployed adults in today's world, and another 900 million are underemployed. Well over 90 percent of the 700 million young people who will join the search for work in the next fifteen years will enter underdeveloped economies—those that can least accommodate them. Classical economic theory assumes that markets expand to employ a growing population, but in the real world, unemployment remains extremely high in all countries with rapid population growth. Sixty million of those looking for jobs around the world are aged fifteen to twenty-four—they are the shoeshine boys of Mexico City, or the young men selling cigarettes three at a time to taxi drivers almost as poor as themselves in Lagos and Manila. In Egypt, where 800,000 young people enter the job market each year, the underemployed are called *hayateen*—"men who lean against walls." Rapid population growth has been accompanied by massive urbanization in the developing world. Unlike their parents, many of tomorrow's youth will be born in cities, and will know—if only via television screens glanced through shop windows—that there is also a world of brand-name sneakers, attractive cars, and pop culture glamour that they cannot access. Indeed, even the most dysfunctional megacities present a study in the contrasts between the globally wealthy and the locally destitute. The dispossessed children of the future will see, as their contemporary equivalents already do, the rich rolling by in one world, while they suffer in another. And as today, they will listen to religious and political militants promising a better life, or preaching hatred of those who already enjoy one. It is a volatile mixture that already has brought us much of modern terrorism, from Black September to the hijackers of September 11. The anger of young men seems to be especially explosive when they cannot marry because they lack resources—and in India, as we have seen, the situation will be exacerbated by a sex ratio skewed against women.

Stumbling Toward Solutions

When Saddam Hussein was born, his mother could not bring herself to look at him. Her husband and twelve-year-old son had just died and she was unable to adjust to another pregnancy. The abandoned Saddam was brought up by an uncle who taught him to trust no one—and the world has seen the result. The infant Saddam's birth was not the geopolitical equivalent of a butterfly flapping its wings in China and causing a hurricane on the other side of the world; innumerable conditions and circumstances led to his violent rule over Iraq, and its bloody, chaotic end. And we certainly can't make policy based on a single case, however dramatic. But Saddam's sad tale points the way to underlying patterns that we ignore at our peril.

Any discussion of an issue as complex as the role of women in war and peace is bound to attract charges of reductionism. But our argument is not that population pressures are the only cause of wars or terrorism, or that a high ratio of young men in the population always triggers conflict, or certainly that all unwanted children turn into tyrants. Just as an evolutionary precedent can't excuse wartime rape any more than it can remove the shine from battlefield heroism, all wars, battles, and terror attacks are the product of myriad individual circumstances and human decisions. But neither does this mean that biological and demographic factors can be dismissed. As we've seen, they are not only worth examining; in some cases they turn out to be significant, decisive factors in the course of human events. And crucially, the social and demographic factors, unlike many others that may be related to warfare and terrorism, are open to change.

Can a lower birth rate bring peace? Certainly, it will be associated with less competition for resources and a lower ratio of young, volatile males in the society. That can't prevent the occurrence of wars and civil disturbance altogether, and there may be any number of unforeseen negative consequences to living in a shrinking population. Still, there are good reasons to believe that competition for scarce resources is at the heart of a great many wars, human or otherwise—and the larger our population, the more intense the resource competition will be. The additional one million people added to the global population every 110 hours are going to grow up needing to eat, expecting to be employed, and eager to enjoy their share of the world's finite riches. There will be a great many angry young men living out frustrating lives in ever-growing mega-cities, and already, they are banding together as gangsters, warriors, and terrorists. Today, one billion people live in urban slums around the world—in the *favelas* of Brazil, the *bustees* of India, or the *katchi* of Pakistan. The slums first grew out of urbanization, but have taken on a reproductive momentum of their own now, and the number of people living in slums will double by 2030 as a result of recent rapid population growth. People who can see their neighbors with air-conditioning and cars, but who themselves live in shacks and often lack sewage disposal or a reliable water supply, have little to lose, and may find violence a rational option.

It is important to emphasize that team aggression, whether expressed as an urban riot, a terrorist plot, or a chimpanzee raid, does not benefit females. Raiding male chimpanzees eliminated rivals and increased the size of their territory, but this did not increase the number of fruiting trees in the forest. The females could have eaten just as much if the raiding males had stayed at home, and indeed they might have been better off if the battles had never taken place. The record of history and the insight of biology is that "we band of brothers" cannot be transformed into "we band of sisters." It is reasonable to assume that under certain circumstances women (perhaps along with older men) will have a moderating effect on male bellicosity, promoting peace and reducing conflict.

The good news is this: Even though most people want more material things, many parents want fewer children. One of the most important things we can do for the safety and peace of the planet is to let women have control over whether and when to have a child. Looked at another way, if males' violent impulses and predisposition to team aggression and warfare are the result of evolution, then what better way to combat those impulses than by committing the most basic reversal of Darwinian logic of which—to use an appropriate word—we can conceive. Darwinian evolution has tailored a woman's reproductive system to produce five to seven children in her lifetime—and up to ten when patterns of breastfeeding change. Empowering women with easy access to contraception and safe abortion, which they do want, is the only way to counter this pattern.

We can only truly win the war of nature when our offspring survive into the future, generation after generation. It is not hyperbole to suggest that this victory will become considerably less likely if current population trends continue unabated. People's access to the means to control their family size, based on informed personal choice and safe, available options, is as essential to modern living as is literacy, and it has the potential to be a potent antidote to war and terrorism. Unfortunately, the ability of women to control their family size is often resisted by men, whether subsistence farmers, factory workers, priests, politicians, or lawyers. We will return to this sad, fascinating, and important issue in chapter 13. But first, let's take a closer look at the practice of warfare and how it has changed over time.

As we consider the role of women in war and peace, we need to ask ourselves: Is it war we need to explain, or is it peace? The standard view of many anthropologists and sociologists, that armed conflict would not arise if only society and politics could be put together in a rational way, is undermined by a great deal of what we now know about biology and history. But if we accept the evolutionary perspective that conflict is the default human behavior, are we condemning ourselves to live out, in ever more horrific technological glory, the rough pattern of sneak raiding that our pre-human ancestors perfected on the African savannah? Indeed, and emphatically, *no*. For history and biology also show that even if we are a violent species, in which males have an inborn predisposition to engage in team aggression, the specific conditions of our environment, culture, economic well-being, and demographic structure can have dramatic impacts on how our most fundamental impulses are expressed. As is so often the case in an instance of an academic dichotomy, it turns out that there are elements of truth on both sides of the nature vs. nurture debate. If we are indeed violent by nature, so too are we subject to the moderating or inflammatory effects of nurture, in all its forms.

7

RAIDS INTO BATTLES

... beyond the limited and temporary relations of the family, the Hobbesian war of each against another was the normal state of existence.

—Thomas Henry Huxley, 1894[197]

I

N September 1991, two hikers in the Ötzal Alps on the Italian-Austrian border found a frozen corpse at the edge of a retreating glacier. It was not, as they first thought, a recent death but the oldest complete human body ever found. The "Iceman," or Ötzi as the media calls him, is now known to be 5,000 years old. He had a dagger in his hand, a bow was found nearby, and he carried a copper war axe. Initially, it was thought he was a hunter or perhaps a gentle shepherd who froze to death. Only ten years after his discovery did x-rays reveal that a flint arrowhead had lodged in his left shoulder blade, almost reaching the lung. When he died he had fresh cuts and bruises on his hands, wrists, and chest, and DNA studies identified the blood of four other individuals on his dagger and woven-grass cloak.[198, 199] Far from being a hunter, it appears that Ötzi himself was the human quarry. A Rousseauean optimist might propose a one-in-a-million fluke as the explanation for why the only body we have from this age happens to show clear signs not just of murder, but a murder that bears the hallmarks of team aggression. A Hobbesian realist, arguing that such violence may have been the norm, would have probability on his side.

Comparison of Raid and War Behavior in Homo Sapiens	
Raid	War
Small group (can be less than 10)	Can involve millions but fighting units still small groups
Young men in the prime of life	Young men in the prime of life
Older men and women give support	Older men may command; women give support
Warriors genetically related and share a common background	Warriors often bonded by training and battle experience
Brief; may last minutes	Can last for years
Stealthy, unexpected	War can be "declared," but battles are still built on surprise
Risk to attackers relatively low	Risk to attackers may be greater than to defenders
No distinction between combatants and civilians; prisoners rare	May exclude attack on civilians; surrender and taking of prisoners formalized
Mutilation of corpses and rape of captured women occurs	Mutilation of corpses and rape of captured women sometimes occurs; often (although not always) condemned by leaders

In the 1940s, the noted human paleontologist Sir Arthur Keith suggested that peace was the order of the day early in human evolution. Wars were aberrant events until the Old Stone Age, he believed, when human beings began to use tools to conquer their environment—and each other.[200] Harry Turney-High, an anthropologist from the University of South Carolina, who along with Quincy Wright, a lawyer from Chicago, was one of the pioneering students of preliterate warfare in the mid-twentieth century, dismissed raids as "reflecting the ways of human infancy."[201] There is perhaps a grain of truth to this observation, in that raiding and warfare appear to go back to our earliest development. But as we'll see in this chapter, even the most primitive human raiding was hardly child's play, and many aspects of the behavior have persisted to this day. These early writers failed to comprehend the ubiquity and antiquity of our same-species killing, and they overlooked the simple fact that when repeated frequently, and with bloody purpose, even small-scale raids can exact a very high mortality rate. Today, we have access to vastly more archaeological evidence and the tools with which to interpret it. If the predisposition for raiding among young men can indeed be traced back to the common ancestor we share with chimpanzees, then we should find an unbroken chain of evidence showing raids and wars throughout the history and prehistory of our species. The fossil, archaeological, and anthropological evidence all suggests that poor Ötzi's death was anything but an anomaly.

Fossil Evidence

We all know the famous image "The March of Progress," or at least one of the many takeoffs and spoofs. From the monkey on the far left, to the ape on its right, up through various stages of caveman and knuckle-dragger, and finally to modern man striding purposefully toward the edge of the page, the illustration captures a view of human evolution as clear, orderly, and almost inevitable. The real story is considerably more complex, and more interesting. Importantly, it is based not on speculation, but on more than one hundred years of evidence gleaned from early fossils, archaeological excavations, and more recently, DNA analysis. Many of the twists and turns of our journey from ape to man remain obscure and heavily disputed by the academics who study them. But the broad picture is becoming clear.

Our forebears separated from the ancestor we share with chimpanzees in Africa at some point five to seven million years ago. The two lineages have continued to grow and diverge ever since, with many new branches appearing on the family tree of related species and most of them disappearing as our evolutionary cousins and ancestors died out. Today, there is only one human species—us, *Homo sapiens*—but throughout history there have been many members of our lineage, starting with *Australopithecus* something like 3.4 million years ago and collectively known as hominids. The Australopithecine fossils, such as the well-known "Lucy" skeleton from Ethiopia, are close cousins to our own line, but probably not direct ancestors. Remains of our earliest direct forebears have not been identified yet, but the first known member of our own genus, *Homo habilis*, arose in Africa perhaps 2.4 million years ago. By two million years ago, creatures with a brain size and body structure intermediate between chimpanzees and modern humans began to migrate out of Africa. These too were our evolutionary cousins, in much the same way that bonobos are cousins to chimpanzees. As for us modern humans, our own species can claim perhaps 200,000 years of history, if that much. We too arose in Africa, and appear to have started our spread around the world some 60,000 years ago.

If our thesis is correct, then not only chimpanzees and their ancestors but also humans and our ancestors should show evidence of a history of violence at each step of our march of progress. It is not always easy to distinguish injuries caused by violence from postmortem damage to a skeleton or injuries from predators such as big cats or bears, however. And when the evidence is sparse and difficult to interpret, as is inevitably the case with the earliest human remains, then discussions often become polarized. Either all our ancestors were helpless weaklings, running from every passing predator while living in harmony with their neighbors, it seems, or they were unstoppable killers of other animals and their own kind. This is another academic dichotomy, and as usual, the truth probably lies somewhere in between.

When Raymond Dart discovered the first Australopithecine skull (two to two-and-a-half million years old)[202] in South Africa in 1924, he argued that these early hominids had been effective hunters.[203] By 1969, Marilyn Keyes Roper interpreted the fact that twenty out of thirty-six known Australopithecine skeletons showed signs of significant injuries as evidence that these creatures also fought with and sometimes killed one another.[204] David Carrier from the University of Utah argues that the relative shortness of the legs of the Lucy fossil provided a mechanical benefit during fights among Australopithecines,[205] rather as pit bull dogs have short stocky legs; other physical anthropologists, however, dismiss this as too fanciful an interpretation. It has been suggested that the massive brow ridges and unusually thick cranium characteristic of the Homo erectus skulls found in China and Java may have evolved in parallel with the propensity of these tool-using hominids to attack one another with blows to the top of the head or face. One of the skulls of Peking Man (Homo erectus, 250,000 to 400,000 years old) does show evidence of an old, healed fracture.[206] (This could have been the result of an accident rather than an assault, of course, but poignantly, the skull itself was destroyed in World War II.)

The base of some early human skulls found in China had been opened to extract the brains, which has been interpreted alternately as evidence of human cannibalism and of attacks by giant hyenas, the fossils of which have been found in the same deposits as the early human bones.[207] Donna Hart from the University of Missouri and Robert Sussman at Washington University in St. Louis have demonstrated that early hominids were indeed hunted by a variety of carnivores and suggest that the violent nature of early hominids may have been overemphasized.[208] However, there is nothing implausible about arguing that both things are true—that early hominids were vulnerable to attacks from powerful wild animals, even as they learned to hunt smaller animals, and each other. Certainly, chimpanzees are killed by leopards, and also hunt and kill small mammals and conduct lethal raids on their neighbors. Armed with a stick or stones, an early hominid might have been an effective killer of his fellows by day but still fall victim to a large predatory cat at night. In Gombe, the chimpanzees are relatively safe, because leopards have been nearly extirpated in the area. But in Senegal, where French primatologist Caroline Tutin followed chimpanzees in the 1970s, leopards remain a real threat to the apes, which in turn behave with appropriate caution, for example nesting high in the forest canopy at night.[209]

Considering how rarely hominid fossils are found, in some ways the evidence of same-species killing is surprisingly strong—and some of it is clear enough to convince a coroner of foul play. Some superbly preserved 780,000-year-old hominid fossils have been excavated from the Gran Dolina site in northern Spain.[210] The fossil remains of six individuals, identified as *Homo antecessor*, show cut marks that indicate "the flesh of at least six people of all ages at Gran Dolina was put to very calculated, meticulous, and efficient use, as if they had been game animals captured for meat," according to a study of early cannibalism by Tim White, a human paleontologist at U.C. Berkeley. A skull of similar age from Ethiopia also shows cut marks.[211] Take a handful of skeletons—or even 6,000—from any contemporary cemetery from our modern, cannibal-free society and the chance of finding evidence of such violence and butchery would be minimal.

Homo antecessor was probably close to a common ancestor for *Homo sapiens* and *Homo neanderthalensis*. Neanderthals had bigger brains than *Homo sapiens*; they were not the dumb brutes of cartoons. Yet they became extinct. Recent DNA evidence suggests that there was little or no interbreeding between Neanderthals and our own ancestors. Given the well-known physical differences between our ancestors and Neanderthals, did they present the ultimate outgroup? Did our ancestors exterminate Neanderthals in our first excursion into "specicide"? One Neanderthal skeleton from France, buried on the floor of a cave and surrounded by a ring of stones, bore evidence of "one or more violent blows on the right temporal region." No doubt Neanderthals also faced attacks from animals and other Neanderthals, as well as from *Homo sapiens*, but a Neanderthal skeleton from Iraq had "scars on the forepart of the head suggest[ing] injuries received in combat," and another had flint weapons stuck between its ribs—whether the attacker was Neanderthal or modern human, it certainly wasn't a big cat. In *Constant Battles: The Myth of the Peaceful Noble Savage*, Harvard archaeologist Steven LeBlanc estimates that 5 to 25 percent of Neanderthal deaths were the result of raids and wars.[212] The fossil evidence strongly suggests that *Homo sapiens* and our hominid predecessors have been killing one another for millions of years—and that there is no broken link in our evolutionary chain of warfare here.

Archaeological Evidence

As we dig our way through human history, the availability of specimens and the richness of information they provide increases substantially the closer we get to our own time. As we move from the scattered, fossilized bones of paleontology to the more complete picture painted by archaeological excavations of burials, settlements, and battlefields, the evidence of warring behavior continues to build. Signs such as a skull shattered by the impact of a flint axe or an arrow point embedded in the spine represent unambiguous proof of human raids or warfare. The reverse, however, need not be true; the absence of signs of violence does not necessarily prove that the society was peaceful. When chimpanzees or people are killed one by one in a dense forest, the chance of the skeleton being found millennia later is remote in the extreme. Among the Yanomamö people of the Amazon, whom we will meet more fully later in this chapter, one third of adults die as the result of team aggression, but given the isolated circumstances under which most killings occur— one at a time and often some distance from any dwelling—there's a good chance that even a crime scene investigator would fail to find physical evidence of most warrior raids. For an archaeologist working hundreds or thousands of years after the fact, the probability of missing subtle clues of violence increases many fold.

Fortunately, archaeologists have many ways of seeking evidence as they revisit the coldest of cold cases. Fortifications indicate warfare, for example, and suggest that raids and battles were quite frequent, but they provide no direct evidence of the lethality of the fighting. Grave goods may tell of a warrior, but the physical evidence will not show how he behaved in battle, or how often he fought. Skeletal remains may show how he died, if it was a skull fracture from a mace attack or a piercing arrow. The cut marks from scalping can survive on the skull, and embedded arrowheads would even outlast the bones they've pierced. But many battlefield wounds involve soft tissue, which with the rare exceptions of Ötzi and a few other similarly preserved corpses, has long since disappeared by the time archaeologists arrive on the scene. A spear thrust to the small intestines or a knockout punch followed by strangulation does not fossilize. Abandoned weapons and corpses may be washed away, removed by animals, dismembered by the victors as trophies, or buried, burned, or otherwise disposed of by the vanquished after defeat. It is reasonable to expect that direct archeological evidence of warfare will be limited, and that it will actually tend to underestimate the frequency and bloody nature of past conflicts.

Given the implausibility of physical evidence surviving, it is an indication of our violent ancestry that so much archaeological evidence of killing does indeed exist. Simply put, Ötzi wasn't alone. The Kennewick skeleton, which is twice as old and among the best-preserved human remains from the early New World, has a flint projectile embedded in his pelvis.*[213] It is thought the flint projectile was fired by a spear thrower from a distance—most likely evidence of a raider's sneak attack rather than a murder executed in a moment of passion. An even older skeleton, found in a cave in San Teodoro, Sicily, reveals that, as with chimpanzees, there were no gender niceties in early human killing—this time it is a woman, from the last Ice Age some 13,000 years ago, who was found with an arrowhead in her pelvis.[214]

Not every individual in our past died a violent death, of course, but the litany of known episodes of ancient violence certainly suggests that our ancestors showed a martial swagger at each step of our march of progress. At Grimaldi in Italy, a child's skeleton 20,000 to 35,000 years old was found with a stone projectile in the spine. The excavation of a 10,000-year-old Mesolithic cemetery in Eastern Europe, Vassil'evka, uncovered skeletons with arrowheads in the ribs and spine. In a 14,000-year-old Stone

---

* Unfortunately, scientific evidence of this type is being threatened by the emotional drive to rebury ancient skeletons under the Native American Graves Protection and Repatriation Act, even when there is no plausible link to any existing group today.

Age cemetery in Nubia, Sudan, 40 percent of the fifty-nine skeletons excavated had projectiles embedded in or associated with the bones—one woman alone had twenty-two distinct projectiles in her body. As further proof that this Stone Age equivalent of "shock and awe" was not an isolated cataclysmic episode, some of the skeletons also showed healed fractures, testimony to prior battles.[215] Cave paintings from Pech Merle and Cougnac in France show men wounded with arrows, as does South African prehistoric rock art depicting Bushmen fighting. The French paintings are approximately 30,000 years old, and they include men wearing masks. They have been interpreted as portraying "magic killing," but with so much other evidence of violence occurring at the time, it seems unnecessary to invoke fantasy rather than explaining the depictions as common events in the lives of the cave painters.*

If the evidence of ancient violence is abundant, excavations around the world have also uncovered compelling evidence of raids, or small wars, in which many people were killed at the same time. At an upper Paleolithic site called Dolni Vestonice in the Czech Republic, archaeologists found a number of men who apparently died together some 26,000 years ago and were buried in the same grave. One of them had a fresh head wound. (Another possible explanation is that their community hunted mammoths—dangerous work, and obviously it is hard to distinguish a tusk wound from a club wound.) At a Mesolithic site at Ofnet in Bavaria, thirty-eight men, women, and children were killed together some 8,500 years ago; one hundred years ago, archeologists found two pits containing their decapitated skulls.[216] At Asparn, a village in Austria, about 150 Neolithic skeletons estimated to be 7,000 years old were uncovered together, some with skull fractures and other evidence of horrific violence.[217] In a feat of ancient forensic sleuthing, a team of Viennese researchers was able to use the naturally occurring element strontium to distinguish locals from outsiders. Strontium is picked up from the diet and transferred into bones

---

* Some authors have suggested that the familiar "hand stencil" style of cave painting indicates that Stone Age people could be right- or left-handed just as we are today. Left-handedness is a partly inherited trait which, for unknown reasons, is associated with a slightly shorter life expectancy. Geneticists have struggled to understand how the trait has survived, assuming there must be some advantage to offset the shortened lifespan. One possibility is that left-handed people are at an advantage in fights with right-handed individuals, perhaps especially when using weapons. Boxing, fencing, and the martial arts all have a disproportionate number of left-handed winners, while in swimming and racing, left-handed people show no advantage. Left-handedness is also particularly common among the Yanomamö (23 percent) and New Guinea Highlanders (27 percent), where raids are common.

and teeth as they grow, and subtle variations in its chemical signature provide evidence of where a person lived. In the case of the Asparn skeletons, strontium analysis of the bones showed that most of the skeletons belonged to the villagers who had been massacred, while a small number of skeletons belonged to assailants from another area who had died in the attack.[218] Taking the analysis one step further, the researchers observed that there were fewer skeletons of young adult females than we would expect, given what we know of the age structure of such populations. Had the attackers killed the others, and carried away the young women?

Tell Brak, an ancient mound near the border between Syria and Iraq, contains the remains of one of the world's first cities, estimated to be almost 6,000 years old. In 2006, a bulldozer digging a trench half a kilometer from Tell Brak unearthed piles of human bones, some with signs of injury. The dead were mainly young or middle aged and a lack of foot and hand bones suggests the bodies had partially decomposed before they were buried. Clearly a costly battle had been fought. Whether the archaeologists had found the victors or the vanquished is uncertain, although it is clear from the broken pottery and cow bones that someone held a feast on top of the mass grave.[219] Almost 4,000 years later, in a New World urban setting, another killing likely took place and left an archaeological record: At Cancuen, a Mayan ruin in present-day Guatemala, thirty-one skeletons have been found of a king, a queen, their retinue, and their children. The bones reveal evidence of axe blows, lance thrusts, and mutilation. The king was treated with more respect after his assassination, however, and buried in his regalia with a necklace bearing his name, Kan Maax.[220]

Evidence of historical battles and warfare can be found throughout the Americas, from before European contact as well as after. Some archaeologists argue that the Native Americans fought each other because the Spanish and Northern European settlers introduced more lethal technologies, from horses to gunpowder, and pushed eastern tribes west, leading to increased conflicts among the tribes.[221] However, there is evidence of massacres long before Europeans arrived on the scene.

Take the Casas Grandes site in Chihuahua, Mexico, a heavily defended community dating from 1300 to 1400 A.D. Steven LeBlanc believes that the entire community of perhaps 2,000 people was slaughtered.[222] Throughout the broader American Southwest, twenty-four of twenty-seven known settlements were abandoned at about the same time, almost certainly because of raids and warfare. Further afield, at a carefully studied site at Crow Creek, South Dakota, dated to approximately 1325 A.D., fortifications enclosed a group of fifty houses. Excavation revealed 486 skeletons that appear to have been hastily interred in one of the defensive ditches shortly after a massacre. Forty percent of the skulls had depressed fractures. Broken teeth testified to blows to the mouth, and nine out of ten skulls had the telltale cut marks of scalping. There was also evidence of severed hands and cut marks inside the jaws, suggesting efforts had been made to cut out the tongues of some victims.[223] As the evidence of historical warfare accumulates, discerning its causes is still a challenge. But it is notable that Crow Creek at the time of the massacre seems to have been in the midst of a phase of rapid population growth, and while it is sited in the midst of a fertile valley, the skeletons show signs of malnutrition including stunted long bone growth, scurvy, and other evidence of vitamin C deficiency.[224] Remembering the demographic factors we discussed in earlier chapters, it is not unreasonable to wonder if this was a case of a community that fell victim to the volatile combination of resource shortfalls and a spike in the proportion of young, aggressive men.

At Larsen, also in South Dakota, near the Missouri River, excavations have uncovered twenty-nine houses protected by two ditches and a wooden palisade—clear signs that attacks were a serious concern. The houses had been burnt and over seventy bodies were found inside and scattered around the site. All the bodies showed signs of mutilation, including scalping, decapitation, removal of the hands or feet, and probably disembowelment. The male genitals or female breasts may also have been removed as trophies, judging by similar mutilations seen elsewhere, but this is a case in which it is impossible to know, since only the skeleton survives.[225] The Larsen site was occupied between 1750 and 1785, and musket balls as well as arrowheads were found. At Crow Creek and the Larsen site, as well as at Ofnet 6,000 years earlier, women aged twenty to twenty-nine were slightly underrepresented among the dead, again suggesting that they may have been captured and carried off as the rightful spoils of a successful raid.

Anthropological Evidence

In 1857, an American stagecoach driver witnessed the last battle between two Native American tribes at Puma Butte on the Gila River in what is now southern Arizona. Isaiah Woods was running the mail coach on the month-long journey from San Antonio, Texas, to San Diego, California, and came upon the fight accidentally.[226] The war, fought without modern weapons, began with a raid on a Maricopa village by Yuma and Mojave warriors, who killed three people, burned the village, and destroyed the crops. In response, the Maricopas mobilized an army of perhaps 200 horsemen armed with bows and arrows. The Yumas and their allies, also numbering about 200, were on foot. The Yuma fought bravely but their line was broken by the Maricopas' arrows, and then the horsemen killed the fleeing Yuma warriors with war clubs, even coming back later to kill the wounded. The main battle lasted about one hour; over 150 Yuma and 25 Maricopas were killed. In relation to the original number of warriors and the total population of the various tribes, the battle Isaiah Woods stumbled upon that September day was an order of magnitude more lethal than any modern war fought with tanks and machine guns.

In the late 1960s, I talked to an old Dyak man in Sarawak, Borneo, who lived in a traditional longhouse with a number of other families. He told me that he had sold the shrunken head of the person he had killed as a young man to a tourist some years previously. The Iban, a group within the Dyaks, believed that collecting and drying an enemy's head transferred the spiritual energy of the dead person to the Dyak warrior, bestowing magical powers to cure all misfortunes, from crop failure to human infertility, or end a period of mourning for a deceased relative. At an ecological level, killing neighbors and also proved an effective way of extending territory and increasing food supply.[227] However, taking heads for their magical powers became a motive in itself. In the eighteenth and nineteenth centuries the Dyaks became pirates along the coast of Borneo, sometimes killing one hundred, or in one case perhaps as many as 400, sailors in order to collect their heads.

The days of tribal raiding were long over on Sarawak by the time I arrived. The granddaughter of the one-time headhunter I spoke with wore a bra, his grandson a T-shirt, and a broken-down automobile stood as the sole reminder of the money the clan chief had acquired selling tribal lands to outsiders. Today I am told that some Dyak long-houses have Internet connections. There is something deeply sad but also inevitable to the way all preliterate societies begin to decay almost immediately upon encountering Western technology and culture. We vaccinate their children but fail to give them control over the number of children they have; we proselytize them to our religions and we stop them fighting one another, but we seem always to bring our guns and alcohol along as well. And then we watch in surprise as their ancient cultures and traditions slip away.

It is hard to know if what we learn from contemporary hunter-gatherer groups truly reflects the state of affairs during the longest phase of modern human evolution. If we want to be sure we understand wars and raids in genuine preliterate societies, then we need to go back two or three generations more, even thousands of years. Julius Caesar opens his history of *The Conquest of Gaul* by emphasizing that Celtic warriors "used to fight offensive and defensive wars almost every year," and he reported that the Helvetti "are in almost daily conflict" with the Germans.[228] Strabo, the Greek writer, considered the Celts to be "war-mad."[229] Invading Britain in 54 B.C., Caesar described his opponent Cassivelaunus as being "continually at war with other tribes." The Roman historian Diodorus Siculus says the Celts in Gaul cut off the heads of their enemies to "fasten upon nails upon their houses, just as men do, in certain kinds of hunting, with the heads of the wild beasts they have mastered."[230] An excavation at Ribemont-sur-Ancre in northern France confirms those writers' tales of constant fighting, with eighty headless human skeletons from that time being found immediately outside a ritual enclosure.

The warring ways of the ancient Celts seem to have been the rule, not the exception, for preliterate humans all over the world. The Hawaiians fought Hawaiians; the New Guinea Highlanders fought New Guinea Highlanders, and sometimes still do; the New Zealand Maoris fought other Maoris; Native Americans fought other Native Americans. Even if the record is not as clear in other places and times, there is no reason to think that things have not been always thus across all the inhabited continents and throughout history. The conflicts were for the most part either stealthy raids or small wars, often lasting less than a single day. But they were repeated so frequently and with such lethal intent that they would have dominated much of human life, for much of human history.

As European explorers fanned out across the world from the fifteenth century onward, they came into contact with many preliterate societies, sometimes fatally so. Both Ferdinand Magellan (d. 1521) and James Cook (d. 1779), two of the greatest navigators and explorers of the Age of Dis-

covery, were killed by the preliterate warriors they encountered.* Typically, the early accounts of preliterate warfare were written by Western ad-
venturers, or by missionaries, but even allowing for the inevitable biases,
the record is starkly bloody. LeBlanc uncovered the tale of William Buckley, a convict who escaped the Australian penal system in the early nineteenth century and lived with the Aborigines. Buckley described raids
such as a night attack on sleeping enemies who fled "leaving their war
implements in the hands of the assailants and their wounded to be beaten
to death by boomerangs."[231]

Appropriately enough as we explore an evolutionary view of human
warfare, one of the most vivid accounts of human aggression comes from
the hand of Charles Darwin. As the *Beagle* rounded Cape Horn at the
southern tip of South America, Darwin had the opportunity to observe
the Fuegians who, despite living in a desperately cold, wet, and windy
corner of the world, built no houses and made no clothes. The young naturalist watched as sleet fell on one naked women and the naked child at
her breast. "The tribes," wrote Darwin, "have no government or head, yet
each is surrounded by other hostile tribes, speaking different dialects; the
cause of their warfare appears to be the means of subsistence."[232] Given
their relentlessly inhospitable environment, the Fuegians might represent
a particularly bleak extreme of the preliterate condition. But even with
better weather, much of our history may well have been spent in a similarly hostile, uncertain world.

In the 1930s, when Europeans first explored the Highlands of New
Guinea, they were astonished to find almost one million people living in
a lost world with a Stone Age culture, totally isolated from external civilization. Like the Fuegians, the New Guinea highlanders were divided into
numerous small clans of a several hundred or a few thousand people, often each with its own language. Almost one million people lived in Stone
Age isolation from the modern world until the first gold miners, missionaries, anthropologists, and administrators penetrated the Highlands
in the 1920s and 1930s. The first human beings had reached the island

---

* At the risk of stating the obvious, many if not most of the early European explorers were also
extremely violent by today's standards, and with the exception of a few officers, most would have
been every bit as illiterate and "premodern" as any tribespeople they encountered. And of course
their famous "voyages of discovery" could easily be seen as the first globalization of team aggression. Among the impressions he recorded of the first people he met in the Caribbean in 1492,
Christopher Columbus speculated that he "could conquer the whole of them with fifty men, and
govern them as I pleased." He had already noted the prevalence of scars among the men, the results, they indicated, of fighting off raiders from neighboring islands.

Credit: © Stapleton Collection/Corbis

The English explorer Captain James Cook was the first to circumnavigate New Zealand and interact with the native Maori population. A brilliant cartographer, Cook also kept careful records of peoples he met on his three voyages around the world. Cook's artist here captures the ostentation of a Maori war canoe as well as how almost-continuous warfare drove the Maori to live in defensible places—such as this fortress built on a natural rock arch above the sea—not unlike the ornate armaments and walled cities of medieval Europe. Cook died in a 1779 fight with Hawaiian natives.

of New Guinea about 40,000 years ago, just 20,000 years after anatomically modern humans first left Africa. By 6,000 years ago, they had established one of the world's earliest agriculture-based civilizations, growing taro and other vegetables.[233] In the Highlands, an irrigation system supported relatively dense populations. The young men were initiated into adulthood in secret, painful, and bloody ceremonies that taught them to be aloof or overtly hostile to women. In some societies men joined homosexual groups shortly before puberty and then reverted to heterosexual relations a few years after. Wife-beating was (and is) common and sometimes extreme.

War between some villages was almost "permanent" and fighting was the "breath of life" for certain clans.[234] There seems to be no obvious relation between the frequency of fighting and social structure or population density. Raids on a neighboring village commonly sparked retaliatory raids, until the two parties became weary of fighting or too many had been killed. Truces involved payments of pigs as reparations for deaths or damage, and peace was often short-lived. The twentieth-century Australian administrators of the Highlands outlawed raids and wars, though insults, fights, and clandestine raids remained common. The advantage of pacification, however, was that a man "could eat without looking over his shoulder and could leave his house in the morning to urinate without fear of being shot [with an arrow]."*

The precarious situation faced by New Guinean people gives us a unique and invaluable look at what life might have been like for most of humanity throughout most of human history. In the Eastern Highlands, wars were so common that villages built palisades for defense and watchtowers to detect attackers. If a village was overrun, families might be trapped and their houses deliberately set afire. When conflict arose between larger groups, several hundred men could be involved. Prior to the Western pacification, a war leader would call out his men and challenge his opponents. As with Greek hoplite battles, a set battleground was chosen. Both sides would attack with barbed, stone-tipped arrows, spears, and sometimes axes. The men carried shields and adopted loose, agile formations. A "war" might last no more than several hours, with men entering into and retiring from the fray more or less at will. Sometimes the battle would be renewed the next day, and sometimes not. These encounters were not games or stylized competitions. "The stated aims of warfare," wrote social anthropologist Lewis Langness in the 1960s, "were the complete and total destruction of the enemy if possible. This included every man, woman and child, whether old, infirm, or pregnant. Although it is true that most raids resulted in only one or a few deaths, cases were known in which entire groups were destroyed."[235] One clan of the Maring tribe lost 5 percent of its population in just two battles, and another 7 to 8 percent of its total number in a single defeat. As with chimpanzee raids,

---

* Across the globe, in West Africa, an old man of the Soma people of Upper Volta had a similar perspective. "Our ancestors were born with their hoe, their axe, their bow and arrow. Without a bow you cannot work in the jungle. With a bow you can acquire the honey, the peanuts, the beasts and then a woman, their children and finally you can buy domestic animals, goats, sheep, donkeys, horses. These were the riches of old. You worked with a bow and arrow in the jungle because there could always be someone who could surprise and kill you."

such losses can accumulate quickly, and if one side is able to score a series of victories, they would soon assume the territory and resources of their opponents. If the two sides were more evenly matched, however, the cycle of warfare and recovery could in theory carry on indefinitely.

Yanomamö

Chronic, costly raids and warfare are found in many preliterate societies and across all continents. But few warring tribes have been studied in as much detail, or with as much controversy, as the Yanomamö people of the Amazon Basin.* The Yanomamö are divided into 200 or so villages distributed over 20,000 square kilometers of tropical forest around the headwaters of the Orinoco River in Brazil and Venezuela. [236] As is the case in New Guinea, some political units are as small as 200 people. In this violent society, as in New Guinea also, love between the sexes is not recognized. Men beat their wives with clubs as a way of warning them to stay away from other men. An errant wife might be punished by being shot with an arrow in a non-vital area, such as the buttocks, or have her ears cut off. Raiders abduct women so often that women take their young children with them into the forest, so that if they are captured, their children will go along with them. Brave and effective Yanomamö warriors are known as *unokais*. *Unokais* are more successful at obtaining wives; on average they have two and a half times as many wives and three times as many children as other men. [237]

Violence is ubiquitous in Yanomamö society, and it takes many forms. In chest-pounding duels, a man throws down a challenge by standing legs slightly apart, arms behind his back, and chest thrust forward. His opponent measures the length of his fist to the man's chest, perhaps makes some trial runs and then starting some distance away rather like a bowler in an English cricket match, winds himself up, runs, and punches forward to deliver the hardest possible blow to the man's rib cage. The victim

---

* American anthropologist Napoleon Chagnon has made twenty-five field trips to the Yanomamö and speaks their language. The emotional and intellectual chasm dividing anthropologists like Chagnon, who believe societies are innately warlike, and those who believe they are innately peaceful, is illustrated by the work of Patrick Tierney, another anthropologist. Tierney also visited the Yanomamö and accused Chagnon and his now-dead colleague James Neel of exaggerating the Yanomamö's warlike tendencies by supplying them with machetes and deliberately infecting the tribe with measles. Chagnon pointed out that measles came to the Yanomamö (who live in an area the size of Texas and are in contact with many traders and missionaries) before he and Neel arrived, and that they worked hard to keep ahead of the epidemic by vaccinating the Yanomamö. Chagnon did give the Yanomamö machetes, but he criticized Catholic missionaries for supplying shotguns.

Credit: Victor Englebert/Time Life Pictures/Getty Images

Warfare and violent conflict are a constant fact of life in many preliterate societies, as they no doubt were throughout most of human history. Many of the Yanomamö people of the northern Amazon still live a largely traditional hunter-gatherer lifestyle. This group, photographed in 1999, is putting on a ritual display of force at a funeral, but war among preliterate peoples is hardly play acting. As many as a third of Yanomamö adults die as a result of raids and battles, a number not unusual for tribal groups.

may be stunned temporarily or even fall, but he commonly accepts four or five blows before it is his turn to pound the chest of the other man. The supporters of each combatant cheer loudly, dancing around the two men. If someone who has delivered his blows tries to escape without standing to receive a return pounding, then a general melee ensues until the deserter can be brought back to take his share of punishment. The only way to stop this tit-for-tat fight is to be injured, or to feign injury. Chest-pounding bouts normally takes place between groups with a history of antagonism, but they can also round off a feast between friendly groups—just to demonstrate no one is a coward.

Side-slapping is a variation on chest-pounding. Opponents in this case deliver the hardest blows possible with the open hand to a man's flank, over his kidneys. Still another variation, the most vicious, involves the use of heavy, flexible wooden clubs, eight to ten feet long. The object is to hit the other man as hard as possible on the top of his head while he struggles to remain standing until his turn comes. Some men display the scars on their heads with pride. Club fights are less orderly than the measured blows in chest pounding, and blows with clubs may strike the shoulders, or break an arm instead. Most club fights are over women and therefore they are most common in large villages where extramarital sexual unions are more frequent. Once a club fight begins, other men may join in until most of the adult men of the village are fighting. Clubbing, side-slapping, and chest-pounding follow the rules of ingroup aggression, and like a boxing match, combatants avoid the genitals or causing severe laceration. (Even when machetes or axes are employed in side-slapping contests, only the flat of the blade is used.) But the restraints on violence are always fragile. During a chest-pounding, some men conceal a rock in their fist to increase the force of the blow.

Raids in preliterate societies are easy to distinguish from this sort of stylized ingroup violence. And they seem to follow the ancestral primate pattern of team aggression closely. Raids involve stealth and surprise. If the village to be attacked is some distance away, the raiding party may travel by a circuitous route and avoid lighting fires, even though the nights are cold. While the goal is to kill men, if a woman can be captured the warriors consider this an added bonus.

We would do well to remember that if these expressions of aggression seem bizarre, we don't have to look terribly deeply into our own experience—small town politics, campus committees, school-ground fights—to find versions of the same type of swaggering, macho, quick-tempered, senselessly competitive behavior. Terms like "preliterate," "tribal," or indeed, "primitive" can create an impression of difference. But the Yanomamö and the New Guinea Highlanders are not only like us—they *are* us. Presumably an American baby raised among the Yanomamö would be as eager and able a chest-pounder as any potential competitor, while a Yanomamö baby raised in San Francisco, say, could be expected to run someday for Congress with every bit as much or as little mudslinging as any other candidate. Our cultures may be dramatically different, but they are each expressions of the same evolutionary history.

Fighting Back

As we look back over our shoulder at the long march of human evolutionary history, it seems increasingly clear that our bizarre predisposition to kill one another was present at every step along the way. But there is no reason to think that our ancestors stood still and defenseless like dueling Yanomamös in the face of an assault. Any military strategist will tell you that you won't last long in hostile territory without a good defense, and given that team aggression has been an inherited trait for literally millions of years, it is to be expected that a series of behaviors also have evolved to respond to such attacks. The surprise is that our ancient inborn responses to team aggression may be every bit as dangerous to our modern welfare as the predisposition to kill itself.

Chief among these defensive responses is an intense alertness when a member of the troop is attacked or killed. Chimpanzees react very strongly when they see the dead body of a troop member, for example. They will stare, seemingly mesmerized, and may approach the body and then pull back. The cause of death doesn't seem to matter. When an adult male fell from a tree and broke his neck in Gombe Park, the rest of the group became intensely excited. They screamed and displayed around a corpse and ended up throwing stones at it.[238] It is interesting to note that predators, such as leopards, tend to drag a corpse away and eat it, so finding a member of your troop mutilated but uneaten corpse would be convincing evidence of aggression by neighboring chimps.

As hominid groups grew larger, new defensive impulses may have evolved. One very significant innate response might have been to react especially strongly to, and be wary of, any situation where more than one corpse was found together. As many raids in preliterate societies were conducted stealthily, it follows that finding one or more corpses would provide a compelling reason to be afraid of attack and extremely alert. Chimpanzees kill one by one, but weapons changed the face of same-species killing, and a time would have arisen in hominid evolution when multiple victims were killed at once. Perhaps this is why in the contemporary world a train wreck or plane crash catches public attention much more than a similar, or greater, number of people dying in isolated incidents. Human beings do seem to react disproportionately to several deaths occurring at the same time, and crucially, this is something that terrorists aim to capitalize on with their bomb attacks and hostage takings. By contrast, when people die one at a time, whether in auto accidents, or of cancer caused by smoking, or even in isolated robberies, muggings, and other attacks, we are less alarmed even though the cumulative burden of death and disease is very much greater.

The corpses of chimpanzees killed in raids make a grisly sight, beaten, torn, and perhaps castrated without the benefit of a sharp blade. Human warriors, too, often mutilate those they kill. In October 1993, when U.S. Marines were attempting to pacify warlords in Somalia, militias disfigured the bodies of Americans they killed and dragged them through the streets of Mogadishu. It horrified the American public, and is something U.S. Marines talk about with anger and disgust to this day. In an effort to kill or capture the Somali commander two U.S. helicopters were shot down over Mogadishu, eighteen Americans were killed and seventy-three injured. (The Americans, however, had much greater firepower, and possibly 500 Somalis were killed.) A great effort was made to recover the corpses of those who had died in the crashes. It was difficult to pull the damaged helicopters apart and in the struggle to free the last body several Americans became otherwise avoidable casualties.[239] In the Iraq War, in 2004, images of the scorched corpses of American security contractors suspended from a bridge in Fallujah, Iraq, had a greater impact on the public than any report of more numerous deaths of soldiers unassociated with such troubling images. There is a strong military ethos about recovering the bodies of fallen comrades—in order to avoid such spectacles of desecration, surely, but also, and perhaps more deeply still, because the obligations of mutual support and sacrifice implicit in team aggression don't end with death.

Perhaps no single mass casualty event in recent history has caused a larger outpouring of fear, anger, and intriguingly, camaraderie, than the September 11, 2001, terrorist attacks on America. For months afterward, in the media and in private conversations, the focus was on the sudden deaths of just under 3,000 people and the desperate, heroic attempts to recover remains. Over the full course of 2001, the average citizen was more than three times as likely to be murdered by a fellow American as to be killed by a foreign terrorist, and over ten times as many people died in road accidents than in the Twin Towers. But cold statistics stand no chance against hot, Stone Age emotions, and it is these emotional responses that rivet our attention and therefore also the attention of the media, elected leaders, and the voting public. (Unfortunately, terrorists too seem to know this, and shape their attacks accordingly.) In the run up to the 2003 invasion of Iraq, members of the U.S. Congress, sensitive to their electorate and no doubt frequently recalling that the Capitol building may also have been a 9/11 target, gave free rein to their Stone Age emotions and a perceived need not to appear weak in the face of an external attack. They avoided a much-needed frank and open discussion of the complex issues at stake as a result, and we have been paying the price ever since.

Our response to evidence—or suspicion—of violent attack from an outgroup is often disproportional to the real level of threat. This reaction tends to come in three distinctive forms: an intense desire for revenge, an elevated sense of camaraderie among the attacked, and widespread fear and anxiety. As a result, the initial enthusiasm for fighting back is often both widespread and heartfelt. Many of the men who volunteered at the beginning of World War I were elated to have the opportunity to fight, responding to propaganda about manhood and patriotism* as well as the pressure of being mocked as cowards by men and women alike if they did not sign up. In the months following the attack on the World Trade Center and the Pentagon, there was an overwhelming desire for some sort of national response, and while the rush to enlist was not nearly as universal as it was during the lead-up to World War I, many young men and women were inspired to join the military.** There was also a brief but memorable unity and friendliness among those closest to the catastrophe; acquaintances who had previously passed one another in the street without so much as a nod, now said "Hello," and "How are you?" for the first time. Still, as the stock of empathy went up, so too did that of fear and anxiety—the stock markets lost $15 billion in value as one consequence of that pervasive sense of insecurity.

Widespread fear in the populace is particularly powerful after an attack, and often out of proportion with the real risks. The anthrax scare that followed September 11 capitalized further on our fear of attack, and provoked a response that resulted in approximately one billion dollars of expenditure for each of the five actual anthrax deaths. During the First

---

* In Britain and her colonies, the call was for defense of the Empire. In France, revenge for the Franco-Prussian War was an added incentive, while in Germany, which had been a united country for just forty-three years, young men responded to a sense that they were surrounded by enemies and had to strike first to avoid being swallowed up. In all three cases, these primal sentiments were nurtured and amplified by the government.

** The most famous example is Pat Tillman, the professional football player who left a lucrative sports career in 2002 to join an elite U.S. Army Ranger regiment. Tillman signed up with his brother Kevin, and was killed in an infamous friendly-fire incident while serving in Afghanistan in 2004. There are many similar stories. Dimitrios Gavriel was a twenty-six-year-old Wall Street analyst when the airplanes hit the World Trade Center towers, and he lost four friends in the attacks. Gavriel had to convince the Marines that he wasn't too old to serve, and he dropped forty pounds in order to get in, after the Iraq war started in 2003. He was killed in the November 2004 Battle of Fallujah, after insisting he be returned to the front following an ankle injury.

Gulf War, more Israelis died of fear-induced heart attacks than were killed by Iraqi Scud missiles. Following 9/11, one in fifteen Manhattan residents were reported to show symptoms of post-traumatic stress—including many who lived miles away from the World Trade Center and never saw Ground Zero.[240] Even more remarkable, 29 percent fewer low-birth weight boys were born in the two months after 9/11 than at other times.[241] My colleague Professor Ray Catalano, who studied birth statistics in New York and across the continent in California, suggests that stress increases the risk of spontaneous abortion and premature labor and that male fetuses are more susceptible than female to this risk. It looks as if we are evolved to respond to unexpected attacks and disaster in ways we never suspected, and the effects can be long lasting. Seven years after 9/11, between one-third and one-half of Americans reported fearing they might be subject to a terrorist attack. Alison Holman and Roxane Cohen Silver of the University of California, Irvine, are conducting a long-term study of heart disease among a representative sample of 2,700 Americans. After 9/11 they added questions about fear of terrorist attacks and found that one in seventeen people remained acutely worried even many years later. This subgroup, even after a careful analysis to exclude other risk factors for heart disease such as obesity or diabetes, were three times as likely to develop cardio-vascular disease as other people.[242] It is scientifically problematic to prove causation based on such correlations, but even if the effect is an order of magnitude weaker than this study suggests, then by far the largest number of people killed by Mohammad Atta and his band of brothers will not have died at the World Trade Center but at home in their beds years later.

For much of human evolution it would have been adaptive to respond to finding a corpse showing signs of violent death with increased alertness, vigilance, and ingroup camaraderie. A powerful, unifying response to unexpected violent death would have reinforced the tribal group's ability to defend against and resist future acts of team aggression. In a complex industrialized society, however, an exaggerated response to violent death can prove maladaptive and lead to counterproductive decisions, such as rushing into an ill-advised war or expensive, meaningless efforts at increased security. When the U.S. Department of Homeland Security issues a terrorist alert and people buy plastic sheets to protect their homes from a poison gas attack, they are responding to a vanishingly small risk. The risk of a cooking stove fire is much greater, but worried citizens are more likely to buy duct tape to confront a perceived terrorist risk than buy a fire extinguisher to confront the vastly more probable danger of domestic fires.

Many of the behaviors discussed in chapter 3, such as conformity to group decisions or a crowd's reaction to an ambiguous situation, could enhance any response to an attack by our own species. Predispositions that would benefit the early hominids response to being attacked may also explain the puzzle of what has been called the "bystander phenomenon." In the 1960s, when a New Yorker called Kitty Genovese was attacked in the middle of the night as she returned home from work, thirty-eight neighbors heard her scream for help for over one hour, but no one came to assist her, or even called the police. Social science experiments conducted in the wake of this event showed that uncertainty about exactly what is occurring makes an individual more likely to conform to the group response rather than to their own personal judgment of the situation.[243] Such behavior can be lethal on a dark night in the modern urban jungle where communication with neighbors has broken down, but in a close-knit clan of early hominids it would have been unwise for an individual to explore a violent situation unless he or she could be certain that the rest of the group was coming along as well. As an analyst studying the Kitty Genovese case wrote, "Each person in an ambiguous and potentially dangerous situation looks to others to gauge their reactions."

Evolutionary psychology may also help explain one of the most important generalizations to come out of modern analyses of genocide: that "he who does the evil is typically convinced that evil is about to be done to him." Evolution has programmed us to fight back when under attack by dehumanizing our enemy and by responding quickly, violently, and without mercy. In the modern world, we certainly do still need to defend ourselves against attacks. But if contemporary leaders are to make optimal decisions then they need to adopt a focused, analytical approach, including recognizing the biases evolution has built into our behavior. The point of understanding our enemies' behavior is not to try to excuse it, but rather ultimately to defuse or destroy it.

The "saddle up and ride" response to a group attack is very understandable. The group rush to aggressive mutual defense and revenge would have been an important survival impulse when a full muster meant perhaps one or two hundred fellows, and the penalty for a false alarm might be no more than a few lost hours and a wasted adrenaline rush. But as our populations have grown and our technology has become ever more lethal, the potential costs of indulging our Stone Age emotions have become prohibitively high.

Looking back on his errors as U.S. secretary of defense during the Vietnam War, Robert McNamara pointed out that America had misjudged the geopolitical intentions of the Vietnamese and exaggerated the danger to the U.S.[244] Overestimating an adversarial threat may have provided an important safety margin during the Stone Age, but it can seriously mislead us in our decision-making today. It can stir the emotions of the professional intelligence services and elected leaders as readily as those of a farm boy or stockbroker volunteering to fight. Prior to the invasion of Iraq, the CIA consistently and uncritically accepted misleading data on the weapons of mass destruction Saddam Hussein was thought to control.

During the 1970s, U.S. politicians claimed that even the CIA was underestimating the Soviet military capacity, and voted for huge increases in the military budget. We now know that the U.S.S.R. was near economic collapse at the time, and far less able to harm us than we thought. The American overreaction continued, however, and may ultimately have helped bring about the implosion of the Soviet Union in the 1980s as its leadership, also driven by misjudgments of the threat they faced from the West, bankrupted their nation in the attempt to keep up in an impossibly expensive arms race. The end of the Soviet Union was certainly to be desired, but the stakes in this nuclear brinksmanship were outrageously high, and the dangerous process of nuclear brinksmanship could easily have ended in a thermonuclear exchange of unimaginable destructiveness. In 1983, for example, when NATO held a fairly routine nuclear launch exercise, the Soviets responded to the simulation by putting their forces and nuclear arsenal on actual alert, in readiness for a real nuclear attack.[245] Had one more misinterpretation occurred, it might have triggered a Soviet pre-emptive strike against NATO. Stone Age passions are simply too dangerous in a world with so many weapons of mass destruction. Of course we have to fight back when we are truly threatened, but if we want the world we're trying to protect to survive the defense, we had better do it with cool, analytical objectivity and the maximum public scrutiny the situation permits.

An Unbroken Chain

The physical evidence provided by fossils and archaeological excavations stands witness to a bloody slaughter of our own species stretching back to our hominid ancestors—and as we argue, beyond that to a common ancestor we share with chimpanzees. There are no broken links in the transition from small raiding parties in preliterate societies to the classical battles of antiquity, and on to the laser-guided missiles and roadside bomb attacks of our own time. In the Middle Ages, one king defined a group of less than seven armed men as *brigands*, seven to thirty-five as an *armed band*, and more than thirty-five as an *army*. He was right in the sense that the transition from a raiding party to full-scale warfare occurs not when groups grow from thousands to tens of thousands, but in the transition from tens to several tens. The evolved predispositions that drive a band of Stone Age warriors setting out to ambush a rival[246] and one million Russian soldiers attacking Berlin at the end of World War II are the same, in the sense that a large army is made up of tiny units, sometimes less than a dozen men, who are intensely loyal to one another and who do indeed share the basic characteristics of raiding parties. *

One of the surprising differences, however, is that war has actually become relatively less lethal as it has become more specialized and regimented. Descriptions of preliterate warfare by anthropologists, missionaries and travelers paint a picture every bit as vicious and—in relation to the population base—even more lethal than the great World Wars of the twentieth century. The population of Native Americans in California, before contact with Europeans (and their diseases), is thought to have been about 250,000. A tribe such as the Mohave may have numbered 5,000. A raiding party might lose seven or eight out of fifty warriors, and if fifteen died it was a disaster.[247] Compared to the size of the population, these were indeed high losses. Clubs and bows and arrows have killed on a greater scale, proportionately to the population, than atomic bombs.

William Eckhardt spent half a century collecting numerical data on armies and their casualty rates for his book *Civilizations, Empires and Wars: A Quantitative History of War*, which was published posthumously in 1992.[248] Quantitative history is an especially challenging undertaking, and Eckhardt's careful work reveals a final important insight into

---

* As our hominid ancestors moved out of the forest and into the savannah, they probably began to live in larger groups. A group which is twice as large as another takes four times the brainpower to track all the possible ingroup interactions. Robin Dunbar of Liverpool University has made the interesting suggestion that because of this, cranial capacity may predict group size. If Dunbar is right, then we seem to have evolved to live in groups of about 150. It is a number that makes intuitive sense: It is the number of people who might come to a wedding, or to whom we might send cards at Christmas.

Deaths from Team Aggression Per Generation, Per Each 100,000 of the Population		
Pan troglodytes		10–15,000
Homo sapiens		
Preliterate group:	Yanomamö, South America	10,000
Classical:	Greece	2,300
	Rome	1,800
U.S. Civil War		1,965
World War I:	Germany	4,400
	Britain	2,300
	U.S.A.	110
World War II:	Russia	8,800–13,000
	Germany	4,750
	Britain	988
	U.S.A.	300
Vietnam War:	U.S.A.	33
9/11:	U.S.A.	1

the nature of human wars through time. The historical record is incomplete; victors often exaggerate the number of those they killed in battle, the wounded may die long after the fight, and diseases associated with war often cause more casualties than the clash itself. There are also problems of definition. Should an analyst consider the ratio of war deaths to the population of the country during a period of war, or instead make the comparison to those who lived during the peace preceding or succeeding the war? Do we try to count war deaths per year, or per generation? But even considering all these problems, careful statistical analysis of historical war data reveals a clear progression, with the percentage of deaths in any population from raids and wars decreasing over time. Among preliterate societies, archaeological evidence and anthropological observation demonstrate that from 5 to more than 30 percent of adults died in raids and warfare.[249] This also seems to be the level among chimpanzees.

Classical and medieval wars could be exceedingly bloody, but considering the total population, there seem to have been fewer deaths than in preliterate raids and wars.[250] For the American Civil War we have accurate statistics, and many battles were extremely lethal. For example, on September 17, 1862, at the Battle of Antietam, one man was killed every two seconds for eleven hours. During the whole of the Civil War there were 618,000 Union and Confederate deaths, from a population of 31.5 million—almost 2 percent of the total population of the country but still far less than even the lowest proportion of war-related deaths seen among preliterate tribes. By the time of World War I, both the numbers of combatants and the lethality of the technology had been ratcheted up again, and the slaughter was even more horrible. On July 1, 1916, the first day of the Battle of the Somme, British Empire troops suffered 19,000 killed and 40,000 casualties; total deaths on both sides would top 300,000 men before the battle finally ended—essentially in a draw—four-and-a-half months later. And yet, for all the extraordinary horror of the First World War, by the end Britain had lost 2,300 out of every 100,000 members of its population, a rate just over that of the U.S. during its Civil War.

In World War II, the losses on the Eastern Front were especially terrible. Germany had deployed the largest army in history[251] to invade Russia in June 1940, destroying 200 Red Army divisions and taking 3.5 million prisoners in one hundred days. The tide turned in the Soviets' favor in February 1943, and Field Marshal General Frederick Paulus surrendered to the Russians outside Stalingrad. Estimates of the number of Germans, Romanians, and Soviet defectors who were captured on that day vary from 200,000 to almost 300,000, but only 5,000 ever returned home. Overall in World War II, Germany suffered an estimated 4,400 military and civilian deaths per 100,000 citizens. Russia paid the highest cost, with an estimated fifteen million deaths in a population of 170 million, or 13,200/100,000 of the adult population. By contrast, less than one-third of 1 percent of the American adult population (300/100,000) was killed in World War II. The courage of those who fight and the pain of those bereaved is the same whatever the ratio of deaths to the total population. At the same time, however, it can be useful to have a sense of proportion and to recognize, for example, that U.S. war deaths in the 1940s were less than one twentieth of the death rate recorded among preliterate tribes in the Highlands of New Guinea at approximately the same time. If we could resurrect Ötzi from his Stone Age slumber in the high Alps and transplant him to America, then it is likely that once he got over the shock of automobiles and canned food, he would be astonished as to how safe the world has become.

The preliterate tribes living in the New Guinea Highlands represent the way of life our own Stone Age ancestors lived for thousands of generations. It may be comforting to distance ourselves from such "uncivilized" behavior, but if we are to understand contemporary wars and terrorist attacks, then we need to recognize that we carry deep within us the same impulses and predispositions. After all, we are by definition the descendents of the victors in this most violent war of nature, and the inheritors of their evolutionary legacy. This doesn't mean that we are destined to be violently aggressive ourselves; violence is not all we have inherited. Indeed, New Guinea Highlanders spent most of their time tending their gardens, worrying over their pigs, nurturing their children, and sitting around the fire gossiping. But raids and warfare ruled their lives in a way that still casts its shadow over our own behavior. Shining the full light of inquiry on our history and nature represent our best hope of emerging, finally, from the lingering darkness.

On September 11, 2001, one American in 100,000 was killed. That year, more Americans were killed by fellow Americans with handguns than died in terrorist attacks. Yet 9/11 brought about a tectonic shift in U.S. politics and global diplomacy, and it led directly to hugely costly, unresolved wars in Afghanistan and Iraq. We were not evolved to handle such numbers dispassionately; had the U.S. been managed by a dispassionate super computer programmed to maximize economic growth and life expectancy, then 9/11 would have produced hardly a blip on the national wealth or death rate. But for better and for worse, we are all governed by Stone Age emotions. Our response to outside threats is colored by adaptations to the hunter-gatherer environment in which our ancestors evolved, where being killed via team aggression by a rival group of hominids was an everyday fear. In a world armed with nuclear warheads, weaponized microbes, and noxious chemical bombs, where a miniscule number of terrorists can always find vulnerable targets to attack in our modern civilization, vigilance is still essential. But so too is sober second thought. If we are to survive our own deep impulses, and perhaps take the next step on the ongoing march of human progress, then the urge for self-defense and revenge must be tempered by cool, objective analysis of risk; measured, productive political responses; and military reaction only when absolutely necessary, appropriate, and effective. Understanding the evolutionary roots of our psychological responses can play a useful role in this difficult task.

8

WAR AND THE STATE

"Empires are primarily organizations of violence."
—QUINCY WRIGHT, 1942[252]

T HE WINDOW FLEW UP and six or seven young men rushed into the room shouting. They snatched the hat of my colleague and co-teacher, history professor Tom Barnes, and swept up a young woman from the class and carried her on their shoulders screaming from the room. Undergraduates at University of California are renowned for creative thinking, and one group of students decided to preface their class presentation with a physical demonstration that team aggression is about sex and resources—not that Tom's hat was of great value. The abducted classmate turned out to be a part of the group, although at the time Tom and I, and the rest of the class, were genuinely startled. The presentation that followed was particularly well-informed, partly because one of the students had been in the U.S. military and under fire in Somalia. But perhaps even more radical than the students' approach was that of their teachers. It is highly unusual for faculty from disciplines as far apart as history and public health to teach a course together.

Tom Barnes has an encyclopedic knowledge of military history, from ancient Sumer to the modern Iraq War, and like me he found that insights from chimpanzee behavior were useful in explaining some of the deeper mysteries of human history and behavior. Halfway through the first course we taught together, he leaned over and said, "This is really working." It became apparent as we taught together that although technology, culture, and the growing human population have changed the face of warfare, they have not buried its origins, nor obscured its twin driving emotions of ingroup reciprocity and outgroup aggression. As we followed the story of human history, we saw very quickly that developments since the origin of city-states, some 5,000 to 8,000 years ago, have profoundly shaped the expression of our evolutionary predispositions, and that these deep impulses have in turn helped shape the character of the civilization we have created. If we are to understand the particular nature of war in our modern world, we next have to take a closer look at two often-contentious cornerstones of our culture: politics and religion. Both are powerful delineators of the line between ingroup and outgroup—it's no wonder they divide us so readily.

Consider the Cuban Missile Crisis in 1962, a moment when the world came close to nuclear holocaust. The United States and its allies had 500 warheads pointed at the Soviet Union, some from as nearby as Turkey, while the U.S.S.R. had perhaps 300 aimed at the U.S., all from a much greater distance. Soviet Premier Nikita Khrushchev, like any competitive primate, wanted to secure an advantage over his adversary, so he decided to deploy thirty-six Russian SS4 missiles with nuclear warheads to Cuba, just ninety miles south of Florida. A U.S. U2 spy plane discovered the missiles much earlier than Khrushchev had expected. The American generals, led by Air Force Chief of Staff Curtis LeMay and neatly displaying the classic primate threat response, wanted to bomb and invade Cuba. Fortunately, in this instance Khrushchev and U.S. President Kennedy behaved more like two mature males vying for the alpha male position in the global troop than raiding parties trying to destroy their rivals and conquer additional territory.[253]

Sometimes our leaders, be they democratically elected or despotic, religious prophets or military commanders, will compete for power under the ingroup rules of alliances, reciprocity, and long-remembered hurts and favors rather than the outgroup rules of swift and vicious attack. Kennedy understood that humiliating Khrushchev might set off just such an outgroup confrontation, and that it could lead to mutual nuclear annihilation. (Khrushchev realized that "all humanity might perish," while Kennedy gauged the risk of total devastation as "somewhere between one-out-of-three and even."[254]) Instead of attacking, Kennedy readied 250,000 troops and 1,000 aircraft in a show of force, but limited his actual intervention to a naval blockade. It was a clever move in that it prevented further missiles arriving, but did not start a war—a little like rival chimpanzees displaying at each other but stopping short of an actual attack. Kennedy agreed with Soviet Ambassador Anatoly Dobrynin that the U.S. would remove its Jupiter missiles bordering the Soviet Union in Turkey in return for Khrushchev removing the SS4 missiles from Cuba. Kennedy also made sure the agreement remained secret so he would not lose face in America—he had his own home-troop alpha male position to think about, after all.

The Cuban Missile Crisis played out in much the same way that chimpanzee power struggles do—there was a lot of hooting, banging about, and displaying, but no real change occurred and no physical violence was done. In the end, however, it was as much luck as good judgment that prevented an all-out atomic war. The Soviet Politburo in Moscow decided to send a public message to Washington that they would remove the missiles, but the man carrying the announcement got stuck in the elevator on his way to Radio Moscow—poor maintenance could have started World War III, if the trapped man had not been able to push the message out of the stalled cage. Meanwhile in the Caribbean and unbeknownst to Kennedy, one Russian F class submarine carrying nuclear-tipped torpedoes narrowly avoided being cut in half by diving just ninety seconds before it would have been hit by an American destroyer. The Soviet crew wanted to launch a torpedo but officer Vasily Arkhipov refused[255]—the fate of the world hung on the quick decision of one professional warrior.

This chapter examines how two of our most important human institutions, the political state and organized religion, have transformed the face of warfare but not buried its evolutionary origins. Both national and religious identity can act as vast, powerful ingroups, but both have also provided the organization and common purpose needed to expand team aggression from a matter of brief, scattered raids, to what are probably the largest, most sustained group efforts in human history—our wars. The first city-states were already characterized by the division of labor, accumulation of wealth, and storage of food and other commodities, which are quite limited in preliterate groups but essential for modern societies and nations. The effects of these societal changes on warfare were profound. The division of labor allowed a specialized, full-time warrior class to emerge for the first time, and the accumulation of reserve resources created conditions that allowed wars to go on for years, decades, or even for over a century—as did the Hundred Years' War (1337–1457) between the royal houses of England and France. A chimpanzee raid may last only minutes, assaults in preliterate societies may end in an hour or a day, and even the Greek hoplites eventually had to put down their lances and return to the fields. Modern wars involving industrialized nations, however, have come to engage the efforts of millions of people and can envelop the whole planet for years on end.

The expanding scale of warfare also served to reverse the equation between the risks and benefits of team aggression. As we have seen, raids developed because they were a low-risk strategy for the attackers, who waited until they found a single foe they could easily overpower. The Yanomamö consider a raid a failure if even a single raider dies, even if the attacking party ends up killing many more of the enemy. Anthropologist Napoleon Chagnon, who studied the Yanomamö, tells of a raid in which the attacking group killed two of another village's men, including its headman, but then was overtaken on the return journey, losing one of their number in the skirmish that followed. The raiding party killed two more of the enemy, but even so Chagnon's informant was explicit that the raid was a disaster.[256] Not only does the raiding party's reaction to a single loss demonstrate just how important mutual protection is in team aggression, it also confirms the power of responding to an attack with an immediate, swift, and vicious counterattack. But the rise of city-states and small empires changed all that. Banded together into a much larger force, controlled by an elite level of society, and battling often against fixed defenses, individual teams of attackers lost heavily. So long as the force as a whole prevailed, however, the state and its elites would benefit, no matter the cost in deaths, pain, and bereavement to the front line troops and their families.

As we look back on the changes in warfare that have occurred during the evolution of our ancestors and on through our more recent history, three profound, potentially disturbing questions emerge. Did our deep-seated predisposition to engage in team aggression also feed back into the evolution of our large brains? Is it possible that warfare played a bloody but essential role in the transition from hunter-gatherer bands into increasingly larger and more tightly organized tribes, city-states, nations, and eventually all the civilizations that followed? And perhaps most challenging of all, did organized religions develop because they helped to define clear ingroup identities, and unite and motivate tribes into becoming ever-more successful warrior units?

Agriculture and Industry

The city-states of the ancient Fertile Crescent, stretching from present-day Syria to Iraq, controlled a few hundred square miles of territory each, with populations of perhaps 30,000 to 100,000 people. Rulers had professional bodyguard corps of 500 to 700 men, and at times of war, 4,000 to 5,000 men might be mustered for battle.[257] Tribal egalitarianism was replaced by increasingly strict hierarchies and top-down rule. Even if poorly led, an army of thousands would almost always defeat the hundred or so warriors a lesser clan might be able to field. The man with the largest army, therefore, was usually victorious. Strong leadership, even if capricious and cruel, was rewarded—as it has continued to be in more recent history.*

Urban living in the earliest cities encouraged a division of labor, which in turn allowed specialization and expertise. Some people worked as farmers, providing food for the entire community. Others became craftspeople, including those who mastered iron and bronze manufacture and leatherwork, which led to sharper, more varied weapons and stronger armor. And a dedicated, full-time military caste was also born, meaning that for the first time, warriors wouldn't have to leave the battlefield for spring planting or fall harvest. On the battlefield and off, opportunities for making and breaking alliances, for espionage, and for decoy feints and entrapping maneuvers arose. Surrender became an option (though not a foolproof one: After the battle of Cynoscephalae in 197 B.C., the defeated Macedonians held up their spears as a gesture of surrender. The Roman commander Flaminos tried to stop his troops from killing the prisoners, but they did so anyway[258]). Warfare took to the seas, and as cities built defensive walls, armies responded with the first siege engines. By the second millennium before Christ, the Hittite capital of Boghazkoy, in what is now Turkey, had imposing double walls of stone and brick.

---

* Evolutionary biologists make the important distinction between adaptation and acclimation, and we must stress it here. Adaptation refers to the biological changes that accumulate as a species evolves to suit its environment more perfectly. Acclimation, meanwhile, refers to the short-term adjustments individuals or groups of individuals make to respond to a changing environment. The first process is slow, largely irreversible, and dominated by chance—in short, it is biological evolution. The second amounts to making the best use of one's evolutionary endowment in a given environment, and it can be rapid, reversible, and at least in humans, intentional. Given the long timescales of evolution, it is safe to assume that any changes in human behavior occurring over the last several tens of thousands of years represent cultural acclimation rather than biological evolution.

Chimpanzees in a raiding party momentarily stop foraging for food, and focus on the stealthy actions needed to outnumber and kill an individual of a neighboring troop. A preliterate human raiding party carries, or collects on the way, the ten to twenty pounds of food and bows, arrows, or spears it needs to mount a raid on a nearby enemy. The Maoris in New Zealand sometimes solved the problem of provisioning their warriors in a way only a person evolved to dehumanize his enemies could—by eating those they killed.[259] One European in the 1820s saw twenty women slaves carrying baskets filled with human meat. Settled agriculture with food surpluses, draft animals, and boats enabled armies to be provisioned for much longer periods and to embark on more distant journeys. Today, a typical soldier needs a minimum of one hundred pounds of supplies a day and an airman 1,000 pounds.[260]

The first historical record of an organized state launching a formal war comes to us from the Stele of the Vultures, a Sumerian monument portraying the defeat of the King of Umma by the King of Lagash in 2525 B.C.[261] The King of Lagash is portrayed in a four-wheeled chariot drawn by four asses, wielding a battleaxe and leading a phalanx of troops six deep—clearly, military technology had advanced far beyond the sticks-and-stones stage. Wars between the first city-states of the Fertile Crescent were frequent. For example, the Akkadian king Sargon the Great records fighting thirty-four wars during his fifty-year reign, starting in about 2325 B.C. The Egyptian pharaoh Amenhophis I (1550–1528 B.C.) led his army to within sight of the snow capped Taurus Mountains in southern Turkey, and Thutmoses III marched 20,000 men and 900 chariots into Palestine for the Battle of Megiddo in 1468 B.C. By the eighth century B.C., the Assyrians were fielding armies of up to 50,000 men, equal to five contemporary American divisions.[262] Three hundred years later, the Persian King Darius the Great sent 300,000 men and 60,000 cavalry 1,500 miles in their attempt to subdue the Greeks. Ancient armies lived off the land they passed through, and women and children often trudged along with the soldiers, serving as scavengers, auxiliaries, and companions. A Carthaginian army 10,000 strong, led by military genius Hannibal (247–183 B.C.), survived in hostile Italy for fifteen years by living off the land. In the fourth century B.C., Alexander the Great of Macedonia forbade women to follow his army, which greatly increased its mobility. Now, however, the soldiers had to find sex as well as food from the lands they marched across. Both were taken by force.

In a parallel of the increased specialization in society as a whole, as armies have grown larger, fewer men have actually fought. In 1066, the 12,000 cavalry and 20,000 Norman foot soldiers who faced King Harold's Saxons at the Battle of Hastings in 1066 represented virtually all those who had crossed the English Channel seventeen days earlier,[263] but in the U.S. Army in World War II, only one in six of those conscripted actually came under fire.[264] In parts of the Pacific theater, it took eighteen men to keep one frontline soldier supplied. In Afghanistan in 2001, only a handful of Americans experienced genuine combat. Much of the ground fighting that did take place was carried out by Afghan warlords, paid off with forty-five million U.S. dollars.

Despots and Democracy

All primate societies, whether built around females as among rhesus monkeys or around males as with chimpanzees, are comprised of competitive individuals jockeying for position within a hierarchical social structure. At any given time, there will be one or two dominant, and many subservient, individuals. But primates, especially chimpanzees, have also evolved behaviors to soften hierarchical structures and permit individual freedom of action. As we saw in chapter 3, a chimpanzee troop has an easily recognized alpha male at the top, along with a more subtle but still important female hierarchy. The alpha male can be a bully, but he is not a true despot and the other males often challenge one another to maintain or improve their place in the overall hierarchy. Males within a community very occasionally kill one another, but for the most part they deal with tension within the group by symbolic displays of power and important rituals of submission and reconciliation, such as mutual grooming.

Human preliterate societies developed even more powerful leveling mechanisms, which partly conceal the hierarchical structure of human, male-dominated society although they never totally eliminate it. In New Guinea, there are village chiefs, or "Big Men," who are expected to be good orators and generous leaders. But genuine authoritarian power is strongly resisted by the remainder of the group. Boasting is one potentially effective means of establishing dominance, but a !Kung hunter who has just killed a large antelope does not rush into camp and brag about his achievement. Rather, he waits for people to ask how his day went, and then makes self-deprecating remarks: "Oh, it was nothing really." Christopher Boehm of the Jane Goodall Research Center, University of Southern California, reports a !Kung man who quipped to an anthropologist, "Of course we have headmen...each of us is headman over himself."[265]

But such egalitarian tendencies have their limitations when it comes to more involved hunting expeditions and extraterritorial raids. In such situations, and especially as group size grows, lack of clear leadership could spell death. When the Mae Eng tribe of Highland New Guinea goes to war, all the warriors meet and discuss tactics at great length. Everyone is heard. On the battlefield, however, a "supreme chieftain" is appointed. His power comes more from knowing the consensus of the group than from central authority, however, and his control evaporates once the battle ends. It's not hard to see how similar wartime arrangements could lead soon enough to permanent authority—indeed, the emergency power grab is a favorite of military strongmen and would-be despots to this day.

As the earliest states began to coalesce from smaller clans, the changes in military organization again paralleled those in the broader society. Just as clear class and social stratifications were appearing in civilian society, as raids turned into wars, a defined chain of command developed among the warrior class. Even as the distinction between officers and enlisted men emerged, the spontaneous responses young men make to opportunities and threats began to be supplemented by leaders giving orders. In the second millennium B.C. armies of the Fertile Crescent, historians suggest, there was one officer wielding a sword for every ten spear-carriers.[266] The platoon or fighting unit, sharing as it does many features in common with a huddle of preliterate warriors raiding a rival clan, or adult male chimpanzees killing a member of a neighboring troop, remained the foundation of every army. But the command structure that makes more complex wars possible has no parallel in biology. Levels of control and coordination above the platoon enable armies to bring together huge numbers of men without losing small-team cohesion. Without a multi-layered chain of command, Darius could not have marched his troops to Marathon, Julius Caesar could not have invaded Gaul, and Red Army Marshall Georgy Zhukov could not have encircled the German Sixth Army at Stalingrad. Interestingly, the military command structure of generals, lieutenants, and many other ranks has also become the foundation of all modern administrative systems. But whether civilian or military, and no matter how large, the chain of command is only as strong as its weakest link—just as it is in a raiding party of chimpanzees. The Iraqi army which faced coalition troops in the two Gulf Wars had literally thousands of generals but no sergeants between the officers and privates. This missing link, all technologies apart, may have played a role in the rapid defeat of Saddam Hussein's troops in both 1991 and 2003.*

As specialization continued to develop, the commander at the battlefront and the political leader no longer needed to be the same person. Kings could remove themselves entirely from the physical dangers of war and eventually, even the general was able to retreat from the frontlines to his bunker, leaving the soldiers to bear all the risk of a new form of team aggression, planned and orchestrated by an elite few and carried out by a powerless many. In the twenty-first century in Afghanistan and Iraq, however, technology has brought this evolution of risk avoidance full circle for some. A high commander's cell phone communications can be intercepted, his car can be tracked by a pilotless spy plane, and his bedroom destroyed by a 2,000-pound bomb guided by global positioning satellites hurled high into space years before any conflict takes place. The commander may now be killed before his troops.

In most cases, when chimpanzees commit as a group to launch a patrol into a neighboring territory, they seem to act opportunistically. Sometimes, however, the alpha male excites the other males by leading a display, or even taking on the role of a genuine battle leader. Boehm describes how Goblin, the Kasakela troop's alpha male at the time Boehm worked in Gombe, was leading a patrol near the no-man's land between two troops of chimpanzees. When the patrol heard the cries of the next troop, a junior chimp began to scream in response, but stopped immediately when he realized that Goblin was silent. Goblin moved to a vantage point, scanned the area where the "enemy" was located, and appeared to contemplate the situation for a full minute. Then he began a fearsome set of hoots and physical displays and this time, all his companions joined in.[267] Goblin demonstrated true leadership, and his party benefited from his analysis that it was a sound strategy to reveal their location and strength by responding with a hostile display.

---

* The unusual, and basically unnatural, structure of the Iraqi army also created problems after the U.S.-led invasion. When the Coalition Provisional Authority disbanded Saddam's army and sent them home without pay, it soon became apparent that building a new army from scratch would be very difficult—there was no tradition in Iraq of loyalty among potential recruits beyond their longstanding tribal obligations and religious affiliations.

Bluff and Blunders

In warfare, each side sets out with the illusion it can win, yet relatively often the losing side seems to misjudge the possibility of victory. Whole books have been written cataloguing military blunders, which generally take the form of vastly overestimating one's own prowess or the adversary's weaknesses. Why do commanders behave in this way? Once again, a hint at least comes from the broader animal kingdom. Higher primates struggling for dominance sometimes display behaviors that can be best described as symbolic. For any animal, the least costly way of winning a confrontation is a successful bluff. The alpha male chimpanzee in a troop will break off and wave tree branches, perhaps plunging into water and turning over stones or drumming the buttress roots of trees, and charge and vocalize with its hair standing on end. It is an impressive display for a human observer, as well as for other chimps. Bluff and bombast may have been selected for by evolution because in certain situations they can establish social order without causing injury. As the ancient Chinese military commentator Sun Tzu wrote, "All warfare is deception."

For the most part, it is reasonable to assume that those who gain positions of leadership, whether chimpanzee or human, are intelligent as well as strong or determined, and also experienced in anticipating the unexpected. Evolution certainly would favor animals that are clever enough to detect a deception, and that creates a challenge for would-be bluffers. As every conman knows, the best way to convince a mark that you're telling the truth is to believe your own lies. When it comes to animal confrontations, what better way to convince your rivals to back down than believing in your own inflated prowess and flamboyantly expressing it? Richard Wrangham has indeed suggested that a big-brained, thoughtful primate may successfully deceive an opponent by bluffing.[268] A warrior group may also exaggerate its own capacity to win in order to defeat a larger group—by scaring it off, or by boosting one's own confidence and sense of invulnerability. This behavior can lead to disaster, especially when the enemy knows what you are doing and understands that you are bluffing, but sometimes in surprise attacks a weaker group does indeed defeat a stronger force. It seems that high morale—which after all is a form of self-deception—can also prove adaptive.[269] Apparently these traits have been with us for so long in evolutionary history that we respond strongly to leaders who are convinced of their own superior strength, and who enjoy an eloquent ability to portray the enemy as demons and cowards.

The trouble is, overconfidence is only a short step from crass stupidity. In August 1915, British, Australian, and New Zealand (ANZAC) forces landed at Suvla Bay in Turkey as part of the Dardanelles offensive designed to open shipping routes to Russia's Black Sea ports. The plan was sound, military surprise was achieved, and airplanes spied on the Turkish reaction. Within twelve hours, 20,000 men had been landed. But the British general in charge, the sixty-one-year-old Sir Frederick Stopford, who was grossly overweight and reputedly too weak even to lift his own dispatch case, somehow convinced himself that he and his troops were so superior to their Ottoman Turk counterparts that neither surprise, nor haste, nor preparation were particularly necessary. Stopford did not come ashore for another forty-eight hours, and instead of capturing the high points overlooking the beaches, the leaderless troops went swimming in the Bosporus Strait. Said one observer, it looked like an "August Bank holiday in England." The Turkish forces dug in and ultimately, 7,000 ANZAC troops were to die capturing the heights overlooking Suvla Bay, which as one officer remarked, could on the first day have been taken "at the cost of walking up them." [270]

At the beginning of World War II, British-held Singapore was perceived to be the key to controlling the sea routes between Europe, Australia, and New Zealand. Winston Churchill wrote in a cabinet minute, "Singapore [can] only be taken after a siege by an enemy of at least 50,000." When Japanese forces began to leapfrog through the Malaysian peninsula, the British sent the battleships *Repulse* (which dated from the First World War) and *Prince of Wales* to defend Singapore. The aircraft carrier *Indomitable,* originally intended as part of the Singapore task force, had run aground in the West Indies, but Admiral Tom Phillips, who commanded the battleships, underestimated the importance of air power and sailed on regardless. Both battleships were sunk by Japanese planes and 840 men drowned. Ultimately, a numerically inferior force of Japanese captured Singapore and 20,000 British, Indian, and Asian troops entered prison camps, where many were to die.[271] Churchill was a brilliant leader and strategist, but this was one case where believing in his own bluster—and in the inherent superiority of British troops over the Japanese—turned into a costly blunder.

It is easy to criticize Churchill and his commanders in the defense of Singapore, and much more so the complacency and prejudice of the commanders at the Dardanelles. But peering more deeply at why they erred reveals evolutionary roots to their behavior. In the case of Singapore, Churchill's overconfidence was clearly detrimental. But might his pugnacious spirit and belief in his own bombast also have been central to his success as a war leader? A relentlessly rational person might have abandoned Singapore, or would have defended it in an entirely different way. It could also be argued, however, that the same rational person faced with the British defeat at Dunkirk in 1940 might have decided that immediate surrender was the only logical course. After that defeat Lord Halifax, a senior member of Churchill's war cabinet, came close to asking Mussolini to help negotiate peace with Hitler at that time.[272] Churchill knew he could not defeat the Nazis alone, but he could have lost the war for all the Western allies if he had not been so determined to fight on against seemingly impossible odds until American help finally arrived. Could Churchill have stirred the British nation with his magnificent speeches if he had not genuinely believed in them? His ultimate role as victor and legendary wartime leader supports Wrangham's hypothesis that evolution may have favored those males who believed their own exaggerated claims. Britain and the world may in fact owe at least part of the triumph over Nazi Germany to Winston Churchill's quite illogical and very ape-like refusal to believe that he could ever be defeated.

In 2004, with the U.S. invasion of Iraq already going seriously wrong and the lack of weapons of mass destruction calling the original justification for the war into question, George W. Bush was elected for a second term. In all likelihood, every politician dissembles or lies outright during a heated election fight. But Bush was unusual in the degree of his own belief in his policies, even though many strategists saw them as deeply flawed and unfolding events proved them so. His opponent, John Kerry, spoke in long nuanced sentences and changed his mind on some issues, only to be slammed as a "flip-flopper." In a complex world, it is often wise to question a strategy and sometimes to change direction. But Bush's unusual certainty seems to have tapped into reassuring Stone Age predispositions, and appealed to many voters: 86 percent of U.S. citizens listed terrorism as the prime issue in the 2004 election, and a little preliterate bombast, swagger, and certainty seems to have been broadly reassuring to the electorate.*

---

* Interestingly, the majority of voters also emphasized the importance of "moral values" in the 2004 election. We'll explore the important role of religion and moral codes in warfare and outgroup dynamics later in this chapter.

Conscription and Coercion

Preliterate raiding parties are made up entirely of volunteers: The only coercion is the fear of being seen as a coward by your peers. Caesar in *The Conquest of Gaul* describes how the Suebi, a pre-state Germanic tribe, raised their army. Each year, every canton had to provide 1,000 warriors. Those left at home supported those fighting by doing double duty in the fields, and the following year they switched places with the men recruited to fight.[273] This kind of system worked well enough for small groups, but when larger, more organized political units arose, volunteerism quickly became supplemented by conscription.

Most states have been able to organize forces on a huge scale because they haven't deviated from the basic fighting unit. The predisposition to engage in team aggression, which we have argued is in the hearts of all men, and especially young men, is built around a few dozen individuals who know and trust one another and who are bonded by a common experience of combat. Conscription changed this dynamic fundamentally. Where once brothers, cousins, and genetic kin had fought side by side for solid (if unconscious) evolutionary reasons, now a hodgepodge of battalions, often including conscripts fighting against their will and hailing from entirely different regions, might enter battle together. Formal military training, stern discipline, and a well-developed command structure became essential to replace the natural cohesion and unity of purpose inherent in a chimp raid or a tribal battle.

Ever since the ancient Assyrian Empire flourished in what is now Iraq some 3,500 years ago, the state has chosen who was to fight and then literally beaten men into shape, adding the whip and the threat of execution to the natural predisposition of young men to join together and mount a raid. As the Persian Emperor Xerxes sat on his gold throne on Mount Aigaleos to watch his navy fight the Greeks in the Gulf of Salamis, on September 21, 480 B.C., he dictated to his scribes the rewards and punishments he intended to mete out to those who fought, though in the end 60,000 Persians were slaughtered before the incentives could be applied.[274] For minor disciplinary offenses Roman soldiers had their pay cut, or they were flogged. For serious offenses, such as sleeping on duty, the *fustuarium*—being beaten to death by cudgels wielded by fellow troops—was imposed. A unit that disobeyed orders or showed cowardice could be *decimated*: One in ten was clubbed to death. Officers as well as the rank and file were subject to decimation—a punishment Crassus

imposed when his legions were defeated by a slave army led by Spartacus in 71 B.C. In the 1640s, during the English Civil War, Lord Hop-
ton maintained that a good general should "command well, pay well and
hang well." [275] Frederick the Great of Prussia ordered that any soldier who
turned round in battle be bayoneted by the man behind.

In the American Civil War minor offenses were punished by assign-
ment to digging latrines or mucking out stables. More serious offenses met with solitary confinement or a "bucking and gagging," where a
man was left for hours with a wooden gag in his mouth and with his an-
kles and wrists tightly bound. Deserters had their heads shaved and were
branded on the hips, or they were executed in front of their comrades.[276]
During the Battle of Stalingrad, the Soviet secret police executed 13,500
men for desertion, "cowardice," and even for not shooting fellow soldiers
if they saw them crossing over to the German lines. The condemned men
were forced to strip naked before being shot, in order not to damage their
clothes, which would be reused. In just one example of how blind adherence to rules triumphed over logic and group interest, a Russian pilot
who was shot down thought he had parachuted behind the German lines
and prudently tore up his Communist identification card. In fact, he fell
in Soviet-controlled territory, and was executed on the spot under Stalin's order 270, which sentenced anyone who removed his insignia during
battle to be shot.[277]

Patriotism

If heavy-handed discipline and top-down control are essential ingredi-
ents in maintaining a large military, they are certainly not the only ways
fighting men are kept in line and on task. Sooner or later, a state usually
finds ways to build on the innate loyalties evolved to support coalitional aggression, just as Henry V called his men together into a "band of
brothers." It is no mistake that our word for ardent love of and loyalty to
one's own country—patriotism—is derived from the Latin for "father,"
and no surprise that leaders appeal to this artificial sense of kinship, especially when they are exhorting us to sacrifice our own interests for a
common goal. Hitler spoke continuously of the Fatherland with himself as the *Führer* or leader, and Austrians in World War I fought for
their "father," Emperor Franz Joseph, at least until his death in 1916.
Joseph Holper of the University of Illinois compared the words used

by Alexander Hamilton and James Madison in their personal letters to those in their political writings. As authors of the *Federalist Papers* (1787–1788), these American "Founding Fathers" helped turn public opinion toward ratifying the U.S. Constitution.[278] When trying to inspire patriotism in their public writings, Madison and Hamilton were ten times as likely to use kin terms such as "father," "brother," and "mother" as in their private business letters. Abraham Lincoln began his Gettysburg address, "Four score and seven years ago our *fathers* brought forth on this continent a new nation" (emphasis added). Five weeks after 9/11, when the U.S. Congress approved a 342-page act giving the government many new powers to tap telephones or restrict private liberties—it was called the Patriot Act.

Credit: National Archives (photo no.

From simple raids on neighboring tribes to the largest and most complex of modern conflicts, all war has its roots in the biology of human behavior. Warriors across cultures and back through time to prehistory would recognize the look on the face of this U.S. Marine. He is returning from two days and nights of continuous fighting against Japanese forces entrenched on Eniwetok Atoll, in the Marshall Islands in 1944.

In World War I, generals on all sides spoke often of *élan*—the energy characteristic of enthusiastic young soldiers. Military leaders may not recognize the evolutionary origins of their policies, but they were struggling to maximize the predisposition for team aggression in their troops. The French Commander-in-Chief, Joseph Joffre, believed all attacks should be "pushed to the extreme with the firm resolution to charge the enemy with the bayonet, in order to destroy him."[279] The German Theodor von Bernhardi dismissed "mechanical appliances, be they ever so excellent" and extolled "boldness" in infantry attack. British textbooks and infantry regulations in World War I asserted, "The main essential to success is to close with the enemy, cost what it may," and suggested that "the most soldierly way out of what looks like an impossible situation is by attack." Countless night raids were ordered into no-man's land as part of trench warfare. Exhausted men risked their lives crawling among decaying corpses killed in previous patrols and all to little or no effect, except to express the general's deep-seated belief that wars were won by morale not machines. Such thinking would have served well as a pep talk to chimpanzees starting out on a raid, or to a band of Yanomamö leaving to attack another village, but it cannot stop a hail of bullets in no-man's land.

For all the triumphs of strict military discipline and high-flung political speeches, it seems that many humans do indeed retain their ability to sense a bluff. Egalitarian societies tend to produce better fighters than those with great disparities in wealth. After examining two dozen recent conflicts researchers found that the nation with the more equitable distribution of wealth defeated the nation with the less equitable distribution in three-quarters of the wars. When they looked at eighty wars going back to Napoleon, they observed that the nation with the least social stratification won in eight out of ten conflicts. Enemy spies and dissidents probably find it easier to work in a country with exaggerated social and economic disparities, and an unjust regime has to spend energy controlling its own people as well as fighting abroad. But it also seems reasonable to suggest that a sense of equality would more closely mimic the dynamics of team aggression, leading to tighter bonds of mutual loyalty and willingness to sacrifice for one's fellows. While chimpanzees are hierarchical, every animal in a troop has strong social ties to every other, and such a structure would also have been typical for most of hominid evolution. Warriors from an egalitarian society are probably more effective precisely because they are more in touch with the evolutionary roots of raiding.

Slavery and Racism

Slavery and racism are the ugly cousins of raids and warfare. Racism remains a powerful force in the contemporary world, even though genetically the human species is amazingly homogenous. There is less genetic variation among human races across the globe than there is between the various groups of chimpanzees in Africa. Like chimpanzees, however, we probably evolved in such a way as to maximize facial differences as it helped us to categorize those around us into three relevant groups: relative, potential mate, and enemy. Added to this original variation, different environments promote certain features, whether changes in skin color (a recent event in evolutionary history) or differently sized noses (a large nose is adapted to hot, dry desert air, a flatter nose is better adapted to the moist tropics). Besides, humans seem to have a knack for inventing differences, where none exist, so perhaps it's not surprising that even evolutionary differences as insignificant as these would serve to strengthen ingroup recognition of kin and friends, and also facilitate outgroup hostility.

For millennia, slavery was as universal, and as unquestioned, as warfare. The predisposition to raid and kill easily becomes converted into a predisposition to raid and enslave, as an even more effective strategy for enlarging the raider's resources than simply killing competitors. Most writers agree that the first slaves were probably prisoners of war. The feminist writer Maria Miles rightly asserts, "Slavery...obviously did not emerge out of trade, but out of the male monopoly of arms."[280] In Classical times, slave traders followed armies, purchasing prisoners in job lots for resale later.[281] The Old Testament Jews purchased slaves from the "heathen" around them, but they also enslaved fellow Hebrews. In the New Testament, St. Paul exhorted Christian slaves "to obey their masters,"[282] and it took almost another two thousand years before Christians universally condemned slavery, despite the central message of equality in their religion.

The uses of war captives are not always purely practical or intentional. Once we understand the evolutionary roots and pragmatic importance of ingroup cohesion, however, a chilling logic lurks beneath some deeply horrifying behaviors, such as the use of slaves not as laborers but as symbols of chiefly power and religious supremacy. Many of the Native peoples of the Pacific Northwest owned slaves, even though they had little need of extra workers.[283] For example, Tlingit chiefs sacrificed a slave when a new house was built, and the Kwakiutl killed slaves in order to use their bodies as rollers to pull an honored visiting chief's canoe on shore. Once individuals are identified as belonging to an outgroup, there seems to be no limit to the human capacity for cruelty. Despots exploiting captured members of outgroups displayed the same extravagant lack of empathy in ancient Persia, the Aztec empire of Mexico, and the West African kingdom of Dahomey, among many others.

Chimpanzees sometimes return from a raid on a neighboring troop with a captured mature female in tow. Surviving females belonging to a troop that is being undermined by raids may well switch their allegiance to the successful troop. In human raids and wars as well, another man is a rival to be killed, but a woman can represent a valuable opportunity to transmit the victor's genes. In preliterate societies, women are sometimes dragged along by force after a raid, raped, and once pregnant, made dependent on their captors. In what is now the West African nation of Burkina Faso, a hundred years ago men formed raiding parties with the specific goal of capturing women who could be used as slaves or exchanged as bride wealth.[284] In the Old Testament, the Jews could free a male slave after seven years of bondage, but female slaves did not enjoy this release.[285] When Alexander the Great stormed the city of Tyre in 332 B.C., his soldiers killed 7,000 to 8,000 men, crucified 2,000 others, and sold 20,000 to 30,000 women and children into slavery.

Slavery taps into the basic predisposition to dehumanize other people, which we have argued is essential for an intelligent, social animal to kill members of its own species. Slave-owning societies maintain over time the same dehumanizing attitude toward slaves that warriors express toward enemies in the heat of battle. In order to coerce adults to be slaves, there must be both a social structure with the power and wealth to acquire slaves and instruments such as manacles, whips, and branding irons to control them. The very word *bondage* reminds us of the role of technology in slavery. The use of wooden restraints or iron chains to secure slaves is particularly important during the transport of slaves, who are often traded over great distances. Only through severe whippings, or the threat of punishments such as crucifixion, can adults be terrorized into life-long submission.

In some circumstances, however, the barriers between ingroup and outgroup can erode with time and proximity can erode outgroup definitions. It can be difficult to maintain the dehumanization associated with warfare when a slave works, eats, and sleeps in your household for several years.* Slave masters sometimes freed favorite slaves as a result. Horace,

---

* The twists and turns of human psychology regarding outgroups can be deeply confounding. Surely Thomas Jefferson, another of America's Founding Fathers and the author of the Declaration of Independence, believed as he wrote, "that all men are created equal." As a Virginia slaveholder, however, he refused to free his own slaves, and despite a few early gestures, never truly pushed for broad emancipation. That he carried on a longtime affair with one of his slaves, Sally Hemings, and fathered several of her children while also maintaining that Africans were "inferior to the whites in the endowments both of body and mind" truly defies explanation.

the Latin lyric poet, and Josephus, the historian, were both the children of freed slaves. When young children were captured and brought into slavery, they might be treated as a slave or as an adopted child, as their captors preferred. Some forms of slavery overlapped with apprenticeship, while some household servants might regard their lives as little different from slavery. The Turkish Sultanate enslaved Christian boys and turned them into an effective army, the Janissaries, eager to lay down their lives in wars against Christendom.

Slavery is now universally condemned, but it has not universally disappeared any more than the behavioral predispositions underlying it could have. For example, young women are purchased from their families in Nepal and sold into prostitution in Mumbai (Bombay), India. Some years ago I gave a lecture in Nepal, at a meeting chaired by the Minister of Health. The Minister told me how he had just come back from visiting some villages in the Himalayas. As he walked along a narrow alley he had heard a mother shouting so loudly at her daughter that her voice carried through the thick stone wall of the house. The girl had broken something, or somehow triggered a family altercation. "If you were not such an ugly bitch," screamed the mother, "we would have sold you to Bombay." Many of the sex workers in Bombay are Nepalese girls who were sold by their families. In a moment of anger, was this mother's brain able to place her own daughter in the evolved framework of an outgroup? Perhaps it was an insult that would not have been carried out, but it saddened the Minister that poverty could give rise to such thinking.

The Cost of War

From Classical times onward, organized military activity, unlike chimpanzee raids or raids in primitive societies, has usually cost much more than it has achieved. Following the Treaty of Westphalia in 1648, which ended the Thirty Years' War (begun as a political and religious conflict in the Holy Roman Empire), the leaders of sovereign states began to recognize that something had to change. Warfare had become so pernicious, painful, and costly that nations began to back down from constant, aggressive competition for territory and influence, defining their borders on maps but pulling back from interfering in each others' internal affairs. Catholic and Protestant states began to live in uneasy peace, even though deeply divided by religious beliefs. From the seventeenth century onward systematic, although often flawed, efforts were made to separate combatants from civilians. By the twentieth century, as we will see in chapter 10, victorious nations actually began to help the vanquished recover after defeat and genuine, although still weak, supranational organizations were developed. Despite, or indeed because of, new and highly destructive methods of killing, humanity began devising policies to limit killing to certain areas, under certain regulations, and to extend humanitarian aid to the suffering.* We are reminded almost daily of the shortcomings of the complex, artificial world of agreement and cooperation between competing sovereign states, but it is worth taking a moment to contemplate the fact that it has literally no precedent in animal behavior.

Still, modern war remains extremely costly for the individual soldier, and for nations as a whole. The U.S. Civil War cost $6.6 billion in 1860 dollars, enough to provide every slave forty acres of land and a life-long income. American military spending now has reached well over one billion dollars a day, more than the eight next-largest national military budgets combined. True, the U.S. is fighting two overseas wars at the moment. But at least one of those is a war of choice; since the implosion of the Soviet Union, there has been no real threat of a large-scale attack by an outside power. Terrorist attacks have become increasingly important, but confronting them does not require large standing armies, and intercontinental ballistic missiles are useless against lone suicide bombers. Even discounting the human and financial costs of active war, maintaining a technologically advanced military comes at the cost of added expense and lost opportunities. The armed services' annual oil consumption would be enough to run all the mass transit systems in the U.S. for fourteen years, for example, and military activities produce more toxic waste than the top five American chemical manufacturers combined.[286] Meanwhile, democratic nations subsidize their armaments industries and often sell arms to unstable and totalitarian states—sometimes only to have those weapons turned back upon their makers or their makers' allies. Even a well-financed terrorist group could never manufacture a supersonic, shoulder-launched missile with an infrared homing device capable

---

* The debacle of America's invasion of Iraq, and the vast civilian suffering it has caused, can obscure the fact that despite a few high-profile instances of abuse and murder, U.S. forces have taken on personal risks and gone to technological lengths to avoid civilian casualties to a degree that was unheard of as recently as during the Vietnam War.

Credit: National Archives (photo no. 165-SB-94)

Our brains evolved to respond to images of war, not statistics. This mental shortcoming is the source of terrorism's power to cause fear even when the real threat is small. Approximately 10,000 Union and Confederate soldiers were killed or wounded at the Civil War Battle of Cold Harbor, Virginia in June 1864 (workers burying bodies in 1865, above), and ultimately the Civil War killed one in fifty of the population. The terrorist attacks of September 11, 2001, also left an indelible image of death, but horrible as that day was, only one in 100,000 Americans died. In 2001, Americans were fourteen times more likely to die in a road accident than in an Al Qaeda attack, and six-and-a-half times more likely to murdered by a fellow citizen.

of detecting and downing a target jet airplane. And they don't have to. An estimated 100,000 of these missiles already exist, and 50,000 have been sold to developing nations. At least thirty insurgent groups globally are thought to possess these most perfect weapons of terror. Already, they have been used against U.S. military aircraft in Iraq and it may be just a matter of time before they are used against civilian airliners as well.

Military expenditures, because they produce nothing lasting, encourage inflationary economic pressures. The defense industry has only one customer—a government spending the money it collects from its citizens—and in a very real sense, those writing the contracts and those competing to fill them belong to the same exclusive ingroup, often linked by the "revolving door" that shuttles military retirees into service with defense contractors and their lobbyists. Before 9/11, the Congressional Budget Office estimated that the U.S. would have a projected surplus of $5.6 trillion by 2011; since 9/11, they've changed that estimate to a $2.3 trillion deficit. Part of that change is driven by shifts in the economy and political decisions about tax cuts, but a considerable portion is the result of the cost of the wars in Afghanistan and Iraq. The U.S. has been spending a quarter of a million dollars a minute in Iraq for several years, and depending on how long we stay, the total financial cost, including billions for lifelong support to injured veterans, could soon top $1 trillion.[287]

The military expenditure of developing countries is about one-fifth that of the developed nations, but in terms of human expenditure their commitment to team aggression is even greater. In the 1990s, while the West saw some reduction in expenditure, the military budgets of Africa rose steeply.[288] Ethiopia spends $16 per capita on their military and $1 on health; in Chad and Angola that ratio reached twenty-five to one. Globally, an area equivalent to the combined size of Indonesia and Turkey is reserved for military use, and spending on weapons research exceeds that for improving health, increasing food production, controlling pollutants, and saving energy combined.

The cry is often heard that all the world's problems could be solved if military budgets were redirected to the common good. It is true that even in the 1990s, the money India spent buying twenty MiG 29 fighters from Russia could have educated fifteen million girls or that when Nigeria chose to purchase eighty tanks from Britain they were in effect opting not to immunize two million of their own children.[289] But such comparisons ultimately serve only to reinforce the oddity of our species' evolutionary history. Even without the military expenditures, it is unlikely that the modern state would invest anywhere near as deeply in more noble, altruistic efforts such as global disease eradication, family planning, environmental conservation, and sustainable economic development. It's a shame, really, as these are exactly the sorts of projects that could actually help to make military force less necessary. But for the present at least, is seems that it is only warfare to which we are committed so unreservedly. Only a virtually universal and deep-seated, inherited drive could explain the expenditure a modern state is willing to invest in the military, even in times of peace.

The Utility of War

In the modern world, the costs of war are obvious and unambiguous. The original survival advantage enjoyed by individual males with a predisposition for team aggression has long since been replaced by a major, verging on suicidal, disadvantage for our species as a whole. But as we conclude our diagnosis of the biological roots of war, we still have to ask if there are benefits of raids and warfare that we have thus far overlooked, in our early history or even in the present day.

Wars are technological pressure cookers and they have accelerated the development of many things other than weapons. Penicillin was introduced in World War II, our microwaves are powered by variations of the vacuum tubes in the radar sets that helped win the Battle of Britain, and the first jet airliners were powered by modifications of military engines. But all these advances would have come in a decade or two without the cost of the Second World War. In a much more profound way, however, warfare may also have driven human evolution itself, and also helped bring about the beginning of civilization. The raids our hominid ancestors seem to have fought so often may have played a role in the enlargement of the brain, which is the prime characteristic of our species. The evolution of a large brain is more of a puzzle than it may at first appear. Rats survive and multiply with quite small brains, for example, and hummingbirds can navigate thousands of miles with a cerebrum the size of a grain of sand. Our vast brains, meanwhile, come at a significant cost. Large even in infants, the human brain and the skull that protects it is almost the entire reason why human childbirth is such a risky, painful experience, endangering the life of every mother in labor. If the child survives the ordeal of parturition, its brain will consume as much as a fifth of all the energy he or she derives from food for a lifetime. In evolutionary terms, anything as metabolically expensive and lethally risky as delivering a baby with a large brain must be very important indeed.

Recall the Kasakela alpha chimpanzee Goblin, and the mental work he did in working out how to answer a challenge from the chimps in the territory his team was attempting to raid. To succeed, he had to assess the strengths and devotion of the other males in his own troop, recall rele-

vant previous experiences with other troops, and make life and death decisions about the possible intentions of his enemy. * Our aggressive urges
may be instinct, but the mental processing that goes into controlling and
shaping them into actions requires a great deal of neural computing power. Our ancient predisposition to attack and kill our neighbors has always been a risky behavior, and it could be that this violence played a key
part in the evolution of our ancestors. In simple terms, natural selection
would have a field day with a big, aggressive, violent ape that also happened to be dumb. Maybe the sharp selection of death in battle created
such a positive bias toward intelligence that it helped drive the evolution
of the size of our brains. Certainly as groups of our ancestors warred back
and forth, any mutation that led to an even marginally larger or more effectively analytical brain would have been a major advantage. It is arrest-
ing to think that the incredible intellectual abilities that distinguish us
from all other animals may have arisen because for millions of years, we
have been trying to kill one another.

What we know for certain is that the human brain did go through
an impressive jump in size as we evolved. Mark Finn, David Geary,
and Carol Ward of the University of Missouri pinpoint this change to
about one-and-a-half to two million years ago. At that time, fossil hominid skulls demonstrate the beginning of a rapid evolution from a modest 600 milliliters of cranial capacity to our current volume of 1,300
milliliters. Our ancestors were already walking on two legs when this
great leap in brain size began, and they had even started making stone
tools. It was also around this time that they began to manipulate fire
for protection and cooking, which together with the first crude weapons placed our human ancestors at the top of the food chain. (Previously inedible grains and woody corms—the bulb-like structures of food
plants such as taro and arrowhead—could be roasted and softened, and
cooked meat keeps much longer than raw before going putrid.) A better diet could support a bigger brain, but this by itself does not explain
why our brains evolved to be so large. The fact is that we already had the
brainpower to kill deer, or even to combat the colder climates we met
on the migration out of Africa. By the time our brains started their rapid increase in size, our worst enemy was ourselves, which would have

---

* Goblin did indeed manage to live to a ripe old age. In fact, Martha and I met up with him on a
path in Gombe forest just a few days before he died. He was too weak to climb a tree to get food,
and from a mere five feet away looked us straight in our eyes, sadly, bewildered, as if to say, "What
is happening to me?"

made outsmarting aggressive neighbors our greatest mental challenge. In the words of Finn and his colleagues,

> The conceptualization of natural selection as a "struggle for existence" of Darwin and Wallace becomes, in addition, a special kind of struggle with other human beings for control of the resources that support life and allow one to reproduce. In this situation, the stage is set for a form of runaway selection, whereby the more cognitive, socially, and behaviorally sophisticated individuals were able to out maneuver and manipulate other individuals to gain control of resources in the local ecology....[290] [emphasis in the original]

In other words, our very viciousness may have been the catalyst for the unique mental capacities that now allow us to contemplate our own mortality and question whether team aggression was such a good idea in the first place.

As we've mentioned, warfare may have helped propel the development of religions and, perhaps even more importantly, it may also have been the stimulus that moved us from life in small bands to working together in increasingly large and powerful groups. Archeological excavation demonstrates that for much of the New Stone Age or Neolithic era (starting about 10,500 years ago) people lived in modest villages, probably representing autonomous clans. By analogy with existing hunter-gatherer bands, when a group was defeated probably either it was eliminated or the survivors fled to some other area. But flight would not always have been possible in certain geographical settings. Robert Carneiro of the American Museum of Natural History argues that "warfare [was] the mechanism by which autonomies were overridden and larger political units established."[291] Perhaps this is why the first Chinese state developed along the length of the Yellow River, where 5,000-foot mountains would have prevented defeated tribes from moving elsewhere, and the Egyptian and Assyrian empires arose on the flood plains of the Nile, Tigris, and Euphrates Rivers, surrounded by inhospitable semi-deserts. In these settings, one clan may have conquered neighbors who had nowhere to flee and absorbed them. Or perhaps the defeated clan was eliminated and the victors gained a larger area without having to defend its perimeter—the mountains or the desert did that for them.

Alternatively, perhaps some small clans simply banded together for mutual protection. When the first European explorers penetrated far up the Amazon they found villages that were not "a crossbow shot" apart. Fighting was frequent and intense, and left to themselves, Amazonian Indians might have developed into a state built of many communities. On his visits to the Yanomamö, Napoleon Chagnon observed that in the central parts of the region they occupied the villages were especially close together, headmen have more authority, and fighting is more intense. Given time, perhaps the Yanomamö would have coalesced into a large unit, just as the Roman Empire arose from scattered tribes in central Italy. Regardless, without the predisposition to conquer their neighbors any human group would likely have remained as small, sometimes mobile groups, each independently eking out a subsistence living. Without warfare and the stronger leadership needed to maintain larger cohesive groups, our species might never have crossed the threshold into specialization and division of labor. We would have been trapped in a dead end, a talking, fire-using hominid forever caught around the campfire, often hungry, frequently dying in childbirth, and forever threatened by violent neighbors, our horizons limited physically and intellectually to a world of imaginary spirits.

The prosperity, scientific insights, art, and all that is glorious about true civilization might never have emerged without the ability of the earliest states to organize armies and impose their will on rivals. That is how they managed, quite accidentally, to reach the critical size and social organization required for civilized living. Perhaps this helps explain why in both wars and religion, courage and cruelty are close cousins, and why we glorify war despite its vast costs. It is a sad, sobering thought that the golden heart of civilization needed to be refined and forged in the furnace of warfare. Even if this was the case, however, it is all too clear that war's usefulness is now long past, and it is time to let the furnace cool.

Holy Wars

If our minor physical differences and shared ingroup, outgroup predispositions provide foundations for warfare and slavery, our cultural and religious identifications have helped to intensify those basic drives. Throughout history and across the globe, military and political leaders have invoked the local religion to support their war aims, even when that religion preaches love and forgiveness. From an evolutionary perspective this is not the paradox it seems. Our ancestors who belonged to a warring band with a strong, charismatic leader would be more likely to win against their enemies. A belief in a supernatural leader, whether a sacrifice-demanding deity of the Aztecs or a "Lord of Hosts [who will] avenge me of mine enemies," helps build group identity, makes fighters bolder, and takes military confidence and bluster to an entirely new plane. When David faced Goliath, he was not alone, crying, "I come to thee in the name of the Lord of Hosts, the God of the Armies of Israel."[292] As with all human behaviors, religious belief is a very complex and often contradictory phenomenon. Certainly, many sincere and dedicated religious adherents have lived peaceful, productive lives, guided by their spiritual beliefs, and it would be a mistake to try to reduce the impulse to believe to a simple survival mechanism, similar to the goose bumps that stood our ape ancestors' hair on end and made them look larger and more powerful than they were. But still, the centrality of religion, and particularly organized religion, in the history of warfare can scarcely be overstated.*

Three and half millennia ago, Thutmosis III of Egypt held a council of war before the Battle of Megiddo in what is now Jordan. His lieutenants said, "Your father, the God Amon, will bring the fulfillment of your Majesty's intentions."[293] The pharaoh was not only god's representative on earth, he was a god. In 2003, House Republican leader Tom DeLay, an evangelical Christian from Texas, asked for "divine assistance" to protect both Israel and the U.S.A. Are DeLay and Thutmosis examples of the same predisposition expressed in different historical contexts?

Religious or supernatural beliefs are surpassingly common human attributes, and it seems obvious that our brains are drawn to and welcome many aspects of religious adherence. But from a purely pragmatic view of survival and reproduction, religions are extravagant and expensive. In Medieval Europe the Church was often the largest land owner and in some ways it was organized rather like a modern corporation.[294] Biology demands explanations for costly or time-consuming structures and behaviors. As we saw in chapter 2, modern behaviors that resonate with our ancient survival impulses are often pleasurable, and it may be that some religious messages and practices appeal because they simulate the

---

* Whether we see organized religion as a sincere expression of belief in a real god or as a psychosocial structure built around an imaginary deity doesn't particularly matter here. The fact is that belief in a god or gods has influenced the human history of warring profoundly, whether or not any particular instance of belief is sincere or substantiated.

feelings of belonging to a tight-knit group with a strong but benevolent leader. Whether it is the comfort and reassurance of a community of fellow believers and knowing that a powerful god is on our side, or the fear of authority and punishment for even minor transgressions, we seem predisposed to welcome religious ideas into our minds, no matter how fantastical they may seem from the outside. Unfortunately, this predisposition can leave us open to beliefs that are not just supernatural, but also extremely dangerous in the here-and-now.

Throughout history, religion has been used to support warfare, to define strategy, and to encourage troops to defeat the enemy. This remains true whether we see such martial uses as a corruption of religion or its central point, and conflicts need not be "about" religion in order for religious belief to impact them profoundly. America's invasion of Iraq in 2003 was largely propelled by Vice President Dick Cheney, Secretary of Defense Donald Rumsfeld, and his deputy Paul Wolfowitz, but the president's religiously grounded sense of destiny helped him misjudge the support of other countries,* and was among the factors preventing him from asking probing questions about the conduct of the invasion or plans for winning the peace. (That the planning document for the post-war administration of Iraq—National Security Presidential Directive #24—was cobbled together in one week[295] is as clear a sign of misplaced belief in miracles and divine intervention as we could ever hope to see.) After he stepped down as British prime minister, Tony Blair spoke publicly about his deep Christian faith. Blair may have been the one person who could have stopped the Bush rush to invade Iraq; instead he saw his religion as a moral imperative to intervene. Blair did not wear religion on his sleeve while in office, but Bush paraded his faith enthusiastically. His religious outlook resonated with many American fundamentalist Christians, whose contrived interpretations of the rambling Book of Revelation have sinister implications for war and violence. In one strain, a belief has emerged that the Temple of Solomon has to be rebuilt in Jerusalem in order for the Second Coming to take place—and that "keeping" Jerusalem Jewish is a necessary step on the way. Beyond being poor theology, this interpretation encourages foolish military action in order to hasten the coming of the end times, but still it finds a receptive audience in the United States. One

---

* This was revealingly if inelegantly expressed, "Well, we're never going to get people all in agreement about the use of force, but action, confident action that will yield positive results, provides a kind of a slipstream into which reluctant nations and leaders can get behind and show themselves that there has been—you know, something positive has happened towards peace."

survey has shown that over half of all Americans believe that Revelation is relevant to current events, and 30 percent believe that 9/11 was predicted in the Bible. Michael Drosnin, who wrote *The Bible Code*, implying extraterrestrial forces embedded a secret code in the Bible only modern computers can unravel, was invited to brief "top military intelligence officials" in the Pentagon following 9/11.[296] Whatever the original evolutionary benefit of blind faith in such patently ridiculous explanations of the world may have been, its application to modern international relations is clearly and wildly maladaptive.

Charles Darwin summarized religious devotion in his usually clear and eloquent way, as "consisting of love, complete submission to an exalted and mysterious superior, a strong sense of dependence, fear, reverence, gratitude, hope for the future, and perhaps some other elements."[297] Until the rise of science and secular education, religious beliefs were a human universal.[298] Darwin thought religion arose partly from an innate desire to explain the world around us, and evolutionary biologists have been struggling to explain religious beliefs ever since. Late in his twentieth-century career, the Oxford marine biologist Sir Alister Hardy speculated that just as domesticated dogs transfer devotion from their mother to a human master, so religious belief reflects a similar mechanism whereby, "This 'new master,' a supposedly invisible being, is imagined by primitive man to account for the 'something' real beyond the self he felt himself to be in touch with."[299] Lee Kilpatrick of the College of William and Mary in Virginia also links religion to attachment theory, arguing that a belief in god may be triggered by the same predispositions that bond a mother and child.[300] He makes the interesting suggestion that adult love may be an extension of the same mechanism, serving as a "commitment device" to bond parents. Kilpatrick metaphorically compares religion to an "evolutionary cheesecake"—that is, just as our enjoyment of cheesecake taps into our evolved preference for fats and sugars, so religion is built on our predisposition to submit to a strong leader. However, Kilpatrick rejects the idea than religion evolved to promote group cohesion, because he does not see how such a predisposition could "translate into inclusive fitness," or an improved ability of the group to survive and pass on its genes. The solution to this problem, as E. O. Wilson suggests, is that religion evolved because those clans that believed in a divine ruler would be more tightly bonded, more confident of their victory, and therefore more likely to win in raids and battles.[301] David Sloan Wilson of Binghamton University, another prominent evolutionary thinker, also proposes that religion helped our hominid ancestors form cohesive groups.[302]

A strong argument can be made that those clans, groups, and nations who venerated some totem or deity that was perceived to reward ingroup behaviors became more cohesive, and more hostile to outsiders. A mythical father in heaven who was ready to smite members of an outgroup, who were especially hateful or subhuman because they worshipped a different god, could be a strong and unifying force. Some of the most persistently violent conflicts, as in Northern Ireland or the Middle East, have been exacerbated by sectarian education and incitement, as well as religion-based communal loyalties. "Holy War" may ultimately be a paradox, but many conflicts have indeed been sparked, driven, or intensified by religious differences. In a study of restrictions on civil liberties in the U.S. following 9/11, four out of ten of those who regarded themselves as "highly religious" believed that American Muslims should register their whereabouts with the government, compared to only 15 percent of those who considered themselves "not very religious."[303]

The idea that religious belief fulfills the demands of a deep evolutionary predisposition is supported by an intriguing series of epidemiological studies. Church-going Christians have been found to have a lower risk of heart disease and death by heart disease, for example, and a survey of African Americans showed that religious commitment provided protection against hypertension,[304] while another study found that African Americans who attend church regularly can expect a remarkable fourteen years more life.[305] Recently Yousif Yagoub in Kuwait found that Muslims who prayed five times a day and read the Koran had lower blood pressure than those who did not.[306] As is always the case when religion enters the equation, the number of possible interpretations becomes almost infinite. But given that the health benefits associated with religious belief seem to apply equally to all faiths, even those which are theologically mutually exclusive, we can conclude that we are seeing the positive effects of fulfilling a universal human predisposition rather than the modest manifestations of various divine powers. Even without a specific impulse to religious belief, such benefits are also explicable because religious groups tend to recreate the trust and support of a hunter-gatherer clan. Speculations about the nature of an evolved predisposition for religion are strengthened by the fact that studies of identical twins indicate that religiosity is partially inherited.[307]

It would be as incorrect to suggest that religion causes wars as it is to say that evolution does. But religious belief and identity certainly do shape the expression of our fundamental predispositions, including those related to war and terrorism, in important and sometimes quite direct ways. Take for example the Islamic tradition of Holy War, which began in January 624 A.D., when the prophet Mohammed sent seven to twelve men on a raid from Medina to capture a caravan coming from Mecca. Two of the Meccans were killed. A second raid turned into a small war when 300 men from Medina met a larger number from Mecca at a small oasis called Badr. Mohammed remained in the rear of the troop praying fervently to Allah and promising that anyone who died fighting would be transported immediately to paradise. Hearing the prophet's words, at least one soldier drew his sword saying, "Perfect! Have I only to get myself killed by these men to enter paradise?" He was cut down instantly.[308] For the sincere believer, the promise of a quick passage to an eternal reward can be a powerful motivator in battle, and from the Vikings to the Crusaders to the Mujahadeen, the idea has been widespread in history. Pope Urban II promised salvation for all who took part in the Crusades. Whether or not there is any truth to the pervasive belief in an afterlife, it has the potential to change the survival equation fundamentally. In the purely temporal plane of the here and now, losing in battle means evolutionary catastrophe, whether through death or the loss of mating opportunities. But once the concept of an eternal reward is added in, the immediate concerns of survival and passing on genes can fade to irrelevance.

Here on Earth, success in battle always increases a man's stature in his society. When success is understood to have divine backing, then the cohesive and status-enhancing effects of victory are even greater. The promise of a supernatural reward for those who die fighting can motivate men to risk more for victory, resulting in less fear for the combatants and potentially very significant benefits to their leaders. Mohammed's victory at Badr strengthened his role in the community and reinforced his own faith in Allah's power—and led to a long series of revenge raids back and forth. Mohammed's followers in Medina resisted a siege army said to number 13,000, and the day this force withdrew, Mohammed attacked a Jewish tribe who had sided with his enemies. Hundreds of men were beheaded and the women and children sold into slavery. Mohammed took the lovely Rayhan as his own concubine after he had her husband executed, exactly as King David took Bethsheba after he had her husband killed by sending him to the front of a battle. Unified purpose and religiously in-spired courage can make for success in military battles, and in the biologi-cal war of competition for mating opportunities as well.

Serving a divine rather than a mortal leader strengthens a fighting force, and the emotions of warriors and the strategies of commanders during the Israelite conquest of Palestine (1200 B.C.), the two centu-ries of maximum spread of Islam (622–800 A.D.), the Crusades (1096 A.D. onward), and the conquering Aztecs under Montezuma (sixteenth century), seem to have been propelled by similar religious enthusiasms to attack and subjugate their neighbors. They were proof that overcon-fidence, in the right circumstances, can be adaptive and enhance suc-cess in team aggression. But Wrangham's hypothesis set out above also suggests that religious fervor, while it has motivated impressive martial triumphs, has led, like overblown personal confidence, to ignomini-ous and unnecessary defeats. In the 1880s in Sudan, Mahdi Mohammed Ahmed of Dongola declared himself a prophet. At the Battle of Omdur-man in 1898, his successor Abdullah al-Taashi led 40,000 dervishes, armed with swords and spears, against the rifles and machine guns of a much smaller British and Egyptian force commanded by Sir Hora-tio Kitchener. Despite the dervishes' deeply religious confidence, Om-durman was a highly uneven engagement. Ninety-seven hundred of the Mahdi's followers were killed and another 16,000 wounded, while Kitchener lost only a handful of his men.[309] Whatever else we might think of the religious impulse, it is clear that for these men at least, it had become decidedly maladaptive.

The most extraordinary and explicit intertwining of religion and war-fare occurred during the maximum growth of the Aztec Empire, imme-diately prior to the arrival of Hernán Cortez and the Spanish conquest in 1519. The Aztec capital of Tenochtitlan, now the center of Mexico City, was amongst the largest metropolises in the world, and the Spanish Con-quistadors were astonished by its magnificence. The power of the Aztec hierarchy was built on warfare and ritualized cruelty on a gigantic and horrible scale.[310] Legions of men were captured in battle and kept for sac-rifice by cardiac evisceration—cutting out their beating hearts—drown-ing, burning alive, and impaling with arrows. Others were slaughtered in gladiatorial combats, with the victims tethered to a huge circular stone altar and attacked by four Aztec warriors armed with clubs and swords edged with sharp obsidian blades. About one-third of the human sacrifices were slaves, women, or children rather than opposing warriors. During the recent archaeological excavation of the Templo Mayor in Mexico City, the skeletons of forty-two infants sacrificed to a rain god were uncovered. Fray Diego Durian, a Catholic priest who accompanied Cortez and his men, described one sacrifice in Tenochtitlan:

> ...the temple was reddened with blood of five hundred men. A fire sacrifice was ordained; this was the most terrible and horrendous sacrifice that can be imagined....A great bonfire was built in a large brazier placed on the floor of the temple. This was called the "divine hearth." Into this great mass of flames men were thrown alive. Before they expired their hearts were torn out and offered to the god. [311]

In other sacrifices, men and women were dressed up and led through the city in boisterous celebrations. The warrior who had captured a sacrificial victim would wear his flayed skin (skin to skin, with the bloody side out) and parade through the streets taking part in ritual battles. Lesser citizens would join together to buy a human being to sacrifice jointly. During certain ceremonies the whole population cut and lacerated themselves, even slashing the ears of their own infants still in their cradles to induce bleeding. It was religious devotion to violence on an astonishing scale.

If the Aztecs' belief in the divine was their ultimate undoing—Cortez credited his improbable conquest to the Aztecs mistaking him for a divine emissary, if not a god himself—it certainly also helped them build their empire in the first place. The endless cycle of cruel rituals built fearless warriors and impressed neighboring tribes with the Aztecs' power. As in classical Greece, all grown men were expected to be soldiers. Promotion was awarded according to the number of enemy each man killed or captured for sacrifice. Priests, for example, had to kill or capture at least four men.[312] One of the Conquistadors, all of whom were almost as brutal and as courageous as the Aztecs they fought, wrote with no small measure of admiration:

> It is one of the most beautiful sights in the world to see [the Aztecs] in their battle array because they keep formation wonderfully and are handsome. Among them are extraordinary brave men who face death with absolute determination....During combat they sing and dance and sometimes give the wildest shouts and whistles imaginable, especially when they know they have the advantage....In warfare they are the cruelest people to be

found, for they spare neither brothers, relatives, friends nor women even if they are beautiful they kill them all. [313]

Some modern historians and anthropologists have indulged in the most fanciful intellectual contortions in order to try and describe the Aztec culture as somehow benign. To quote two among many, Pierre and Janine Soisson describe the Aztec "war of the flowers," in which a battle was agreed to in advance, as a "game, played with courtesy and fair play, to decide the best amongst those competing." They frame religious wars as "a sort of collective tournament, in which participants merely captured prisoners, who were then to be kept in reserve for use in sacrifices."[314] The desire to avoid confronting our essential warlike nature gives rise to an extremely unrealistic perspective. Perhaps it is exactly this sort of willful self-deception that ultimately disproves the same Rousseauean belief in the natural "goodness" of humanity that it seeks to prop up. Admittedly unpleasant as it can be to contemplate the reality of human brutality, wishful thinking will not help us to find our better natures. The perspective of evolutionary psychology suggests that only an animal burdened with the most terrible predisposition to engage in team aggression, to dehumanize other members of its own species, to systematically misinterpret the surrounding world around them, to worship symbols, and to invent gods could have produced civilizations capable of generating such horrors.

The French Wars of Religion, the *Reconquista* of Spain and Portugal, the Thirty Years' War in central Europe, and certainly the Crusades confirm that Islam has no monopoly on the intermingling of warfare, identity, and religion. But given the particular relevance of fervent Islamist belief in contemporary wars and terror campaigns, the Islamic martial tradition merits closer examination. One tradition among several in Islam is that there are two types of *jihad*, or Holy War. The first is an offensive war aimed at destroying or converting infidels. It is conducted by the state and led by its ruler or caliph. The second is a defensive war responding to an attack on Islam, where each individual has an obligation to fight, even in the absence of centralized leadership. This view was put forward by Iba Taymiyya in the thirteenth century and by Abd al-Wahhab in the eighteenth century, whose fundamentalist teachings form the basis of Wahhabism, the conservative strain of Islam upheld by the contemporary Saudi Arabian royal family. Osama bin Laden, Mohammed Atta, and others perceived Islam as under attack from American secularism and so they perceived an obligation to fight back even if they died in the process.[315]

Importantly, perhaps this mix of resentment, anger, and outgroup hostility is just as possible without religion. But religious differences have often served to intensify hatred, excuse killing, and promise rewards. In short, religious belief may or may not be the ultimate cause of global jihad and Islamic terrorism, but it is hard to argue that it does not amount to a particularly combustible fuel thrown onto whatever historical, environmental, and biological fires are already burning.

It should be said that religious belief and practice are not incompatible with intelligent, critical thinking and a scientific outlook. My father's family were Congregationalists and my mother's uncle a minister in the Methodist Church. My wife's grandfather was a minister in the American Episcopal Church. At my English grammar school, in Anglican tradition we began every day with a hymn and a reading from the Bible—something that I still value for the beauty of the language and the insight it gave me into my own culture. Like most boys of that age, I read the religiously themed works of C. S. Lewis, such as *The Screwtape Letters*,[316] and though I struggled with the contradictions in religion then, I later attended Lewis's lectures at Cambridge University.

For brains that evolved to solve problems, as ours surely must have, the shifting cycles of the seasons, droughts and storms, and above all the deaths of loved ones can seem mysterious and unpredictable. The religious beliefs of preliterate societies accommodate the arbitrary ways of the world, and perceive spirits behind thunder or hiding under every stone. When skepticism emerges religions demand the believers have "faith." As societies became more complex through human history, so one god (or a pantheon of gods) was perceived to decree moral codes and to relay laws relating to the behavior of believers. Especially as children and young adults, it is surprisingly easy for us to believe in the supernatural and the transcendent. My co-author and I can both look back on a phase of ardent adolescent belief. He remains a practicing, if sometimes skeptical, Catholic and I became what my wife calls a "spiritual atheist."

The predisposition toward religious belief does seem to intensify for a time in adolescence or early youth. Is it coincidental that this is also the age group providing most of the world's warriors? As we have said, belief in a divine ally can strengthen the fighting powers of a social group and reinforce the authority of a leader who can claim some version of rule by divine right. We might embrace religious belief as appropriate devotion to a real god, or we might dismiss it as a bug in our brains, perhaps an unintended and unavoidable evolutionary consequence of our creative intelligence. In either case, we can see its powerful impact on the development of human societies and their armies. As larger groups began to muster more powerful military forces, belief in a powerful god who teaches a uniform morality and "protects" all those in his devotional ingroup would certainly help overcome the tendency for larger societies to dissolve into competing smaller groups and clans.

All religions seek to explain pain and death, and many promise a heavenly hereafter. But it seems to me that evolution can offer as much or more comfort for the truly bereaved. C.S. Lewis was unusual in that he was an atheist as a youth, and later turned to religion as an adult. He fought in the trenches in World War I, where his best friend was killed. He converted to Christianity in his thirties. This writer of children's stories was also a theologian, and like all mature theologians he struggled with the question of why a benevolent god permits so much suffering in the world. A quintessential Englishman, Lewis married Joy Gresham, an intelligent, lively American woman who later died of bone cancer.[317] His faith did nothing to assuage his deep sadness. In a way, that Darwinian evolution is driven by death and suffering needs no explanation, as each generation must die to make room for the next as we adapt to a changing world. I was also an Englishman married to an intelligent, lively American. When she died at age fifty-two, I didn't know whether I was on my head or my heels. But at least I had a mental framework that made sense of my grief.[318] Death is as much part of biological evolution as reproduction, and I took solace in that fact. Ending The Origin of Species, Charles Darwin wrote, "Thus, from the war of nature, from famine and death, the most exalted object which we are capable of conceiving, namely, the production of higher animals, directly follows." I was fortunate to remarry Martha, but now I am seventy-three and at the age when cancer strikes, strokes happen, and hearts stop suddenly. Few people welcome death, but understanding its place in the natural order takes the fear away. For me at least, understanding the place of death in evolution provides more comfort than the promise of an afterlife offered in exchange for following some particular code of behavior and belief. While I live, I will remain passionate in trying to understand how I evolved. And I will continue to share Martha's commitment to Horace Mann's aphorism, "Be ashamed to die until you've won some victory for humanity." That is a drive I think Darwin would also have understood.

9

WAR AND TECHNOLOGY

Almighty father, be with those who brave the heights of Thy heaven and who carry the battle to our enemies.

—prayer written by an army chaplain and given to Captain Paul Tibbets, who dropped the atomic bomb on Hiroshima, August 6, 1945[319]

WULF SCHIENFENHÖVEL, a German anthropologist, was studying a preliterate tribe in the Highlands of New Guinea. He asked for their help to build a small airstrip, and when it was completed, he offered two leaders of the tribe a flight. The men had hardly seen an airplane before and certainly never flown in one. They came naked except for their penis sheaths and then surprised Schienfenhövel asking if they could collect two large stones. "Why would you want to do that?" asked the anthropologist. The men explained that if he would be kind enough to leave the cabin door open and circle over an enemy village, they could drop the stones on it. Frans de Waal, a Dutch scientist who is now Professor of Primate Behavior at Emory University in Atlanta, Georgia, in retelling this story, writes, "The scientist could note in his diary that he had witnessed the invention of the bomb by Neolithic man." Every new technology, it seems, is applied to warfare with alacrity.[320]

Credit: Courtesy of the Jane Goodall Institute

Mike was a low-ranking male at Gombe Stream National Park in Tanzania, until he learned to collect empty kerosene cans from outside the tent of Jane Goodall. He would then charge the larger, more dominant males, hair erect, screaming and clashing the cans together. This mixture of bluff and "weapons" won him the top rank in the troop.

Competing chimpanzees pull off tree branches and drag them along the ground or wave them about in excited displays. For ingroup aggression, noise and bombast are chimps' weapons of choice. Mike, a Gombe male, achieved the alpha spot in his troop because he had the intelligence to clash together some kerosene cans, which Jane Goodall had brought into her camp, in order to scare the other males. Outgroup confrontations are more physical, but the value of technological innovation and arms development is no less apparent. Chimpanzees have been seen to pick up stones and hurl them at baboons, for example. In a field experiment, one group of chimps was exposed to a stuffed leopard with a chimpanzee doll between its front paws. They attacked vigorously, and one female armed herself with a long stick, the tip of which hit the stuffed animal at an estimated fifty miles per hour.[321] Still, while chimpanzees may throw objects during a raid, the attacking males inflict their lethal wounds primarily by tearing flesh with their large canine teeth or by pounding an enemy with their hands and feet. As we described earlier, during a raid the clumsy, almost comic bluster that adult males display when showing off within the troop is replaced by a cold, deliberate viciousness.

Interestingly, given the masculine monopoly on team aggression behaviors, male chimpanzees have much larger canine teeth than females. As our human ancestors evolved, their canine teeth became relatively smaller, probably in parallel with the ability to use tools to grub up vegetables, carry food, construct shelters, butcher meat, hunt game—and kill other human beings. At some point in hominid evolution the calculus of weaponry changed the nature of raids, and attacks began to be conducted primarily by hurling, flinging, or shooting wood and stone—and later metal—objects at other individuals.

Slings, Swords, and Barbed Wire

The primary weapons of early warfare—the spear and the club or mace—were only a short step away from the chimpanzee's favored tree limbs. Spears are merely long, straight branches with whatever stone or metal tip is available affixed at the end.* A mace is just a club with a heavy head, but appropriately hafted and wielded, it can exert over one hundred foot-pounds of force per square inch on impact. As it takes just two foot-pounds to penetrate the human abdominal wall and not much more to facture the skull, the mace is a very powerful invention indeed.[322] For all its crude origins, this glorified cudgel's lethality has been used as a symbol of political power from the time of the Egyptian Pharaohs to the current day. In twenty-first century Britain, whenever parliament is in session an ornate mace is carried in a procession and laid on the table in front of the Speaker of the House of Commons—a lethal, apish symbol of authority in a modern democracy.

If clubs and spears enhanced our ancestors' ability to kill members of their own species, the invention of the sling, the throwing stick, the jav-

---

* Chimpanzees living in the savannah-woodland environment of eastern Senegal have recently been observed using sticks as spears by anthropologist Jill Pruetz of Iowa State University. The animals sharpen the tips of branches with their teeth and use the weapons to jab bush babies in their tree-hole burrows. That the first evidence of animals other than humans fashioning weapons for hunting should involve chimps is unsurprising perhaps, but Pruetz also notes that most spear hunting is done by females and juveniles. Chimpanzee hunting had previously been thought to be a male pursuit, but Pruetz suggests that females and the young have a particularly sharp incentive to devise new feeding strategies when food is short and foraging males are not sharing their finds.

elin, and the bow and arrow enabled them to do so from a distance. And that changed the risk calculus of team aggression once and for all. Raiding had always been a risky strategy, which as we've seen is the whole reason for banding together in a group to carry it out. Hitting and jabbing weapons reduced the risk to attackers considerably, but weapons that killed at a distance brought that risk down to almost zero—an assailant using such weapons might even go unseen. The sling was effective at up to sixty-five meters and used in vast numbers by cultures ranging from the first city-states of Mesopotamia to the Aztec Empire. Archaeologist Mortimer Wheeler found 22,260 sling stones stored near the east gateway of the Iron Age hill fort of Maiden Castle in Dorset, England, for example. Simple bows, made of wood strung with animal gut, and arrows were in existence as early as the ninth millennium B.C., and composite bows, which used wood layered with sinew or horn to increase flexibility and power, first appeared in the third millennium B.C.[323] At each step of development, raiders and warriors became more lethal, and new tactics were developed—composite bows can be shorter than the all-wood equivalent, and can be fired by a man on horseback as a result. These simple weapons were so effective that in Britain, for instance, bows and arrows didn't disappear from the field of battle until after 1644,[324] and even today Palestinian youths use slings and stones to attack Israeli soldiers.

The lethal goals of warfare among preliterate groups are illustrated by their technological innovations. Stone Age flint arrowheads were of two types. One design could be hafted tightly to the arrow shaft, making it suitable for hunting small animals, where immediate death was intended. The second was only weakly attached to the shaft, so that it would detach and remain buried in human flesh, causing infection should an opponent survive the initial attack. Stone blades can be used to stab and cut (the Aztecs used wooden swords with obsidian blades inserted razor-like along the cutting edge), but to achieve maximum lethality, swords need to be made out of a metal. As the Bronze Age dawned, one of the first uses of its namesake metal was to fashion swords and daggers, as far back as 4000 B.C.[325] The first metal blades were short and stout in deference to the malleable metals available. The Celts who fought Caesar in Gaul made iron swords over two feet long, but the increased reach came at a cost. One Classical writer says such swords bent easily and the warrior who used them had to stop and straighten the blade with his foot before making a second blow.[326] A study of wounds described in Homer's *Iliad* found that while swords caused fewer total deaths than spears, they were more likely to kill rather than maim whenever they were used. [327]

Technological advances were also applied to defense. Leather coats and wood and leather covered shields provided some protection on the ancient battlefield. Copper helmets were found in graves at Ur (approximately 2500 B.C.) and metal breastplates also date back to the Bronze Age. The Celts perfected chain mail in the first or second century before Christ.[328] However, there was always a tension between the weight and inflexibility of armor and the need for mobility on the battlefield.* A Classical writer describes the Celts at the battle of Telamon in 225 B.C. as wearing just trousers and cloaks, which allowed easy movement. And the utility of protective gear depends greatly on what class of weapon your foe is using. The quilted cotton armor of the Aztecs, designed to counter obsidian blades and sling stones, proved useless against the conquistadors' steel swords, crossbows, and rudimentary firearms.

As soon as human societies domesticated draught animals, they too were introduced to warfare. By 1800 B.C., the Sumerians had perfected the war chariot. Initially it was a slow heavy object with four solid wheels, drawn by four asses. Later it was replaced by the two-wheeled, fast, sophisticated chariot of the pharaohs, or that used by Boadicea, the Queen of the Icini, in a revolt against the Romans in 61 A.D. Such Celtic chariots were used to frighten the enemy, but when it came to close order fighting, the warrior dismounted and attacked on foot while the chariot with its driver waited some way off. Stirrups spread though Europe from about 850 A.D., and they permitted the spectacles of knights jousting with heavy lances and coordinated cavalry charges with flashing sabers. The warrior on horseback came to be celebrated in statuary, painting, and literature, but mounted troops could never penetrate a wall of infantry holding pikes—a horse, lacking the impulses of team aggression, is simply too sensible and always stops short of such an obstacle. At the 1415 A.D. Battle of Agincourt, knights dismounted and fought on foot once their steeds had brought them to the battleground. Still, horses were used

---

* Powerful firearms made the steel plate armor of the Middle Ages obsolete, but the recent evolution of ballistic fiber and ceramic materials has returned protective clothing to the list of standard military gear. There is no doubt that advanced body armor systems have saved many military lives in the wars in Iraq and Afghanistan, but the tradeoff between protection and mobility remains. As one U.S. veteran of the 2004 Battle of Fallujah commented to my co-author, "We're human beings, and when you're wearing fifty pounds of armor, you're going to get fatigued much faster. To me, we've crossed the threshold where we are defeating ourselves by wearing so much armor. We can't be a mobile force anymore."

in vast numbers from their first introduction up until they were replaced by mechanized cavalry units in the first part of the twentieth century.* In the seventh century before Christ, the Assyrian king Ashurbanipal could muster almost 3,000 horses, about two-thirds of which pulled chariots while the rest were reserved for mounted cavalry.[329] At Waterloo, Napoleon collected nearly 20,000 horses, drawing on a horse-using agricultural society the Assyrians had lacked.

Throughout the history of military innovation, new technologies have often given their early adopters a competitive edge as their adversaries sought new strategies to counter them. Building earthen banks, palisades, or stone walls is one of the oldest methods of defense (the Chinese character *cheng* means both city and walls). The pre-pottery Neolithic site of Jericho (8350–7350 B.C.) had walls ten feet thick and thirteen feet high, enclosing about ten acres, but even that wasn't enough to protect the inhabitants—the walls were destroyed and rebuilt nineteen times. Stone walls began to become obsolete when the Chinese invented gunpowder in the tenth century A.D. By 1248 gunpowder had reached Europe and Roger Bacon mentioned it in his writings.[330] But the introduction of gunpowder didn't really start to change tactics for another century. Edward I used crude bottle-shaped cannons (*roundelades* or *pots de fer*) at the Battle of Crecy in 1346, but they had no decisive effect on the outcome. By the siege of Constantinople in 1453, the Turks had developed a cannon capable of firing a 1,500-pound shot a mile—an impressive but limited power, given that these cumbersome behemoths could only be fired seven times in a day. The real improvements came from France, where improved field artillery played a role in driving out the English at the end of the Hundred Years' War (1457). During the fifteenth and sixteenth centuries, the cannon changed the face of naval warfare as the effective range of ship-born artillery began to exceed one mile. The arquebus, a precursor of the musket, was less accurate than longbows, but had the power to penetrate metal, and so helped speed plate armor's demise.

Noise too is part of warfare: It frightens and disorients the enemy and signals the attacker's cohesion. Once a raid is underway, chimpanzees pant-hoot, scream, and drum on the ground. Alexander the Great's

---

* Especially when their adversaries are afoot, warriors and raiders can still gain a devastatingly lethal edge by taking to horseback today. In the Darfur region of western Sudan, the mounted *Janjaweed* militia is notorious for its genocidal campaign of massacre and rape against all-but-defenseless civilians. Their group name has been translated, appropriately enough, as "a man with a gun on a horse."

phalanxes screamed *"alala, alala"* just before they clashed into the enemy[331]—and U.S. "psy-ops" (psychological operations) forces blasted the streets of Fallujah with amplified banshee screams and heavy metal music before Marines stormed the insurgent-held city in November 2004. As battles have become more complex, command and communications have developed in new ways also. Metal trumpets, illustrated in Egyptian and Greek wall paintings and excavated at Iron Age Celtic sites, were probably used to inspire confidence, to scare the enemy, and perhaps to issue signals above the noise of battle. The Hawaiian chief Kiha had a war trumpet made of a nautilus shell that could be heard ten miles away (it was decorated with the teeth of slain enemies).[332] At the Battle of Loos in September 1915, a Piper led the 7th King's Own Scottish Borderers into battle playing *Blue Bonnets on the Border*.

Signal companies and communications technology also proved crucially important as armies grew in size and forayed further from their bases. In the absence of adequate communications, it was possible for whole armies to get lost, as did the English and French at the battle of Poitiers in 1306. Using messengers on foot or horseback, the Peruvian Incas and the Japanese Samurai were able to transmit messages at the rate of about 150 to 200 kilometers a day. In 1794 Claud Chappe invented a visual telegraph that could send messages about 400 kilometers a day.[333] War has always been a promoter of new technologies. Conversely, new technologies such as the Internet (originally developed by the U.S. military) can be exploited to lethal effect. The four suicide bombers who killed fifty-two other people in London on July 7, 2005, were not linked to Al Qaeda or any other terrorist organization, but were driven by religious motives after one of them visited Pakistan. They obtained all the information they needed to make their explosives on the Internet.

If an army marches on its stomach, then its victories depend on its logistics support and supply lines as well as its weapons. The Macedonians of classical times could bring some 20,000 men into battle, thanks to their superb logistical organization. By the eighteenth century the maximum size of an army had increased only to perhaps 100,000,[334] limited by the fact that approximately five to ten kilos of food, fodder, and firewood had to be brought to the front each day for each man. In the second half of the nineteenth century, however, warfare began to change at an accelerating rate. With the industrial revolution came a boost in the availability of provisions, along with the explosion, to use an apt term, in many forms of armament and ammunition (thanks to increasingly powerful firearms and artillery, the lethal zone in which the enemy could be killed expanded by several orders of magnitude, just as it had with the invention of the first projectiles some 11,000 or 12,000 years earlier). Railways and steam ships were ideally suited to moving troops and supplies, and once the electric telegraph was perfected, it could relay intelligence reports and commands over great distances at unprecedented speeds. Over time, the production of weapons shifted from hand crafting to standardized mass production, and the rifling of large and small caliber weapon barrels greatly improved accuracy. Bullets and their explosive charges were joined in a single round of ammunition for ease and speed of use. In 1862, Dr. Richard Jordan Gatling invented a hand-cranked machine gun with rotating barrels that could fire 300 rounds a minute. Gatling sincerely believed that his gun would have such a devastating effect that nations "would see the futility of war and solve differences peacefully," but of course that has not been the case. Even explicitly peaceful inventions were put to dramatically lethal effect. Barbed wire was designed to corral cattle, and patented in 1883 by an Illinois farmer, Joseph Farwell Glidden.[335] Before long, it was to prove horribly effective in slowing and snaring men who tried to cross in front of machine gun fire.

Symmetrical Wars

The ongoing military cycle of technological advances and strategic adaptations has at times led to stalemates—the most notorious being the trench warfare of World War I. The tension between offense and defense makes true warfare different from raiding in several important ways, not least because the attackers in a genuine war often suffer a higher casualty rate than the defenders. As long as the technologies of attack and defense develop in parallel, whether it is stone walls and ballistae (a kind of giant crossbow designed to breach those walls), or machine guns and deep trenches, then warfare often becomes increasingly costly for both the attackers and the defenders.

In the two World Wars of the twentieth century, who possessed the advantage, defender or attacker, swung back and forth, creating prolonged and painful contests. The submarine, the airplane, and a large-scale chemical industry all matured in the years immediately prior to 1914, helping to generate the Great War's exceedingly bloody battles. Aerial bombardment was aptly called an air *raid*. Initially, both submarine warfare and aerial bombardment promised to be decisive, asymmetrical weapons. The bomber, it was felt, would "always get through." John P. Holland, who married the gas engine and electric propulsion to perfect the first successful submarine in the 1890s; the Swedish chemist Alfred Nobel who invented TNT in 1866; and Dr. Richard Gatling who invented an early machine gun in 1862, were all certain that because their inventions were so destructive and defense seemed so difficult, each invention would bring peace. They were wrong, and it remains an open question whether even the destructive horror of atomic weapons will deter warfare in the long term.

I have had a life-long interest in aerial warfare. My father was an engineer on a 560-foot-long hydrogen-filled airship at the end of World War I. In retrospect airships seem one of most improbable military technologies, but at the time aerial bombardment appeared to be a seductively simple way of carrying the battle to the enemy. On January 19, 1915, two German Zeppelins dropped six 110-pound bombs on the seaside town of Yarmouth, Norfolk. The Kaiser specifically forbade bombing civilians,[336] but even from a stable, slow-moving dirigible flying in daylight it proved impossible to bomb accurately. Houses, rather than the railway lines targeted, were destroyed. Soon the German press began to advocate bombing cities. The German newspaper *Kolnische Zeitung* in January 1915, proposed, for example, that, "An eye for an eye and a tooth for a tooth is the only way we can treat the enemy. The best way to shorten the war, and thereby in the end the most humane."[337] In Britain, First Sea Lord John Fisher proposed taking hostages from among the German population living in Britain and "executing one of them for every civilian killed by bombs from aircraft." Winston Churchill, who was First Lord of the Admiralty at the time, overrode this desperate suggestion.[338] Luckily for the British, airships filled with highly flammable hydrogen proved vulnerable to antiaircraft fire, fighter planes, and bad weather. Forty percent of those who volunteered as crew were killed. Yet the primate drive to raid is powerful and the crews who flew these improbable craft persisted, showing great courage.

In 1917, Churchill wrote, "It is improbable that any terrorization of the civil population which could be achieved by air attack would compel the government of a great nation to surrender."[339] But by the outbreak of World War II, the technology had changed drastically. Aerial warfare both engendered unprecedented fear and also held the promise of preempting the type of trench warfare that had killed millions of soldiers only twenty-one years earlier. Along with many Civil War battles, First World War trench warfare stands as a classic example of symmetrical warfare, in which both sides are so evenly matched that neither can gain an advantage and the slaughter continues indefinitely with no clear winner. Military theorists were convinced that strategic bombing, with pinpoint accuracy, could destroy the industrial power of the enemy state. In reality, the strategy of precision bombing during daylight raids failed almost as soon as the war began—it was simply too dangerous for the bombers. Three months after Britain declared war on Germany in 1939, twenty-four top of the line, twin-engined Wellington bombers led by Wing Commander Richard Kellet attacked German ships at anchor in Wilhelmshaven harbor in broad daylight. The formation was broken up by heavy antiaircraft fire. Judging his targets to be too near civilian shore bases, Kellet did not drop his bombs and turned for home still loaded with 1,500 pounds of explosives.[340] Half the planes were destroyed. The bomber did not always get through as the strategists had predicted.

War Without Limits

Unlike the chimpanzees and our ape ancestors, human beings have developed a complex, and often contradictory morality of warfare. The role of non-combatants in conflicts has swung dramatically through the years. The nineteenth century saw increased efforts to spare civilians from the ravages of war, and the 1899 Hague Conference and 1922 Washington Treaty forbade the use of bombing to terrorize civilians or to attack non-military targets. But the limits of technology undercut these protections: Zeppelins and Wellington bombers were terribly inaccurate and could not hit small military targets. Antiaircraft guns and defensive fighters forced the bombers to attack at night, and a bomber flying in the dark was lucky to hit the right city, let alone a targeted factory. Strategically, however, this wasn't a problem. Lord Trenchard, who developed the strategic thinking of the British Royal Air Force (RAF), believed it was easier to overcome the enemy's will to resist than to destroy the means to resist. "Morale effect," Trenchard claimed, "was to material effect as twenty to one."

A similar balance of technology, culture, and biological predispositions drove submarine commanders and crews along a similar downward spiral from controlled attacks to widespread slaughter. Krupp engineers in Germany improved on John Holland's submarine by introducing safer diesel engines. By 1914 the world's navies had 400 submarines, although commanders were divided on how they should be used. In Britain, even before the war, Lord Fisher had concluded, "However inhuman and barbarous it may appear there is nothing else the submarine can do except sink her captives." First Admiralty Lord Churchill replied, "I do not believe this would ever be done by a civilized power."

Initially, the rules of war did indeed forbid the unprovoked slaughter of civilians, requiring submarines to surface, stop an unarmed merchant ship on the open seas, and search the vessel. If the ships were secretly armed, or carried contraband relevant to the war effort, then it could be sunk or taken as a prize of war, but the crew had to be rescued. However, submarines were vulnerable on the surface, the crews were too small to mount a search party, and the vessels too crowded to carry large numbers of prisoners. Still, for the first seven months of the Great War at least, German submariners observed the constraints of surface action and sank only ten merchant ships.

The German submarine blockade was nowhere near as effective as that imposed by the powerful British surface fleet against the Germans—a blockade that included turning back neutral ships carrying food. As was to happen with aerial warfare, technical realities drove policies away from the ingroup rules of submission and back to the no-holds-barred approach of the primate raid. Technologically, the submarine was ideally suited to destroying civilian merchant ships supplying a nation at war, just as the bomber was best suited to destroying cities and the people who lived in them. In February 1915, Germany's Kaiser Wilhelm authorized, with some reluctance, a new policy by which the waters around the British Isles were declared a "war zone," and submariners were given the authority to torpedo merchant ships without surfacing. Within two months 150,000 tons of British shipping had been torpedoed and by April 1917 over two million tons had been sunk.[341]

As we'll see in the coming chapter, Western nations had made some attempts to control the worst excesses of war during the nineteenth century, at least when they fought each other. But World War I brought the tit-for-tat primate mode of fighting back to human warfare with brutal force. The British tightened the surface blockade of Germany, and the Germans responded by removing all remaining restrictions on submarine warfare. In May 1915 they warned American passengers joining the British luxury liner *Lusitania* bound for Liverpool of the risk of unrestricted submarine warfare, but only one person decided not to board. The *Lusitania* sank twenty minutes after it was torpedoed in British territorial waters off the coast of Ireland and over 1,000 people were drowned, including 128 Americans. Many in America called for a declaration of war against Germany in revenge, but President Woodrow Wilson maintained neutrality for the moment. By the time of World War II, there was no hesitation in launching unrestricted submarine warfare from the very beginning.

In the case of the air war, it took slightly longer for technology to erode the restraints on killing civilians—one wonders if this had something to do with chivalry and aviators' status as "knights of the air." The first German bombing raid on London during the Battle of Britain, on August 25, 1940, was apparently a mistake committed by a frightened aircrew, and they were disciplined by the Nazis for their error. But Churchill, by that time Britain's Prime Minister, did not know this and, primate revenge instinct in full force, sent eighty-one bombers to attack Berlin the next night. For a while, he insisted that the RAF should attempt "accurate bombing on military objectives," but on dark nights they should "discharge bombs from a considerable and safe height upon the nearest built-up areas of Germany which contained military targets in abundance."[342]

On September 15, 1940, Germany's air force, the Luftwaffe, sent 1,000 planes in a series of daylight bombing raids against London. But even with fighter escorts, the bombers did not always get through. As Germany recognized they had lost the Battle of Britain for daylight air supremacy, the Luftwaffe switched to night raids. On the night of November 14, 1940, Luftwaffe commander Hermann Göring sent 437 bombers to attack Coventry in the British Midlands, destroying 20,000 houses, gutting the medieval cathedral, and killing 600 people.[343] Britain was outraged, and Churchill ordered a revenge raid on Mannheim. But while aerial bombing could cause great suffering, it could not cripple a nation's industrial capacity. The railways of Coventry were up and functioning within a week, and rather than break a nation's morale, aerial bombing seemed to strengthen it. The American military attaché in London wrote,

> . . . the Luftwaffe under the most favorable conditions had failed to paralyze
> the British or reduce the country to impotence in over a year of attacks at
> very short range, and when its energies were often engaged elsewhere. So

why, I asked, should the RAF believe they could bring down Germany at a greater range and with its targets very much more dispersed than those in England and protected by much better anti-aircraft defenses now than Britain had last year.[344]

In November 1941, at the request of Lord Cherwell, Churchill's personal scientific adviser Mr. D. M. Butt inspected over 4,000 photographs of bombing targets taken on one hundred British raids. For most raids only one in four bombs fell within five miles of the designated aiming point, and for the more difficult to hit Ruhr industrial area, only one in ten came even that close. Results were better during the full moon, but losses from the German night fighters were also worse on such nights. For a while the British halted all bombing activity. A computer fed this information might have ended aerial bombing in 1941, and put the resources into other aspects of the war. According to one calculation, bomber production consumed a full 7 percent of British industrial production. Others put the figure higher still, and A. J. Taylor suggests the bombing offensive accounted for one-third of the domestic war effort.[345] Consequently, the U.K. had to purchase tanks, transport aircraft, and much of its munitions from the U.S., one of the reasons Great Britain ended the war bankrupt. But emotions evolved in the Stone Age interact with new technologies to lead fighting men along strange and costly paths. Churchill, at least partly out of a desire for revenge, switched from the sensible skepticism he had shown about the utility of airpower in World War I to the belief that, "The bombers alone provide the means of victory."

At the January 1943 Casablanca conference of Allied leaders, it was agreed that they would carry out a bombing campaign with the goal of "the progressive destruction and dislocation of the German military, industrial and economic system, and the undermining of the morale of the German people to a point where their capacity for armed resistance is fatally weakened." The words chosen are revealing; it was already apparent that accurate bombing was an illusion, and the euphemisms "dislocation" and "economic system" did nothing to obscure the fact that civilians were now direct targets. The strategic bombers had been built, and they were ready to be flown by crews whose bravery, like the strategy itself, was grounded in the primate predisposition to raid without fear or mercy. After the British army was driven back across the English Channel from Dunkirk in June 1940, Britain was alone, and it had no other plausible way of fighting the Germans than to develop its bomber force. In the House of Lords, Bishop Bell of Chichester challenged the strategic bombing policy on moral grounds,[346] but he was a minority voice and among human beings, as with chimpanzees, the will to fight and a sense of revenge are all but unstoppable.

On St. Valentine's Day 1942, Bomber Command was authorized to bomb "without restriction." The first 1,000-bomber raid was made on Cologne, and the sheer scale of the attack overwhelmed the German defenses.[347] On July 24, 1943, RAF bombers attacked Hamburg in a foray suitably code-named Operation Gomorrah, followed by a U.S. Air Force raid the next day. A firestorm one-and-a-half miles in diameter raged through the city. People were not only burnt to death but also suffocated for lack of oxygen. The wind reached 150 miles per hour, sucking everything into the inferno as the flames rose three miles into the sky. Glass bottles melted, kitchen utensils became puddles of molten metal, and human bodies turned to piles of soft, black ash. The death toll was estimated at between 60,000 and 100,000 civilians.[348] The devastation caused on those July nights was not repeated until the very end of the war when, in February 1945, RAF Lancasters dropped over 2,500 tons of bombs on Dresden. The next day the USAF attacked in daylight again, dropping another 1,000 tons of bombs. Thirty thousand to 100,000 people died, some of them refugees fleeing fighting on the Eastern front.

But as always, new offensive technologies and tactics were soon countered by new defensive measures. Elsewhere, the bombing campaign and particularly the massive air raids mounted by the RAF against Berlin turned into prolonged slogging matches. German night fighter tactics improved and RAF losses became unsustainable: It was symmetrical warfare at its worst. The commanders hoped to keep losses to less than 5 percent per raid—statistically this gave a one in five chance of completing thirty ops (the crew's slang for flight operations) and less than one in twenty for two tours of duty—sixty ops. Of the 125,000 aircrew who volunteered for Bomber Command in the war,[349] 55,435 or nearly half, died in combat or in training.[350] The U.S. Eighth Air Force daylight raids also failed to achieve their goals, at least until early 1944 when long-range Mustang fighters were able to escort the bombers over Germany. Forty-seven thousand American fliers were killed, and the strain on the crews was continuous.

German Zeppelins in the Great War carried sixteen to twenty-two men. Allied bombers in World War II needed crews of up to eleven men. The German Type VII U-boat had a crew of forty-four. The closely bonded group of men exposed to danger in a British Lancaster bomber over Cologne, or a German submarine in the cold Atlantic, were manipulating the most complex and advanced technologies of their time, but they probably also shared many of the same emotions of a troop of fighting men in preliterate societies—or even a chimpanzee raiding party. As an individual member of a bomber crew, each man depended on every other. "In a crew you are all in it together in a sink or swim situation," said one adjutant. English Flight Sergeant Arthur Aaron provides a dramatic example of the extreme effort one member of a fighting team may go to in order to save his fellows. Aaron had his jaw blown off, his arm severed, and his chest battered by flak while attacking Turin, Italy. Unable to talk, the pilot wrote instructions insisting that he be carried back to the pilot's seat. He tried three times to land the plane with one arm before finally bringing it down wheels up, to avoid exploding the bombs which were still onboard. The cockpit windscreen was blown out and he faced a one hundred mile per hour airstream. He died shortly afterward and was awarded a posthumous Victoria Cross for bravery.[351]

The strategic bombing campaign can be seen as an extended series of primate raids, but when the predisposition for team aggression is combined with modern technology, and backed up by the power of the state to control the lives of the raiding parties, and when attack and defense are symmetrical, then battles can become enormously prolonged, be immensely costly, and stretch human courage to the breaking point.

Our earliest ancestors evolved a psychological switch that allowed them to dehumanize and kill members of their own species, and our modern weapons help too, by depersonalizing killing. If an assailant with a bow and arrow was able to kill unseen, a modern warrior with long-range weapons is often able to kill without *seeing*—without having to look at his or her victims. The eye-to-eye contact that allowed a WWI ace to let an adversary escape when his guns jammed is largely a thing of the past. Instead, the military equipment stands in for the fellow humans who man it, so that it is the fighter plane that is "bagged," not the pilot, and the merchant ship that is sunk rather than the sailors who drown. This effect has been intensified dramatically with the advent, starting during the First Gulf War in 1991, of video- and laser-guided "smart bombs." Video-game-like images from nose-mounted cameras showed the bombs' approach to their targets visible to the world, but not the aftermath for the civilians and troops on the ground. There, death and destruction were appallingly real, but for those of us watching on TV at home, the ferocious aerial bombardments of both Gulf Wars took on the appearance, quite literally, of a child's game.

Killing without contact can be dehumanizing for the attackers, in addition to being horrifically fatal for the attacked. During the WWII Allied strategic bombing campaign, men who demonstrated great technical competence and astonishing courage often ended up crushing or burning to death a disproportionate number of women and children. And strangely enough, it proved almost *humanizing* for the technology itself. Airplanes could be imbued with personality—the feminine grace of a Spitfire, the thundering masculine power of a Lancaster bomber. Peter Hinchliffe wrote a history of the German night fighter battles with Bomber Command. "Young men," said one German ace,

> fought fiercely against each other. But in this night-time battle it was never the individual man who was the target, it was the airplane that was trying to drop its load of bombs on the night-fighters' homeland, particularly when the crew of the fighter could see the burning cities, perhaps where relatives lived....The full realization that these aircraft contained human beings like you and like me only came when crew members were taken prisoner and were fortunate enough to find themselves in the custody of an operational flying unit.[352]

This counterintuitive reversal of hatred was proved true when a four-engined Stirling bomber was shot down during the 1943 Gomorrah raid on Hamburg. Three of the four crewmembers who parachuted to earth were captured by a civilian mob and lynched from lampposts. The bomb aimer, who landed by accident on a Luftwaffe airfield, was taken prisoner and treated well.[353]

The WWII air war in the Pacific paralleled the developments in Europe. American industry turned out B-27 Liberator bombers as the same way it had mass produced cars, from a mile-long assembly line employing 18,000 people. As with the RAF, the USAAF turned to area bombing by night. Japanese cities were particularly vulnerable, consisting of densely populated collections of wooden houses. The American commander, General Curtis LeMay, mixed fragmentation bombs with incendiaries to make sure firefighters could not stop the blazes before they took hold. On

March 9, 1945, 349 B-29 Superfortress bombers attacked Tokyo. A million people were rendered homeless in a single night, and an estimated 136,000 were killed—more than died in either of the atomic bomb attacks that would follow that summer.* [354]

As in Europe, the destruction did not weaken the resolve of those bombed, and Japanese industrial output continued. A month later Japan unleashed its largest-ever *kamikaze* attack on the U.S. forces landing at Okinawa, and the U.S. suffered its highest causalities in an amphibious landing. Two thousand Japanese *kamikaze* pilots died.[355] Warfare here, too, had entered a costly, tit-for-tat stage. Commanders including Churchill, Arthur Harris of RAF Bomber Command, and Ira Eaker of the U.S. Eighth Air Force in Europe and LeMay in the Pacific were driven, as indeed we expect commanders to be, by a strong fighting spirit. Above all there was an ample supply of young men whose age-old predisposition to band together in loyal groups provided the raw courage to set out night after night with a one in twenty chance of dying in an inferno, torn apart by shrapnel and literally falling out of the sky. They rained death, injury, and terror on cities whose young men were fighting on other fronts and where women, children, and the old were disproportionately represented among those crushed to death or burnt alive.

Asymmetrical Wars

The cycle of technological development and strategic innovation has led to periods of immense destruction in the history of warfare as evenly matched foes exhausted their resources in futile attempts to gain an advantage. But history also provides many examples of asymmetrical battles between wildly unmatched rivals. And sometimes, again thanks to tactics and technology, it is the few who kill the many.

Even allowing for biases in reporting battlefield casualties, some of history's most famous battles have ended in highly asymmetrical death rates. At Marathon, the Greeks suffered 192 dead and claimed to have killed 10,000 Persians. When Alexander the Great's army fought the soldiers of Darius III at Issus in 332 B.C., several hundred Persian soldiers were slaughtered a minute for eight hours,[356] whereas only 450 of Al-

---

* The firebombing of sixty-seven Japanese cities is thought to have killed one million people in the closing stage of the war. Robert McNamara helped LeMay plan the attacks. Twenty years later, McNamara was a restraining influence during the Cuban missile crisis and in his eighties he called for the abolition of all nuclear weapons.

exander's troops lost their lives thanks to their better armor, better tactics, and greater self-confidence. When Hannibal's elite troops attacked the flank of the Roman legions at Cannae in Southern Italy they had only one goal—to eliminate the Roman army. They slaughtered an incredible 70,000 Romans in four hours.*[357] In 100 B.C. at Aix-en-Provence, the Roman general Marius used his better weapons and superior tactics to kill 90,000 Teutons in a single day.

The Mongol armies of Genghis Khan probably never exceeded 20,000 warriors, but they conquered most of the world's largest continent. The Mongols' ability to wage asymmetrical warfare depended on horsemen and the composite bows they fired from horseback. The Mongols used excellent saddles and stirrups, and learned to ride from an early age. Armies traveled with four or five sturdy horses for each warrior, so they could cover huge distances without stopping to rest. They traveled light, and if food ran short they cut the neck veins of their mounts and drank the blood.[358] At the Battle of Kalka River in 1223 A.D., 20,000 Mongol warriors annihilated a Russian army of 80,000 with few losses to their own troops. Six Russian princes were executed by being bound, placed under heavy boards, and suffocated by the weight of victorious Mongols feasting atop the bodies.[359] A similar asymmetry in weaponry permitted the steel swords and iron pikes of the Spanish conquistadors to decapitate and eviscerate poorly armed Aztecs with brutal efficiency. At one battle, Pedro de Alvarado and a few hundred men slew 8,000 Aztec warriors. In April 1940, the German *blitzkrieg,* combining tanks with mechanized infantry and air support, swept through Yugoslavia, capturing 300,000 troops and killing perhaps 100,000 defenders, while losing only 558 Germans. In the First Gulf War, air superiority and a mixture of smart weapons and gravity bombs killed perhaps 20,000 Iraqi troops, for the loss of 148 Allied troops killed in action and 458 wounded.[360] In the Second Gulf War, sixty Abrams tanks and Bradley fighting vehicles of the Third Infantry division rode through the streets of Baghdad and may have killed 1,000 Iraqi defenders on their home turf without a single U.S. soldier dying.

Asymmetrical warfare arises often, but it rarely survives for long before those attacked figure out an effective response. By 2004, it was abundantly clear to Iraqi insurgent groups that they could not survive American

---

* Those they could not kill immediately, they incapacitated by cutting their leg tendons and then came back later to finish the kill and then loot. The Roman author Livy recounts how some victims, in agony and unable to crawl any further, "had dug holes for themselves and then, by smothering their mouths in the dirt, had choked themselves to death."

firepower in open battle. They turned to guerilla-style tactics, and with roadside explosions and suicide bombers killed more American troops than had died reaching Baghdad in 2003. Several preliterate societies acquired Western guns and used them to deadly effect against neighbors still armed with bows and arrows. In the 1780s, Kamehameha was one petty chief among many in the Hawaiian Islands. Then in January 1789 a supporter of Kamehameha captured a schooner called *Fair America*, which was carrying guns and powder. With two prisoners to show him how to use the new weaponry, Kamehameha became the first king to the rule the entire Hawaiian island chain.[361] In the early decades of the nineteenth century, in New Zealand, the Maoris, who had been fighting with each other for hundreds of years with bows, arrows, and clubs, traded food for guns from European traders and whaling ships. Hongi Hika, a prominent warrior, was invited to England to help translate the Bible into Te Reo, the Maori language. But he jumped ship in Australia instead, sold the gifts missionaries had given him, and returned to New Zealand with more muskets. The Ngapuhi and rival Ngati Whatua tribes almost annihilated one another in what came to be known as the "Musket Wars."[362]

Since the beginning of the Cold War, American and Western European armies have striven to replace manpower with firepower. Since the Cold War ended, U.S. power and precision in bombs, shells, and missiles has put it in a position where it seems able to defeat any other nation in the world in a conventional battle. But given time, enemies adapt. Just as the Iraqi insurgents have, so Somali warlords did in Mogadishu in 1993. Al Qaeda did the same, when it turned to blowing up commuter trains in Madrid in 2004, and public transport in London in 2005 after airport security was tightened post-9/11. Indeed, the overwhelming superiority of American, Israeli, and other modern militaries is exactly what stimulated the development of one the simplest weapons of all—the suicide bomber. "We don't have tanks or rockets," said one Hamas leader, referring to the firepower of the Israeli army. "But we have something better; our exploding Islamic bombs. All they cost is our lives, but nothing can beat them—not even nuclear weapons."[363]

The reality, of course, is that nuclear weapons do trump every other weapon. They represent the ultimate asymmetry in warfare. In the late nineteenth and first half of the twentieth century, fundamental discoveries in physics laid the foundation for the Manhattan Project to develop the atom bomb—an American effort spurred on by fears that Nazi Germany would achieve that terrible feat first. On August 6, 1945, a B-29 bomber called *Enola Gay* dropped an atomic bomb on Hiroshima. The previous symmetrical technologies of bomber, night fighter, and flak that had characterized aerial warfare were suddenly replaced by a single atomic weapon. After Hiroshima, one Japanese eyewitness wrote,

> I came onto I don't know how many, burned from the hips up; and where the skin had peeled, their flesh was wet and mushy....And they had no faces! Their eyes, noses and mouths had been burned away, and it looked like their ears had melted off. It was hard to tell front from back....I saw a man whose eye had been torn out by an injury, and there he stood with his eye resting in the palm of his hand. What made my blood run cold was that it looked like the eye was staring at me.[364]

Unbeknownst to the Japanese high command, after the second atomic bomb was dropped on Nagasaki there were no more bombs available. However, within little over a decade, not only were large stockpiles available in the United States and Soviet Union, but nuclear weapons were married to captured German rocket technology to produce the unstoppable intercontinental ballistic missiles (ICBM)—a union which was consummated on September 9, 1959, when the first batch of U.S. Atlas missiles became operational. Even before that date, in the mid-1950s, long-range bombers such as the American B52, the British Vulcan, and the Russian Tupolev 16 Badger had for the first time in history created a symmetrical threat so devastating that massive retaliation and mutually assured destruction (MAD) became the national polices of the superpowers. By 1964 the U.S. had over 1,000 ICBMs and had begun to mount shorter-range missiles on nuclear powered submarines. Following the leapfrog logic of military arms races, fanciful—and often extremely expensive—efforts have been made to develop anti-ballistic missile systems. After the Soviet Union launched Sputnik, the first artificial satellite, the United States first considered putting hundreds of battle stations into orbit to smash Soviet missiles in flight. The enterprise was called Ballistic Missile Boost Intercept, or BAMBI. In 1972 the U.S. and U.S.S.R. signed the Anti-Ballistic Missile Treaty (ABM), but the primate urge to get the upper hand is strong, no matter the real or potential costs: A decade later, U.S. President Reagan spent over $100 billion on so-called Star Wars projects, including laser weapons.[365] A handful of carefully staged attempts to intercept a missile in flight have worked, but most knowledgeable scientists think any defensive system could be readily outwitted by placing multiple decoy warheads on an enemy missile. Ballistic missiles with nuclear warheads may yet fulfill Holland's and Gatlin's nineteenth-century dream of a weapon that would end warfare because there was no defense against it. We have only had five decades' experience with this particularly sinister symmetry, however, and it may be too soon to tell whether our species will continue to obey the rules of the game we rightly call MAD.

Machines and Men

It is perhaps particularly unfortunate that one of the only two species of social mammals to engage in lethal same-species raids—namely ourselves—is also the only species that evolved the cerebral and manual skills to devise such complex and ultimately world-conquering technologies. Wars, and the threat of wars, stimulate technological development, from the sophisticated Greek warships called triremes to the development of aeronautics in the two World Wars. Technology in turn shapes warfare. Galileo may have been most interested in the solar system, but when he built his first telescope, what impressed the Venetian senate most was its implications for observing enemy ships at sea.[366]

Not all weaponry has been optimized just for killing. The primate drive is more complex than that. Did Mike the chimpanzee, who stole Jane Goodall's kerosene cans, not only scare his rivals but actually perceive himself to be stronger because of his new technology? Many weapons, from daggers and armor to sailing ships and airplanes, have been patiently and artistically decorated, often in ways that demonstrate power. A crest on a helmet makes a warrior appear more frightening to his enemy. Manfred van Richthofen painted his Fokker tri-plane red for the same reason that Julius Caesar wore a red coat in battle—to make himself visible and formidable, in this case as the Red Baron. Such displays can cause serious blunders however, and the drive for size for the sake of size has been the downfall of many a weapons system. In 1628, the Swedish Warship *Vasa* capsized on its maiden voyage in Stockholm harbor, rolling over because the hull was too tall, heavy with gilt-decorated carvings and excess cannons.

In World War II German engineers developed the first guided missile, which they called the V-1 and the British nicknamed the Doodlebug, and also the much more complex V-2 rocket.[367] The V-1 was robust, cheap to make, and delivered a ton of explosives. The V-2 was immensely complicated, unreliable, expensive, and consumed huge reserves of labor to build. It also delivered a one-ton warhead. Yet when Hitler saw the V-2 fired and sensed the raw power of its rocket engine, he made the disastrous decision to build the weapon in large numbers. A barrage of V-1s, even though a few would have been intercepted by aircraft or shot down by anti-aircraft fire, would have been a much more cost-effective means of bombarding London.

Sometimes technology moves so rapidly that commanders can't keep up. Eight years after the end of the Great War, Field Marshall Sir Douglas Haig, who had command of the British forces in France, wrote, "…aeroplanes and tanks…are only accessories to the man and the horse, and I feel sure that as time goes on you will find just as much use for the horse—the well-bred horse—as you have ever done in the past."[368] Haig's belief in a cavalry breakthrough dominated his planning for the Battle of the Somme in July 1916 and Passchendaele a year later. But in every case, barbed wire and machine guns proved a devastatingly lethal combination—whether men were mounted or not. On the first day of the Somme, two brigades of the British Eighth Division lost 218 out of 300 officers and 5,274 out of 8,500 men were killed. A German machine gunner wrote, "We were very surprised to see them walking.… You didn't have to aim, we just fired into them." Haig, however, kept up the attack and in the second week of the Somme the British suffered 10,000 casualties—equivalent to a whole division—every day. At Passechendaele, Haig's forces suffered almost a quarter of a million casualties for an advance of five miles. The fact that these tragedies were possible is testament to the strength of the evolved predisposition of young men to fight. Basic training and the harsh realities of trench life turned those who had been drafted as well as those who had volunteered into "bands of brothers," but in this case their loyalty, courage, and cohesion served only to get them killed. Again and again, the generals ordered offensive actions appropriate to a raid by chimpanzees or preliterate tribesmen, and wildly inappropriate to modern conditions. The soldiers at the front responded every time, sometimes reluctantly, but usually with astonishing courage.

Our Stone Age perspectives continue to mismatch technology and battlefield reality. In 1981, during the Cold War, the U.S. began the development of a new stealth fighter designed to outclass anything the Soviet Union might design. The first flight of the F-22 Raptor took place in 1990, but by this time the Berlin Wall had fallen and Russia was having trouble keeping its aging MiG fighters in the air, let alone developing new ones. Meanwhile Raptor costs rose from $35 million for each plane to an astounding $258 million. By the time of the Iraq invasion of 2003, U.S. forces needed reliable armor to protect its Humvees from improvised roadside booby traps, not jet fighters able to fly at twice the speed of sound.

War and Disease

One of the people I was privileged to know in the 1960s was Lord Florey, the brilliant pharmacologist who shared a Nobel Prize with Ernst Chain and Alexander Fleming for the production of penicillin during the Second World War. As president of the Royal Society, London, he invited me to join a committee focused on population and international family planning. Perhaps he was kind to me because he saw contraception not as a shameful taboo, but as a subject of valid scientific research. Lord Florey, born Howard Florey in Adelaide, Australia, had come to Oxford in the 1920s. At that time, one scientist had been physically thrown out of the zoology laboratory because he tried to study spermicides, and Florey met him trundling his laboratory apparatus on a handcart through the streets of Oxford, looking for a scientific home. Florey provided space in his own laboratory, and he remained convinced of the validity and importance of contraception and family planning until the end of his life. In the 1960s, while family planning was still controversial, Florey wanted to use his prestige to organize a top-level meeting between the Royal Society and the American Academy of Sciences to discuss the issue and lend the credibility of these two great scientific organizations to the subject. Sadly, he died in 1968 before he could do so.

Prior to penicillin and the other antibiotics which followed, many wounded soldiers who were not killed on the battlefield died shortly afterward from their wounds and infections—just as the victims of detachable-headed Neolithic arrows had. And battlefield medicine could be as lethal as the enemy's firepower. In the U.S. Civil War, half those treated surgically died. By World War II that number was down to 5 percent, in no small part thanks to penicillin, plus anesthetics. Charles Bell, a re-

nowned nineteenth-century surgeon, described treating the wounded in
the Crimean War:

> At six o'clock [in the morning] I took up the knife in my hand, and con-
> tinued incessantly at work til seven in the evening, and so the second and
> third day. All the decencies of performing surgical operations were soon
> neglected; while I amputate one man's thigh there lay next to him at one
> time thirteen, all beseeching to be taken next....It was a strange thing to
> feel my clothes stiff with blood, and my arms powerless with exertion us-
> ing the knife.[369]

As terrible as the deaths and wounds caused by swords and cannons
can be, wars before the twentieth century usually killed more people because of the disease and famine that followed the fighting. Disease itself
quite often determined the outcome of a war, in fact. According to the Old
Testament, when Sennacherib was attacking Israel an infectious disease of
some sort "slew in the camp of the Assyrians one hundred and eighty-five
thousand."[370] The power of Classical Athens was broken as much by disease as by losses on the battlefield. According to Thucydides, a plague in
430–429 B.C. killed one quarter of the army.[371] Battlefield archaeologists
in Vilnius, Lithuania, excavated the mass grave of 3,000 of Napoleon's
troops who had died retreating from Moscow in 1812. The researchers
first identified the DNA of body lice, and then found that the vermin were
carrying the bacteria that cause trench fever and typhus. They estimated
that lice-born diseases were responsible for a third of the deaths.[372]

When Europeans reached the New World, they brought the curs-
es of smallpox and measles, and returned to Europe with syphilis in
exchange. If Cortez's 1519 victory over Montezuma and his Aztecs in
Mexico was aided by superior European weaponry and the Aztecs' spiritual beliefs, he received an extra boost from the fact that a devastating
smallpox epidemic hit Montezuma's capital of Tenochtitlan at the same
time the Spaniards did. In 1531, when Francisco Pizarro reached the
Inca capital of Cuzco in Peru, the Inca himself had just died of smallpox and been replaced by a new and inexperienced ruler, Atahualpa.
At the time of the Spanish conquest there may have been one hundred
million people in the Americas as a whole—more than lived in Europe
at the time[373]—and twenty-five to thirty million people in what is now
Mexico. Within fifty years of Columbus, the population of Mexico had
collapsed to an estimated three million.

The Pain of War

The random process of evolution originally fashioned the predisposition that carried young men on raids to kill isolated members of neighboring troops. Human culture, as expressed in increasingly powerful societies and a burgeoning, science-based technology, has prolonged and transformed this instinct in increasingly destructive ways.

But whether the technology of killing is simple or sophisticated, it is equally painful. In Africa, the machete has been used so widely in warfare that it is called a "weapon of mass destruction," even though it takes considerable effort and incredible brutality to use one to kill or maim another human being. In the early 1990s, a colleague of mine in Sierra Leone visited a refugee camp there, and spoke with a little girl of about eleven years of age. Both her hands had been chopped off by a soldier (probably only a few years older than she was) wielding a machete. With the innocence of childhood, the girl looked into the eyes of the nurse dressing her wounds and asked, "Nurse, will my hands grow again?" This is not an unreasonable question if you live in the tropical rain forest where everything cut off grows back rapidly, but the wounds of war do not heal so readily. In the Congo and Rwanda, machetes may have killed and maimed more people than the atomic bombs dropped on Hiroshima and Nagasaki.

World War II was the war of my childhood, and such wars seem to leave particularly deep impressions with us. For myself, I retain the utmost respect for the courage of the airmen I met as a child, and the contribution they made to winning a brutal war against a brutal enemy. But because the costs were so great, I believe it is also important to stand back and look for any perspective that may help analyze the pain of modern warfare. It may seem contrived, perhaps even insulting or unpatriotic, to look to evolutionary biology to "explain" warfare. Yet if we were from another planet and saw an enormously inventive, highly social, commonly empathetic creature bent on killing its own species, surely we would look for some type of explanation. Do young men volunteer for elite units and serve bravely through a living hell in response to Stone Age emotions? Do civilians pay steep taxes and toil day and night to manufacture bombs, bullets, and the means to deliver them because they are caught up in evolved behaviors of self-defense and hatred of an outgroup, even though it no longer makes evolutionary sense? Have the power of the state and the development of technology turned the evolutionary predisposition for team aggression into an immensely costly boomerang situation? Are

Credit: © Donna DeCesare

The violence and brutality of war often extend far beyond the battlefield, and can persist for years after formal declarations of peace. Guatemala's long civil war ended in 1996, but its effects still reverberate throughout Guatemalan society. Rosario's parents beat her and she began living on the streets of Guatemala City at age seven. This photograph was taken in 2001 when she was nineteen, about nine months after she and a fifteen-year-old friend were raped by police. Rape is common during war, and in war-torn societies. The perpetrators almost always go free.

the underestimation of the power of machine guns at Passechendaele and the overestimation of the power of the Raptor jet fighter at the opening of the twenty-first century examples of how ancient warrior emotions cloud objectivity? In the trenches of World War I, hundreds of thousands of lives were wasted to gain a few miles of mud, and even if the Raptor program is curtailed, over $40 billion will have been wasted combating an enemy that no longer exists. The Iraq War may end up costing well over $1 trillion, along with hundreds of thousands of mostly civilian lives, all in service of removing a single dictator from power.

In the history of warfare, the advantage has passed from attack to defense and back again many times. Two atomic bombs ended World War II. On September 11, 2001, nineteen young men shook the country that produced those bombs to its foundations. They established a new form of asymmetrical warfare, in which a weak enemy can never win but neither can they be clearly defeated, fighting on and draining resources from their foe indefinitely. When the technologies of attack and defense are symmetrically balanced, then courage, revenge, and the ruthless application of those technologies transform warfare into a human meat grinder, and the basic primate raid advantages of superior numbers and surprise attacks are reversed. The tragically indecisive slaughter in the trenches of World War I is only the most concrete example of this warfare of attrition. It would be prudent to examine whether we might have entered into just such a state of balance with the forces of global terrorism. Perhaps we have not been "civilized" long enough to shed our inherited predispositions for attack and defense. But whether one's philosophical outlook runs to pacifism or military realism, we all need to grapple with our origins if we are to understand our limitations and begin to build the future we seek.

10

WAR AND THE LAW

All war is immoral, and if you let that bother you, you're not a good soldier.
—GENERAL CURTIS LeMAY (1909–1990)

I N THE EARLY 1990S, when fighting broke out in Sierra Leone,
I flew into a tiny airstrip carved out of the jungle in northern
Liberia. I was working with the organization Marie Stopes International out of London to discover ways of helping Liberian
women plan their families after their own devastating war. Now
tens of thousands of Sierra Leonean refugees had crossed the Liberian
border to escape yet another conflict. The white UN World Food Programme trucks parked in the small compound where I stayed were spat-
tered with mud. Liberia has one of the highest rainfalls in the world and
the dirt roads were a mess. Every time another food truck made the long
journey to the most distant refugee camp, the potholes got bigger and
bigger. Eventually they could have swallowed a Volkswagen.

The World Food Programme ran like a military operation. The conversation was all about logistics, broken shock absorbers, recruiting staff,
bringing in fuel, and above all else, carrying food to refugees. As we sat
around a table on a damp tropical night, I suddenly realized every person in the room came from a different country—Germany, Ethiopia, the
Netherlands, Liberia, the U.S., Kenya, Italy, and Britain. We were a collection of outgroups, working together as an ingroup to save lives in a war

that had nothing to do with any of us directly. We were breaking all the rules of team aggression, and succeeding. Like warriors, like any band of brothers, the staff suffered casualties entering conflict zones, they took risks to protect each other in pursuit of a common goal, and they were bonded in their work.

We have argued that our species has inherited an evolved, biological predisposition for young males to band together and attack their neighbors. The Civil War in Sierra Leone, with its millions of displaced people, thousands of conscripted child soldiers and sex slaves, and eventual breakdown of civil society certainly supplied grim evidence of that. But it also provided clear proof that those same impulses can be shaped and applied to entirely different purposes. Biology may be our starting point, in other words, but it need not be our destiny. If, as biology suggests and history attests, the goal of warfare is to destroy your enemy either by killing him or driving him from his land, then for a supranational organization to come together to feed the dispossessed when they are hungry and to provide medical supplies when they are sick is truly a strange and wonderful phenomenon. I was proud to play my tiny part in this triumph of ingroup empathy over outgroup viciousness—heartened to realize that even if our warlike impulses are inevitable, war and its worst excesses are not.

We have surveyed in some detail the long, grim, vicious history of humanity's war against itself. But there are reasons for hope. The gory catalogue of human warfare has on occasion been interrupted by long intervals of peace, most notably in recent decades amongst the world's most developed nations. World War II is a fading memory for an aging generation in Europe and North America, where most adults have grown to maturity in an era of unprecedented peace and prosperity. There are more than enough potential crises to keep a reasonable person awake and worrying through the night, of course, from the prospect of a nuclear North Korea to the fact that our international treaties and military might count for very little when small bands of young men can highjack modern technology and use it against us in spectacular ways. Yet for all the tensions in today's world, it is difficult to envisage a new war between, say, France and Germany, or the U.S. and Japan—and that is a very significant development indeed. We've looked at the reasons to worry about humanity's deep connection to warfare. Let's take a moment now to look at our reasons to hope.

"Just" Wars

The moral code of war in the ancient world could also describe male chimpanzee behavior during a raid: Eliminate rivals and take their resources for your own. Julius Caesar, like so many commanders before and after, was unambiguously committed to genocide. In 53 B.C., he attacked the Eburones, a Belgic tribe living along the Rhine, and he wrote with pride of his soldiers,

> every village, every building they saw was set on fire; all over the country the cattle were either slaughtered or driven off as booty; and the crops, a part of which had already been laid flat by the autumnal rains, were consumed by the great numbers of horses and men. It seemed certain, therefore, that even if some of the inhabitants had escaped for the moment hiding, they must die of starvation after the retirement of the troops. [374]

Ancient warfare, like chimpanzee raiding, was explicitly about resources. When a city surrendered the inhabitants became slaves, but if a besieged city resisted, then in the words of Deuteronomy,

> when the Lord thy God has delivered it into thine hands, thou shalt smite every male thereof with the edge of the sword. But the women, and the little ones, and the cattle, and all that is in the city even all the spoil thereof, thou shalt take unto thyself; and thou shalt eat the spoil of thine enemies.[375]

With such sentiments written into the founding documents of our civilization, both secular and religious, is it any wonder we haven't yet quite managed to leave our chimp-like brutality behind?

Still, the repudiation of war and violence also has deep roots in our cultural history. Christ certainly went beyond the Old Testament ethos of revenge—"eye for eye, tooth for tooth, hand for hand, foot for foot, burning for burning and wound for wound"[376]—when he exhorted his followers, "unto him that smiteth thee on one cheek offer also the other."[377] For a century or more, the early Christians refused military service and as late as the end of the second century A.D., the early Christian writer Tertullian was arguing that God "in disarming Peter, unbelted every soldier."[378] But our deep, inherited predispositions have a way of resurfacing. When the Roman Emperor Constantine the Great converted to Christianity in 312 A.D., it was not because of theological insight or a turn toward pacifism. Rather, Constantine had a pre-battle vision of the "Chi-Rho"—the first two letters of Christ's name in Greek—with the promise, "In this sign you shall conquer," and went on to devastate his rival's superior force. The Christian monogram was painted on the shields of Constantine's troops before it decorated the churches he built.

It was Aristotle who coined the term "just war,"[379] but he meant "just" as in "justification"—he was justifying the extermination of barbarians by civilized nations. The Christian tradition of just war was not radically different, however. Soon after Constantine, Saint Ambrose (circa 339–397) extended the traditional definition of just wars to include attacks on heretics, who were defined as traitors and therefore subject to offensive warfare.*[380] The *bellum justum*, or just war, concept was most fully developed by St. Augustine of Hippo (born 354). Building on Ambrose, he saw wars as part of God's divine plan for humankind. He developed the concept of a world society (*Civitas terrena*) and a heavenly society, the City of God (*Civitas Dei*). Within such a framework, Augustine argued that it was natural for people to form hierarchical social groups and that therefore they were morally obliged to show "obedience to princes." He wrote in the context of a collapsing Roman Empire, living to see the Sack of Rome by the Visigoths in 411, and his own North African city of Hippo was destroyed the year after his death in 430. In part, Augustine interpreted heaven in terms of the corrupt and decaying empire in which he lived. He gave the theory of just war a new and cruel twist when he argued that it was right to punish sinners to stop them sinning, thereby making war an act of love. If anything, the ferocity of war intensified as a result, but for the first time it came wrapped in a hypocritical cloak of compassion. Although counseling those involved not to take pleasure in the suffering they caused, nor to seek blind revenge, in the last analysis Augustine's writings set no limits to the use of violence and did not distinguish between offensive and defensive warfare. This manner of thought helped pave the road to the Crusades and the Spanish and Portuguese conquests in the Americas, among other wars we would today consider deeply unjust, if not genocidal.**

---

* Ambrose's father was a soldier, and before becoming bishop of Milan in 374, Ambrose himself had been the military governor of the surrounding region.

** These campaigns had more pragmatic roots as well. One of the reasons Pope Urban II organized the first Crusade was to get warring nobles out of Europe, on the theory that it was better for them to go fight the Muslims than stay home killing Christians. There was a similar impetus behind Portuguese and Spanish empire-building in the Americas. After the *Reconquista*, in which the Muslim rulers were driven out of the Iberian Peninsula, there was a surfeit of heavily armed nobles with a centuries-old tradition of fighting. The Popes sought to gain peace in Europe by exporting local violence to distant places occupied by pagans—the ultimate outgroup.

Still, equality and peace are implicit in the Christian message, and official Christianity did make a number of attempts to rein in the horrors of war. In 1027 A.D., the Church under *The Truce of God* attempted to forbid warfare from Saturday night until Monday morning.[381] When the crossbow was invented it was considered a particularly horrific weapon, and the Church tried to outlaw its use—although theologians argued crossbows might still be used to kill unbelievers. In his massive compilation of legal precedents, published in about 1140 A.D., Gratian went even further, holding that the punishment of an evildoer was a benevolent act performed in his or her best interest and that, as non-believers lacked the fundamental spirit of justice, Christian armies were justified in confiscating their property. Winding up a scholarly study of *The Just War in the Middle Ages*, Frederick Russell writes with some truth, "It remains an open question whether just war theories have limited more wars than they have encouraged."[382]

A profoundly different way of looking at war arose in the seventeenth century, in the midst of the pain and destruction of the Thirty Years' War.[383] After the sack of Magdeburg in 1631, in which only the cathedral and 140 houses were left standing, some 30,000 men, women, and children had been slaughtered. Between 1550 and 1650 the population of central Europe dropped by perhaps 30–35 percent.[384] In 1625, Hugo Grotius (1583–1645) published *The Law of War and Peace*.[385] "Wars were begun on trifling pretexts, or none at all," he commented, "and carried out without reference to law, Divine or human." Grotius reasoned, without the benefit of comparative biology, that human beings needed to reverse the basic martial drives of surprise, viciousness, and annihilation of an enemy. "Kings," he wrote, "who measure up to the rule of wisdom take into account not only of the nation which has been entrusted to them, but of the whole human race." He went on to argue that wars should be preceded by a formal declaration of hostilities by civil rather than military authorities, and followed by well-defined treaties. Grotius also tried to reverse the unthinking, passionate hatred of enemies that had characterized raids and wars since our earliest days. Wounded and captive adversaries, he argued, should be treated humanely, and conquered populations should retain their autonomy. "One must take care, so far as possible, to prevent the death of innocent persons, even by accident," he wrote. The idea has stuck with us ever since, though it has often been pushed to the back of our minds. Interestingly, Grotius, born a Calvinist, conceived his arguments while in prison, in part for challenging his sect's belief in pre-destination. By advocating for humane treatment of conquered enemies in war, he helps us make our case that there is an element of free choice in biology as well.

At the beginning of the U.S. Civil War, the Union regarded captured Confederates as traitors undeserving of humane treatment, much as the Bush administration has treated suspects since 9/11. Then in 1863, President Abraham Lincoln asked the political philosopher Francis Lieber (1798–1872) to draw up a military code for the Union Army. A Prussian by birth, Lieber had fought at Waterloo and been injured and left for dead at the battle of Namura. He immigrated to America where he became the first professor of History and Political Economy at South Carolina College in Columbia. When the Civil War broke out, two of his sons joined the Union forces. A third joined the Confederates, only to die at Williamsburg. Lieber's code of 157 articles made clear the distinction between civilians and combatants and forbade torture—as clear a signature as there is of the dehumanization of an enemy. This Lieber Code was widely adopted, and today the military law of all contemporary democracies punishes soldiers who abuse civilians.

Attempts to internationalize laws of war began with the 1899 Hague Convention for the Pacific Settlement of International Disputes. The Conference was organized by Tsar Nicholas II of Russia. Nicholas was a truly kind man, a loving husband and father, but his idealism caused him to lose control of his empire and ultimately led to his assassination after the Russian Revolution of 1917. The Hague conference sought to limit armaments and arbitrate international disputes,[386] but it ended in a clash between idealism and *realpolitik*, a push-and-pull argument that continues to this day. Nicholas's cousin, King Edward VII of Britain, thought the meeting was "the greatest nonsense and rubbish I ever heard." The British government sent Sir Jack Fisher, who was already dreaming of the Dreadnought class of battleships, with 12-inch guns and steam turbines to render every other warship in the world obsolete. "If you rub it in both at home and abroad that you are ready for instant war," said Fisher, "and intend to be first to hit your enemy in the belly and kick when he is down and boil your prisoners in oil (if you take any)...then people will keep clear of you."[387] The German representative at The Hague got Fisher's message and commented that England was firmly committed "to the principle 'might is right.'" A second Peace Conference was held in 1907, but neither conference did much to stop the drift toward war. They did outlaw aerial bombardment from balloons and the use of poison gas, but only because in 1899 these were weapons which the military deemed unlikely to be useful, and therefore cynically appropriate as bargaining chips. In World War I both would be used on a significant scale despite the Hague agreements.

For all that we still have savage civil wars, genocidal attacks, and abusive, negligent, and ill-advised use of military power, humanity has come a long way in the application of moral, ethical, and legal principles to the conduct of war—at least in theory. The sad fact is that we continue to backslide, and to find ready excuses when it suits our needs. The recent invasions of Afghanistan and Iraq were labeled part of a "war" on terror, but when enemy combatants were captured they were not accorded the rights of prisoners of war. Instead, the U.S. Central Intelligence Agency has interrogated about 3,000 prisoners without any legal oversight and the military has held hundreds of "unlawful combatants" in Guantanamo Bay, Cuba, largely incommunicado and without access to a fair trial. There is also always a temptation, when the lives of your own citizens are on the line, to torture prisoners in order to obtain key information. Some suspected Al Qaeda members were deliberately shipped to other countries known to torture prisoners, such as Egypt or Morocco. In the words of one officer, "We don't kick the shit out of them. We send them to other countries so they can kick the shit out of them." John Yoo, a law professor who joined the Berkeley campus the same year I did, has argued that the U.S. president has the constitutional power to interpret treaties and therefore can interpret the Geneva Conventions any way he wishes. All other considerations apart, a person being tortured will invent any fiction to end the pain. I'm not a legal scholar, but I am a biologist, and I recognize the symptoms of team aggression when I see them.

In 2006 the Supreme Court ruled that all prisoners, including suspected Al Qaeda operatives, are protected from "cruel treatments and torture" by Article 3 of the Geneva Convention adopted by the world community in 1949. Senator John McCain, who was tortured as a prisoner of war in Vietnam, resisted legislative efforts to reinterpret the Convention, but in the end Congress passed an ugly compromise that robbed those accused of terrorism of the right of *habeas corpus*—essentially, the right to a fair trial—and left the president to interpret the Geneva Conventions as he pleased.* It might be argued that terrorists do not declare war and that they deliberately kill civilians, so they don't deserve the status of prisoners of war. But torture is almost always counter-productive—even if it does, rarely, lead to real and actionable intelligence, the risk and consequences of false information can be so severe as to make any possible gain meaningless. Ibn al-Shaykh al-Libi, a Libyan who ran the Al Qaeda camp in Afghanistan where two of the 9/11 terrorists trained, told interrogators under torture that Saddam Hussein had trained Al Qaeda operatives in the use of biochemical weapons. This misinformation, invented to avoid further pain, became part of the public "justification" for invading Iraq.[388]

Equally important, torture is as harmful to the torturer as it should be repugnant to us all. Nothing better epitomizes, nor more effectively reinforces, the dehumanization that lies at the core of our same-species killing. Allowing prisoners to be tortured in the fight against terrorism erases our moral authority, and jeopardizes our connections to those, such as moderate Muslims, who would otherwise ally themselves with the West in condemning acts of terrorism. Perhaps most importantly, torture dehumanizes its practitioners as well as its victims, and condemns us to life at the amoral, soulless level of our ape ancestors. The urge to take any and all possible measure to secure information and protect our allies and loved ones is powerful and understandable. The true test of our ability to overcome the predisposition to dehumanize our enemies, which is the evil heart of team aggression, is our ability to treat those we hate, especially with good reason, justly. At the beginning of the twenty-first century, U.S. legislators failed that test and slipped back in to a set of evolved predispositions stretching back to our shared ancestry with chimpanzees.

Civilians and Combatants

Sometimes, the ingroup mentality of honor and chivalry we encountered in chapter 3 tempers the outgroup dehumanization of the enemy. In 690 B.C., after capturing Jerusalem, the Babylonian king Sennacherib records how he "killed the officials and patricians who had committed the crime [of resisting the Babylonian attack] and hung their bodies on poles outside the city."[389] Those guilty of lesser crimes were imprisoned,

---

* In June 2008, the U.S. Supreme Court decided 5 to 4 to extend the constitutional right of *habeas corpus* to prisons in Guantanamo Bay.

but Sennacherib eschewed broader vengeance and released the rest. The Hindu *Book of Manu,* written in the fourth century B.C., prohibited killing those who were asleep, naked, disarmed, grievously wounded, or fleeing "with hair flying." In *De Officiis,* the great Roman orator Cicero wrote, "we must also ensure protection of those who lay down their arms and throw themselves upon the mercy of generals, even though the battering ram has hammered at their walls."[390] Even before *The Truce of God,* a series of Church decrees in the late ninth and early tenth centuries A.D., collectively called *The Peace of God,* prohibited attacks on churches, fruit trees, and animals, and decreed excommunication for soldiers who attacked women, merchants, pilgrims, clerics, and even shepherds.[391] All these laws, decrees, and opinions did not stop rape, murder, theft, and vengeance in warfare, of course, but the impulse to limit these normal excesses of war clearly has deep roots.

The success or failure of efforts to separate civilians from combatants is also partially linked to the technology used in killing. At the Battle of Gettysburg, there were over 50,000 casualties, but only one woman was killed. When firing muskets and cannons, you can select your enemy. By contrast, the Hiroshima and Nagasaki atomic bombs could not distinguish between the sexes or between military and purely domestic targets, and no such selectivity was possible. (Obviously, there is also a choice of whether to target civilians or not, and the important fact that Gettysburg was a battle between armies composed entirely of men.) The development of smart, or guided, bombs, which are extraordinarily expensive, was of course driven mainly by the need for more accurate destruction of specific targets. But the desire for pinpoint destruction was also in no small part focused on reducing civilian casualties. Whether the ultimate motivation was to avoid negative public perceptions or not, it represented a step toward more ethical conduct of war and away from the ape-like impulse to seek revenge, and to destroy everything in our path.

Capturing enemy soldiers rather than killing them became more common as modern warfare developed. But its roots likely go back all the way to the rules of ape ingroup dominance struggles. The archetypal surrender posture, in which a man walks forward with his hands in the air, may well derive from patterns of submission found among other species of social animals. In a pack of wolves, the submissive animal will roll on its back to expose its vulnerable belly as a sign of submission. Hyenas, which like wolves live in matriarchal groups, exhibit an even more extreme show of surrender and vulnerability; any individual re-entering the pack after a brief absence undergoes a "greeting ceremony," in which the animals demonstrate mutual loyalty and trust by deliberately exposing their pendulous genitals to each other's teeth—and hyenas have the most powerful jaws of any mammalian species, capable of crunching the marrow bones of large animals.

Taking prisoners is different from absorbing members of a vanquished tribe, and it may well have begun as an expression of self-interest. It is almost certainly the starting point for slavery. All raids are ultimately about capturing resources, after all, and with the rise of agriculture and stratified societies, human effort became a very valuable commodity indeed. But taking military or civilian prisoners and keeping them captive for their labor, or even just long enough to sell them to others, requires not just the strength to prevent escapes and revolts, but also a particularly deep level of dehumanization. Compared to the long-standing tradition of slaughtering vanquished warriors, however, taking prisoners of war could actually be a sign of ingroup ethical thinking—so long as POWs are kept under humane conditions and repatriated when the conflict ends. In modern times, the victor can also claim the moral high ground, which may increase the morale of his own fighting troops and the support of the population at home. Ethical treatment of prisoners may contain an element of self-interest as well, in that every soldier knows he too may be captured someday, and therefore hopes for reciprocity in treatment.

The ill-treatment of prisoners (many of them civilians) by American forces after the U.S.-led invasion of Iraq in 2003 shocked many people. In so doing it became a vivid, and in many ways encouraging, illustration of the spread of basic humanitarian rules of warfare. At the same time, it remains difficult to apportion blame. The soldiers who humiliated Iraqi prisoners acted exactly as the Stanford students we met in chapter 3 did in Zimbardo's famous experiment, when undergraduates were given power over fellow students. Such behavior possibly should have been predicted—a few reservists with inadequate training were put in charge of thousands of prisoners at Abu Ghraib, the most infamous prison in Iraq.[392] But should some soldier who has never read the Geneva Conventions be punished for behaving as human beings have always behaved, or should a commander or politician with ultimate responsibility for the military be charged with the crime?

War Crimes

As social groups grew larger with settled agriculture, so the power of imposing legal sanctions shifted from the individual and his or her kin to the larger social group. Among tribes such as the Nuer of the Sudan, or the Pueblo Indians of the American Southwest, for example, the chief or the high priest adjudicated disputes, but if the defendant did not agree to the punishment then a blood feud could result. As societies grew larger, the division of labor increased, technology improved, and written precedents could be kept and studied. As part of these innovations, the state took over the role of both prosecution and punishment, which could range from a fine or restitution to torture, banishment, and execution. Modern legal systems are inherently disorganized, having grown out of what came before, and despite regular attempts at updating, reform, and rationalization—they all contain a multitude of imperfections, contradictions, and ambiguities. Still, the best of them handle disputes over property, theft, homicide, adultery, and other behaviors a certain society wants to discourage in a pragmatic, transparent way, and keep up with society as mores change and new technology develops. But in all these cases, no matter how sophisticated, the civil and criminal laws deal exclusively with what can be accurately described as "ingroup" offenses.

This is one of the major challenges in developing international law, and specifically, laws of warfare. How do you apply an ingroup solution (laws, and their enforcement) to an outgroup problem (wars, and the atrocities they generate)? Legal approaches to regulating violence outside the troop face two problems: First, acts that would be called homicide inside the troop become acts of courage to be socially rewarded when committed outside the troop; and second, there is no universally accepted authority to enforce a system of punishments on the armed forces or civilian commanders of independent countries.

Some of the Germans who guarded the concentration camps were soldiers who in many cases could have had themselves transferred to other duties. But in their own minds and emotions, these men saw themselves as no different from any other band of warriors. Johannes Hasselbroeck was a Nazi Party member who found military camaraderie changed him from an atheist to "a believer in God." He was wounded at the battlefront and transferred to the Gross Rosen concentration camp in the Netherlands, where he was promoted to commandant. He described his bond with his fellow prison guards in exactly the same terms we heard men describe combat in chapter 4:

> Even the ties of love between a man and a woman are not stronger than that same friendship that there was among us. This friendship was all. It gave us strength and held us together in a covenant of blood. It was worth living for; it was worth dying for. This was what gave us the physical strength and courage to do what others dare not do because they were too weak.[393]

It is a somber insight into human behavior that the same predisposition to form a band of brothers can lead both to acts of high courage, to the tenacity of the workers in the World Food Programme in Liberia, and to the utmost cruelty. Hasselbroeck was not under fire and what "others dare not do" was to oversee the slaughter of 100,000 unarmed prisoners.* Yet there is reason to believe that the powerful predisposition at work in each situation is the same. Such crimes should be punished but the issue is not as straightforward as convicting someone of murder in an ingroup setting.

The idea of a "war crime," like the concept of human rights, is a welcome and important cultural development, but also a recent one. Slobodan Milosevic of Serbia, Idi Amin of Uganda, Joseph Stalin, Adolf Hitler, and Saddam Hussein merely did what Julius Caesar, Darius of Persia, Ramses II of Egypt, the Assyrian Ashurbanipal, the Greek Alexander the Great, or Moses and Joshua of the Old Testament, and untold thousands of others did before them.

In 1900 in Britain, at the time of the Boer War in South Africa, the *St. James's Gazette* felt able to publish, "We have undertaken to conquer the Transvaal and if nothing will make sure except the removal of the Dutch inhabitants, they must be removed, men, women and children."[394] In the U.S. at about the same time, Lieutenant Preston Brown was only fined two weeks' pay for his murder of an unarmed prisoner during the Philippine War.

Attempts to punish blatant transgressions of the military code were, however, made after World War I. The Allies occupying Germany in 1918 tried and executed a number of German soldiers and government officials, but a high-profile attempt to convict almost 1,000 Germans for war crimes collapsed when responsibility for the trials was handed back to the Ger-

---

* Hasselbroeck was convicted as a war criminal and sentenced to death, but the warrant was reduced and he was released after serving ten years.

mans themselves. The stiffest sentence passed when the vanquished were asked to try their own servicemen was four years' imprisonment for two submarine commanders who had coldly machine-gunned the crew of a British merchant ship they had torpedoed. In between the two World Wars, the Permanent Court of International Justice was set up as an attempt at an impartial, international legal authority. It functioned reasonably well when it tackled cases where both parties agreed in advance to accept the Court's ruling, and where the adjudication of disputes was limited to states rather than individuals. The Court, however, could not stop Japanese aggression in Manchuria and China, the Italian invasion of Ethiopia, or Nazi support for General Francisco Franco during the Spanish Civil War.

The failure of the First World War trials was remembered after the Second. The International Military Tribunals set up at Nuremberg and Tokyo used judges drawn from the victors, and defined certain behaviors amongst the defeated as "crimes against humanity." Of the 185 people tried at Nuremberg, thirty-five were acquitted, twenty-five were executed, twenty were sentenced to life imprisonment, and the rest received lesser sentences, were discharged on health grounds, or like Hermann Göring, committed suicide. The legal processes were somewhat arbitrary, in the sense that at the ingroup level, any war could be defined as a crime against humanity. "Crimes against international law," the Tribunal concluded, "are committed by men, not by abstract entities, and only by punishing individuals who commit such crimes can the provisions of international law be enforced." But these assertions, however eloquent, fail to provide the kinds of philosophical foundation and internal logic that characterize civil law within a single country. Telford Taylor, the chief American counsel, said at Nuremberg, "War consists largely of acts that would be criminal if performed in the time of peace—killing, wounding, kidnapping, destroying or carrying off other people's property. Such conduct is not regarded as criminal if it takes place in the course of war, because the state of war lays a blanket of immunity over the warrior." The trend, starting in the twentieth century, has been to try to extend the application of ingroup morality to outgroup behaviors. At Nuremberg, Taylor answered his own paradox by saying, "But the area of immunity is not unlimited, and its boundaries are marked by the laws of war." However, unlike the "International" Military Trials of WWI, which were ultimately convened and conducted by the victors, the Nuremberg Trials were accepted as "just" by the majority of the Germans, as well as virtually all the Allies. The egregious nature of Nazi crimes, the solemnity of the proceedings and the fairness demonstrated by acquittals, the 1948 Universal Declaration of Human Rights, and the UN Charter all reinforced the perceived justice of the yearlong trials. If a computer were programmed to dispense justice, though, it might have categorized the crushing and burning to death of civilians in Allied bombing raids on German and Japanese cities as war crimes as well.

In the Japanese war trials, which paralleled those at Nuremberg, the immunity Telford spoke of came close indeed to being "unlimited," largely for expedient reasons. Only seven men, including General Hideki Tojo, who had been Japanese Prime Minister through much of the war, were hanged out of twenty-five accused. Shiro Ishii, who had tested bacteriological weapons on live prisoners, was let off the hook in exchange for knowledge about germ warfare that the American military establishment felt it needed. Most important of all, a political decision was made to exclude the Emperor from prosecution. Clearly, all efforts to regulate war run into the basic paradox bequeathed us by biological evolution, namely that the purpose of war—when measured by ingroup standards—is mass homicide. There are no absolute standards for defining a war crime and human judgments are fickle.

While the Nuremberg trials succeeded in bringing closure to a terrible war, there are also a number of reasons why not all war crimes prosecutions lead to reconciliation. Our predisposition for team aggression means that the "other" person's actions are easier to brand as "crimes against humanity" than our own. When a leader is indicted for a war crime, his supporters may interpret any punishment as evidence that they are the real victims, who were merely defending themselves against attack or injustice. Such trials generally fail to indict those members of the local community who may have been most intimately involved, and they can do nothing about those who stood by passively when "crimes" were committed. Can we hold individual members of a killer mob accountable for his (it is almost always a man) actions? In the late 1970s Pol Pot's Khmer Rouge killed an estimated 1.7 million Cambodians, or 20 percent of the country's population. But should the prison guards of the notorious S-21 interrogation camp, where 14,000 prisoners were incarcerated in one small set of school buildings, be held responsible? Many were only fifteen or sixteen when they killed prisoners and each knew he would be killed if he did not carry out orders.[395]

Attempts to convict individuals accused of war crimes are likely to remain confusing and uncertain in their outcomes. In the late 1990s, a Spanish magistrate sought the extradition from London of General Augusto Pinochet, former dictator of Chile. The case reached the British House of Lords, where it was concluded that Pinochet's regime had been guilty of conduct that was "not acceptable on the part of anyone." Eventually, Pinochet returned to Chile but he escaped prosecution when the Appeals Court ruled he was "demented," and unable to stand trial. In 1998, the Statute of Rome, which binds the members of the European Union, established the International Criminal Court (ICC). This court began to push the envelope of international justice in two new ways: It continued the Nuremberg tradition of focusing on individuals rather than the states that are party to treaties, and it set out to impose its judgments both on countries subscribing to the treaty of Rome and on non-signatories as well. As a result it has run into a slew of problems. The case against the Serb leader Slobadan Milosevic, accused of genocide in the Balkans, was so complex and prolonged that Milosevic died before his case came to judgment in The Hague.

Still, the ICC has been ratified by sixty nations, and American President Bill Clinton signed the Treaty for the U.S. in 2000. He didn't submit it for Senate approval, however, and the Bush administration rejected the Court on the grounds that American politicians and commanders might be indicted. The British derided the Americans for their fears, but setting rational limits to the jurisdiction of courts attempting to define war crimes rapidly proved a difficult and contentious task. In 1993, the Belgian parliament went where the ICC was afraid to tread, claiming "universal jurisdiction" over war crimes. The supreme court of Belgium upheld the right of national courts to try, *in absentia*, Ariel Sharon (for his part in the massacres of Palestinians in Beirut in 1982) and Georges Henri Beauthier (the former dictator of Chad). Following the initiation of the 2003 Iraq war, a complaint was filed in the Belgian courts against British Prime Minister Tony Blair for crimes against humanity. Donald Rumsfeld, the U.S. Secretary of Defense, countered by threatening to remove NATO headquarters from Brussels and the Belgium parliament backed down.

Collateral Damage

U.S. Army Field Manual 27-10, the comprehensive guide to the laws of war for the American military, sets out to apply ingroup ethics to the conduct of war, stating, for example, that, "A commander may not put his prisoners to death because their presence restricts his movements or diminishes his power of resistance by necessitating a large guard." In March 1971 Lieutenant "Rusty" Calley was convicted by a military court of leading the massacre of men women and children in the village of My Lai 4, in South Vietnam. He based his defense on outgroup standards, arguing that he had no way to distinguish Viet Cong from harmless villagers, that leaving the villagers alive endangered his own troops, and that he had been ordered by his superior, Captain Medina, to "Kill everything." Calley wrote,

> Our mission in My Lai wasn't perverted, though. It was simply "Go and destroy it." Remember the Bible; the Amalekites? God said to Saul, Now go…and utterly destroy all that they have, and spare them not; but slay both man and woman, infant and suckling, ox and sheep, camel and ass. But people took the spoil—and God punished them. [396]

A couple of years before My Lai, another massacre occurred in a small hamlet called Thanh Phong, also in the delta of the Mekong River. A group of seven elite Navy Seal commandos under the command of Lieutenant Robert Kerrey* landed at night with orders to kidnap or kill a suspected Viet Cong leader. At least thirteen unarmed women and children were killed as the Seals poured 1,200 bullets into the village. A Vietnamese survivor described how the Americans knifed a woman and three children and then shot fifteen women and children at close range. One old man had his throat cut. Some of the Seals corroborated this account, although Kerrey and others maintained that they returned fire into the darkness from about one hundred yards away. Like Calley, Kerrey reported to a swaggering superior, Captain Roy Hoffman in this case, whose rules of engagement were "if you don't kill or capture them [Viet Cong suspects] you'd hear from me."[397]

Another person who used the concept of "collateral damage" to justify killing was Timothy McVeigh, who blew up the Muir Federal in building Oklahoma City in 1995, killing more Americans than died in the First Gulf War. He was an intelligent young man and a decorated army combat veteran of that war, a loner who became sympathetic to the anti-govern-

---

* Not to be confused with 2004 presidential candidate John F. Kerry, who also served in Vietnam and like Kerrey would go on the serve in the U.S. Senate.

ment militia movement. When questioned about the children he killed, McVeigh showed no remorse and labeled their deaths "collateral damage." In some crazy way, he saw himself fighting to avenge the injustice he perceived to have occurred during the government "siege" of the Waco, Texas, compound of the Branch Davidian cult, including the death of eighty-two of its members. Timothy McVeigh was certainly a mass murderer, and perhaps a mentally ill one at that. But from his perspective, his plea of unavoidable collateral damage in a just and necessary operation was no different from the justification given for killing civilians by Calley and Kerrey, to say nothing of Coalition forces in the two Gulf Wars, NATO forces in the Balkans in the 1990s, or Israel in its 2006 attacks on Lebanon.

Whatever the exact details of the various Vietnam massacres, it seems that young males raiding deep in the territory of a pursued enemy can very easily slip back to the basic primate drive of exterminating everything that moves. There were 122 convictions by military courts for war crimes by U.S. troops in Vietnam, and many more that were never prosecuted, as was the case with Thanh Phong. Looking back at Vietnam, Bob Kerrey wrote,

> It's far more than guilt, it's shame. You can never get away from it. It darkens your day. I thought dying for your country was the worst thing that could happen to you, and I don't think that it is. I think killing for your country can be a lot worse. Because that's the memory that haunts.[398]

Once we remember that evolution has given us two opposing moral frameworks, one in which we call killing murder and one in which we call it warfare, then perhaps we can begin to understand why decision-making in relation to "collateral damage" is so intrinsically difficult. Perhaps the complex, morally confusing stories of individuals such as McVeigh, Calley, and Kerrey can be best understood by regarding them as manifestations of the basic primate predisposition for team aggression, with all the frenzy and destruction that implies. They all killed civilians during prototypical primate raids, as men have done for the great majority of human history. Timothy McVeigh was executed, Calley was court-martialed,* and Bob Kerrey was elected to the U.S. Senate. By their own

---

* Calley was convicted of premeditated murder of twenty-two civilians at My Lai and sentenced to life imprisonment at hard labor. He served just three-and-a-half years at the Fort Benning Army base in Georgia.

standards, each was a brave man doing his duty; by some legal interpretations of warfare they all deserved to be punished.

Reconciliation and Rebuilding

Some of the most important attempts to ameliorate the horrors of war arose in the twentieth century. They were largely without precedent, and although not yet universally successful, they have helped to widen the gap between the behavior of partially civilized *Homo sapiens* and the behavior of fully wild *Pan troglodytes*. At a global level, by far the most important twentieth-century innovation has been a series of humanitarian and economic measures to assist the defeated and those displaced by warfare. The League of Nations undertook a limited humanitarian effort after World War I, but following World War II, the United Nations set new standards for a more humane postwar world. As efforts to divide civilians from combatants have increased, so civilian suffering after a war has come to be perceived as involving "innocent" people. International agencies such as the World Food Programme play an immediate life-saving role, carrying food and shelter to communities displaced by war. It is not something Julius Caesar would have endorsed, and the very opposite of what a band of chimpanzees would undertake. Low-profile, politically acceptable activities of this type have changed the face of war in a more profound way than most people realize. The activity enjoys broad support in developed nations; indeed it can be politically dangerous for a leader of a democratic country not to intervene in humanitarian crises created by warfare. Media accounts and TV images, especially of starving children, can arouse powerful ingroup empathy.

To function well, however, relief organizations have to be non-partisan, and that can be difficult: To be non-partisan means to put one's own ingroup affiliations aside in order to embrace a larger, global tribe. In a sense that we sometimes still feel at a deep level, embracing a global community could even be seen as a form of betrayal against our more narrow ingroups, in that we replace the ethos "my country, right or wrong," with "right over wrong, whether it is my country or not." In recent years, independent aid groups such as Médecins sans Frontières have become vocal proponents of human rights. These groups have indeed extended the reach of their ingroup ethics to the entire human population, and have done a great deal of good and important work. Unfortunately, there remains much of that work to be done, for as the journalist and Bard College professor David Rieff writes, in war situations, "The dictates of the Universal Declaration of Human Rights are unlikely to translate into consequences on the ground." Human rights are an ingroup concept, but for the most part we still find it difficult to consider all humans part of our ingroup.

Still, it is important to remember that just as chimpanzees have a passion for revenge, they also have a number of social mechanisms that aid forgiveness, bonding, and reintegration of transgressors.* The late twentieth century saw a remarkable example of how this other primate predisposition can play out in human affairs, if we can find the courage, wisdom, and strong leadership needed to deploy it. The South African Truth and Reconciliation Commission (TRC) has been a unique attempt to heal the wounds of apartheid in South Africa, a conflict which had many of the characteristics of war. It is a clever alternative to the intrinsically difficult task of bringing "war criminals" to "justice." Under the TRC, those who had killed and tortured during apartheid days were confronted by their victims. The aim was not punishment but to secure acknowledgment of what had happened. In the model of ingroup/outgroup behavior used so often in this book, the TRC tried to get those involved in the cruelty of team aggression to review their actions by ingroup standards. Importantly, the TRC was also not expecting those Black South Africans who had suffered abuse from the agents of apartheid to simply forgive and forget. Rather, the formal structure of disclosure, admissions, and expression of anger, pain, and other reactions allowed people on all sides of the conflict to start experiencing their one-time adversaries as human beings—that is, the process was a start toward building a larger, more inclusive ingroup. It did not work in every case, but the TRC was a profound and fascinating attempt to square the human circle of empathy and aggression. It remains an important

---

* There is a continual pressure for individuals of any group of social animals to buck the system, by attempting to climb the hierarchy or by free-riding, which is to say taking advantage of group benefits without contributing to group welfare. For group cohesion and cooperation to survive, evolution has built in counter-strategies. In chimpanzees, grooming reinforces bonds between individuals and helps to reintegrate the troop after ingroup altercations. This sort of social glue works because it is pleasurable but also requires an investment of time and effort that might otherwise have been devoted to finding food or some other task. Anthropologist Robin Dunbar from Liverpool University speculates that speech is the human analogue of chimpanzee grooming. Gossip, story telling, and discussion are also time-consuming processes and, like grooming, they help knit people together.

model for other countries that have undergone civil war, genocide, and
other catastrophic failures of ingroup cohesion.

Supranational Forces

If nations are in some way equivalent to chimpanzee troops—admittedly,
this is speaking quite loosely—then our fellow countrymen comprise our
most important ingroup. The TRC worked by expanding the national ingroup to include all South Africans, but what happens when we try to extend that crucial boundary further still? Many thinkers have argued that
human rights can be upheld only by some sort of supranational agency.
William Penn in the seventeenth century and Immanuel Kant in the eighteenth promoted the idea of a union of states to prevent war. Kant believed "perpetual peace" could be achieved through a league of nations,
and he proposed that the "false pride of sovereignty" could be overcome
by educating children in the need for world government. The League of
Nations was created after World War I "to promote international cooperation and to achieve international peace and security." American President
Woodrow Wilson was a major supporter of the system, but the U.S. Senate failed to muster the two-thirds majority needed to ratify the Covenant
establishing the League and let the U.S. join.

The phrase "United Nations" was first used by President Franklin D.
Roosevelt in January 1942 to refer to the twenty-six nations allied against
the Nazi war machine. The United Nations assumed its present structure
at the United Nations Conference of International Organizations meeting
in San Francisco, which stretched from April to June 1945. The Charter
signed at the end of that meeting grew explicitly out of the experience of
World War II:

> We the Peoples of the United Nations determined to save succeeding gen-
> erations from the scourge of war, which twice in our lifetime has brought
> untold sorrow to mankind, and to reaffirm faith in fundamental human
> rights, in the dignity and worth of the human person....[399]

The isolationist tendencies that blocked U.S. participation in the League
of Nations were replaced by active support of the UN after World War II.
The U.S. Secretary of State called the UN a "new era in global understanding." Over time the UN, while manifestly imperfect, has had more successes

keeping the peace than did the League, largely by providing a venue for international discussion, and perhaps more importantly, through the work of its various agencies, including the World Food Programme, the High Commission for Refugees, and the Department of Peacekeeping Operations.

But talk cannot solve all human problems, and of course it is the rogues, despots, rebels, and general bad actors most likely to start conflicts who are least likely to respond to the encouragement or badgering of the international community. Any legal system needs an enforcement branch, and the United Nations remains a very long way from becoming a global policeman. The UN is built around the concept of national sovereignty. The organization normally acts by consensus in international debates, and even when the Security Council votes, its five permanent members retain a veto. The UN did take notional control of the Korean conflict in 1950, but only because of a diplomatic accident in which the Soviet Union absented itself from the Security Council chamber at the time the key vote was taken. When it came to the more recent fighting in Kuwait and Kosovo, for example, a supranational group of powerful nations was assembled outside the UN system. For the time being, UN peacekeepers remain the armed equivalent of aid organizations—they try to protect civilians and maintain peace, but do not actively engage in conflicts. The European Union is discussing a common EU peacekeeping force, and other regional organizations such as the African Union may be heading in the same direction. The challenges are immense, from issues of command structure to the crucial need to be seen as neutral, but the potential rewards are also substantial. Perhaps the concept of a truly international police force will need to go through several iterations, beginning with relatively small groups of like-minded nations working together, before it can become a reality.

Other elements of international military law include efforts to restrict certain kinds of particularly destructive weapons or those that have a high likelihood of harming civilians. Recent efforts by the UN to control the global armaments trade have largely failed. In 1997 the Ottawa Convention to ban landmines succeeded because anti-landmine groups were able to circumvent the normal UN system of vetoes and stalling. But small arms remain the major killing weapon in contemporary wars from Bosnia to Liberia, and the number of pistols, rifles, and machine guns grows by six to eight million *annually*. Small arms are responsible for half a million deaths per year and their control would reduce deaths from aggression.*[400] Unfortunately, China, Russia, and the U.S. cynically block any agreement on small arms in order to appease their own manufacturers and political lobbying groups. The 2001 UN Conference on the Illicit Trade in Small Arms and Light Weapons achieved nothing except to mobilize the U.S. National Rifle Association, which now outguns the international anti-arms trade lobbyists. A second attempt to limit this trade in 2006 was defeated by a minority of nations including Cuba and the U.S.—a rare and unfortunate case of agreement between these two countries.

It is generally concluded that if any family of nations is to prevent wars, then it must be able to field an army of its own. No matter what international statutes are agreed to, there will always be rogues and bad actors, and when individual nations or coalitions seek to enforce UN resolutions, as in the First Gulf War, they can too easily be seen as aggressors themselves. But any effort to recruit, command, and deploy an international military force cuts across the predispositions underlying effective fighting units. Asking armed forces to "keep the peace" is a contradiction in terms. The basic biological drives that make men fight are undermined in the disorientating world facing the modern peacekeeper. There is no clear enemy. The soldiers and commanders may be fighting with multinational colleagues they find difficult to trust, and the application of force central to warfare would be seriously constrained by complex Rules of Engagement (ROE). Such rules are the military orders that prevent the use of "excessive force" by peacekeeping troops, and are necessary and humane, but that does not make them any less of an oxymoron biologically. Wars can only be fought to be won.

Families at home are less likely to accept casualties in a distant peacekeeping mission they do not understand than when their soldiers are fighting for a nation's immediate interests. And in the absence of clear orders on the ground, soldiers are likely to take the path of least resistance and not intervene. This was one reason the UN failed to intervene effectively in the 1994 genocide in Rwanda. In June 2002, the U.S. unilaterally pulled out of their commitment to provide troops and to pay for part of the United Nations peacekeeping effort taking place in Bosnia. It was a move strongly condemned by internationalists, but strongly supported by some military

---

* In 1999 there were 28,874 gun-related deaths in the U.S., many the result of gang warfare. The intentional gun death rate in the U.S., where some states do not even require handguns to be registered, is 13.47 per 100,000. In Singapore, where there is strict gun control, it is 0.24 per 100,000—fifty times less.

and civil leaders in the United States. It highlighted the profound philosophical and practical problems that arise when attempts are made to link war and law. As we've seen all too clearly in the aftermath of the 2003 invasion of Iraq, police work and soldiering are very different occupations, and soldiers and police are often ill-suited to perform each others' duties.

Ambivalence and Progress

Many challenges and steep biological, political, and practical obstacles remain before we can realistically start to talk about legislating war into obsolescence. Still, humanity has made considerable progress over the centuries by using laws to limit some of the worst excesses of war. In some cases, war has been transformed from an exercise in extermination—the widespread approach of the ancient world, summarized by the Biblical "neither [leave] any to breathe"—into a more complex process, where an army made up of basic fighting units maneuvers to become, in Prussian general Carl von Clausewitz's famous words, "a continuation of politics by other means." The goal of Grotius's *The Law of War and Peace* and Lieber's Code is to reverse the biological drives of surprise, viciousness, and elimination of enemies inherited from our evolutionary past. We are making progress, but slowly and with considerable ambivalence. The laws of warfare are caught perpetually between the immovable commandments of ingroup obedience and the irresistible forces of outgroup destruction. The transition from the viciousness of the team raid to some sort of restraint on killing is necessarily and intrinsically unstable and uncertain.

The more we appreciate from whence we came biologically, the better placed we will be to build up those cultural and environmental influences that help ameliorate the worst aspects of war. The naïveté of attempting to prosecute Blair as a war criminal and the school-yard bully politics of Rumsfeld to remove NATO Headquarters from Brussels tell us that if we are going to make genuine progress toward limiting war's worst excesses, we have to begin by confronting the universal human (or at least male) schizophrenia over killing members of our own species. We may well make the most progress if we build on the progress to date in incremental, pragmatic steps rather than trying to make extreme—if justifiable— claims for human rights that cannot command universal support.

In the meantime, it's a good idea to keep in mind just how much progress we have made, both in limiting the total number of wars and in supporting those who have been displaced, starved, or made homeless by fighting. The Assyrian King Ashurbanipal (668–625 B.C.) proudly decorated the public rooms of his palace with colored bas-reliefs of grotesquely cruel treatments that were deliberately meted out to prisoners of war. Some were skinned alive; others had their tongues pulled out or were beaten with iron rods. Such carvings were meant to scare, impress, and be emulated. The much milder, but still horrible humiliation and beatings meted out by an occupier of the same country—Iraq—in 2004 A.D. produced shame, rather than pride, in many Americans. We cannot claim to have controlled war, but we have begun to channel it in more humane directions, away from genocidal attacks on foreign populations. In other words, we have started to shift war away from its starting point. International laws and their effective enforcement will be an important part of moving that process forward. But we can't ever hope to succeed fully unless we address the war's biological roots much more directly.

We cannot remind ourselves too often of the ubiquitous nature of our Stone Age behaviors. On the same day in 2006, President Bush announced he would veto a Senate Bill loosening restrictions on stem cell research and permit the export of bombs to Israel to use in its war against Hezbollah in Lebanon, where collateral killing of civilians was certain. When I was a laboratory researcher, I needed a powerful microscope to even see a bunch of stem cells, and personally I would have been much less troubled by flushing stem cells down the sink than dropping a bomb on a house full of women and children. Yet our ingrained ability to dehumanize others is so strong, and our ability to "justify" war so facile, that intelligent and well-intentioned people spend more time worrying about embryos than children or adults—provided of course that those children and adults live somewhere else and are not part of our ingroup.

11

EVIL

It is not the murderers, the criminals, the delinquent and the wildly non-conformist who have embarked on the really significant rampages of killing, torture and mayhem. Rather it is the conformists, virtuous citizens acting in the name of righteous causes and intensely held beliefs who throughout history have perpetrated the fiery holocausts of war, religious persecutions, the sacks of cities, the wholesale rape of women.

—ARTHUR KOESTLER, 1967[401]

BEFORE WE LOOK MORE CLOSELY at how we can rein in our warring impulses, we have first to understand the nature of what it is we are confronting. In English, we have one simple word that expresses it perfectly: evil.

All of us have some concept of evil, but we interpret its origins in a variety of ways. For the Christian it is evidence of original sin; in modern sociology it is a behavioral pathology. Should we regard the September 11 attacks on New York and Washington as a quintessential act of evil or as an explicable response to a chain of historical events? The weapons were different, and the numbers much greater, but the attacks were something that a raiding party in Highland New Guinea making a surprise attack on a neighboring village would recognize. It is what Joshua did to the citizens of Jericho (except that the population of Jericho knew they were being besieged). It is what the Japanese did at Pearl Harbor.

To the extent that the people killed were unknown to an attacker who struck without warning, it is what the United States did at Hiroshima. Nor were the 9/11 attacks fundamentally unlike the behavior of a group of chimpanzees killing an individual of the same species they had never met before. Evil, in the human mind, is relative: *They* are evil, but *we* are innocent—or at least vindicated by circumstances. For the other side the descriptions are reversed. In all cases, the worst killings and atrocities happen when we perceive the people we are fighting to be evil. Could evil in fact be understood as exactly that dehumanization that allows us to kill members of our own species? It is a definition worth exploring.

In the case of homicide, there are usually two individuals—the murderer and his victim—and they commonly know one another. The crime writer Agatha Christie loved the fiction of the cold-blooded murderer who surprises his victim and sometimes kills with exotic poison. In the real world, however, most murders are more emotional than intellectual: There is an altercation, tempers flare, the quarrel escalates, and a knife is thrust or a gun is fired.

Team aggression is very different from homicide, though both are primarily male behaviors. Women can kill and torture, but as we argued in chapter 6, women have never been observed to initiate the teamwork central to raids, war, or slavery. Unlike homicide, team aggression permits us to hate another member of our own species and to want to kill them without having engaged in a conventional dispute—it makes us cold-blooded killers. In instances of team aggression, the attackers may never have seen the victim before. There has been no prior quarrel. And the victim presents no direct or immediate threat. It is enough that they live in Jericho, Nanking, Dresden, or New York City. Empathy evaporates. Just as the September 11 terrorists did, every team of lethal aggressors carries its own engine of hatred into the attack. We may disagree with a murderer's motives, but we can often at least understand them. We call murder "senseless" only when the attacker doesn't know the victim, or in cases where the murdered is obviously innocent. When the victim is known, passion is evident, or affiliation with a rival group is clear, we often do in fact see the "sense" in a particular act of killing, even if we abhor the act itself.

Whether we approach the issue as a theologian, lawyer, or barroom philosopher, this thing called evil seems to boil down to gratuitous acts of cruelty against one's own species—and particularly against those one does not know or has not met before. Even our legal system makes distinctions between killing in self-defense, killing as a crime of passion, and the more severely punished cold-blooded, premeditated murder. In this sense, a snake, hawk, or even a shark cannot be "evil," as it has no evolved capacity to empathize with its prey, let alone de-individualize and "de-specize" its own kith and kin. For good reason, Jane Goodall called a book she wrote about packs of wild dogs in Africa *Innocent Killers.*[402] *Pan troglodytes* and *Homo sapiens* have both the ability to feel empathy for a victim and the ability to turn it off, and our killing is not always so innocent as a result.

If, as we have argued, young male chimps and humans both have an evolved predisposition to attack and kill members of outgroups, then cruelty takes on a new face. Horrible as it may seem, such "evil" acts have made evolutionary sense for males for many millions of years. There is a reason why males, sometimes egged on by females, can cause pain without immediate cause. Extreme cruelty is more than the cat playing with a half-dead mouse to teach its kittens how to hunt. Once chimpanzees or human beings identify another individual as belonging to an outgroup, there appear to be no limits to their enthusiasm for literally tearing that other apart. A cat lacks the empathy that would make its drawn-out attacks cruel; with humans, cruelty sometimes is the whole point.

The Fall

As we humans approach adulthood and our judgment matures, most of us are puzzled by the origins of human evil. How could clerics organize the Inquisition, or Germans not protest their own concentration camps? How could an independent United States of America, founded on principles of equality and individual liberty, tolerate slavery? The Aztecs' human sacrifices, Pol Pot's Cambodian killing fields, Tutsis and Hutus in Rwanda murdering their neighbors, the Arab *janjaweed* militia killing non-Arabs in Darfur—they all boggle the mind. These acts of mass cruelty and indifference to suffering can only make sense, in fact, when we start to see them as the symptoms of evolutionary impulses that have long outlasted their usefulness. They are the signs of team aggression gone horribly wrong.

The history of warfare is a consistent, continuous record of courageous altruism combined with breathtaking cruelty. Philosophers and theologians have long puzzled over this dilemma. Saint Augustine asked, if God is good, "Where is evil then, and whence, and how crept it in hither? What is its root, and what its seed?"[403] Muslim theologians, such as the philosopher Al-Nazzam, who died in 845 A.D., believed God could do no evil.[404] Evangelical Protestants emphasize that "all have sinned," and that Christ's death provides universal redemption. In the Judeo-Christian interpretation of the human condition, a perfect Adam is tempted by Eve, who had been led by Satan to eat of the Tree of Knowledge. Theologically, altruism is the natural position and sin needs to be explained. This understanding of human nature, you will recall, is also expressed in Rousseau's conception of humans as innately good, so long as they remained in a state of nature. The Darwinian perspective shows that the reverse is true: Competition and strife are the norm, and while altruism exists, it is a more limited and fragile behavior which evolved relatively recently in biological history. Remembering that evolution adds its innovations on top of what was already there, it holds that altruism would be something of a behavioral veneer over a thick core of bloody competition. It's no wonder we backslide so often.

Jung-Kyoo Choi of Kyungpook National University in Korea and Samuel Bowles of the Santa Fe Institute, New Mexico, use game theory to explore human evolution. Their simulations offer an interesting insight into how competition, empathy, and altruism interact; the research suggests that the predisposition for ingroup altruism and the predisposition to hate an outgroup are mutually reinforcing. Bowles and Choi designed computer models of evolution in which some individuals carried an altruistic gene which encouraged them to help their ingroup, and others had no such gene. Then the scientists added a second variable, making some individuals hostile to outgroups, while others remained tolerant and traded with their neighbors. Tolerant, selfish, non-altruistic individuals who trade with their neighbors form a stable system in Choi and Bowles' simulations. But a society with a high proportion of individuals who are altruistic to their own ingroup yet hate their neighbors consistently attacked and defeated neighboring societies with fewer such parochial altruists in a few hundred generations of simulation.[405] Evil, apparently, needs no explanation beyond the biological.

Even if the roots of evil can be found in our biology, that does not make its expression any more acceptable or easier to witness. And evil has been expressed with sobering regularity through history and across cultures. When evolutionary biologists study behavior, their attention is drawn to those behaviors that are universal in a species, and the human propensity to dehumanize and de-individualize an enemy is just such a behavior. We see its traces everywhere, from the careful impaling of human beings so that they lived as long as possible on their stakes in Assyria twenty centuries before Christ to the genocides that scar our own time, twenty centuries after Christ. Underneath it all runs a deep current of dehumanization—we simply could not do these things to one another if we considered our victims to be truly human.

In 1980 I visited a refugee camp in Thailand for Cambodians fleeing Pol Pot's takeover of their country. One young woman, called Mavany, had seen her husband shot, her grandfather commit suicide, and her baby die of diarrhea. She wrote an account of her suffering in fractured English:

> Every morning at 5 o'clock we must go to the rice field to cultivate from 5 o'clock to 11 o'clock. We stop one hour to eat rice water, then go back to work until 6 o'clock in the evening. We go home take three spoons of rice and at 7 o'clock sometimes we had [to go] to meetings....Khmer Red say in force everyday [we] must go to work. No work—no rice....They were cruel. They thunder. They arrogant. No sympathy.[406]

Cruelty takes may forms, but all of those forms depend on the human ability to dehumanize and de-individualize another person. "They arrogant. No sympathy," said Mavany, capturing perfectly what happens when our empathy switches off. When the home video of Osama bin Laden describing the September 11 attacks was released, viewing audiences were amazed, angry, or let down by its banality. Surely, a man convinced he was working to bring about God's will would express remorse at the loss of life? Or if not remorse, then at least an animating hatred? But no, bin Laden delivered a rambling, disjointed conversation interrupted by occasional guffaws. It certainly didn't fit our image of the devil incarnate, which is exactly what bin Laden became the moment the Twin Towers fell. And yet, banality is what evil is all about. Once an adult human brain has categorized a group as "the enemy," as bin Laden and Al Qaeda have categorized all Americans, then members of that group become no more than a target to be eliminated—like ants from a pantry or weeds from a lawn. The more who die at one time, the better. Chimpanzees kill one at a time because that is all they can manage with their bare hands, feet, and long canine teeth. Although we cannot get into the mindset of another species,

I suspect that if chimpanzees had the technology, they would be pleased to escalate the scale of their attacks on neighboring troops.

Whether we want to or not, we all distinguish between our ingroup and various outgroups, though we may belong to more than one ingroup at a time. We change our perceptions, and ultimately our moral standards, as easily as vaudeville artists change their costumes. We identify with the group we belong to at any particular moment, and all too easily diminish others. An individual who is outraged when a referee unfairly penalizes his favorite soccer team may very well support the carpet-bombing of "terrorists" in some distant clime, even when innocent civilians are killed in the process. But this is not just something that others do; we all do it, all too easily. It is part of being human.

The drive to attack others, or the ability to impassively allow others to be attacked, can be switched on or off by training and the need for self-defense. Circumstances can make most any man into a terrorist, a disciplined soldier, or a soccer hooligan. Some people, however, seem to find it unusually easy to dehumanize others. The journalist Bill Buford describes an English soccer fan called Harry. Buford can't help liking Harry: He is witty, a loving husband with an affectionate, optimistic wife and a young family, and has a strong sense of justice. Once, a policeman trying to break up a rowdy party insulted Harry's friend, ordering, "Now hurry up, you cunt, and get the fuck out of here." Incensed, Harry smashed a window. The policeman fought him to the ground, but Harry bashed the man's head against the floor, rendering him temporarily unconscious. He put his mouth over the policeman's face, and sucked on his eye "until he felt it pop behind his teeth. Then he bit it off." Harry, like warrior males throughout history, was so able to dehumanize his enemy that he felt no guilt. After biting off the policeman's eye, he went home, picked up his wife, and went to the local Kentucky Fried Chicken.[407] While sitting in the window he was arrested, still in his blood soaked T-shirt. In a war, Harry's loyalty to his fellows would have been rewarded; in Britain in the 1980s, his behavior earned him a long prison sentence.

It's easy enough to find examples of particularly heinous violence, but human cruelty is not just an isolated aberration in some individuals. Given the right circumstances, it can involve whole communities. History shows that those circumstances have not been particularly rare. The Jewish historian Josephus described how famine spread through a besieged

Jerusalem in 69 A.D., rendering the survivors too weak to bury those already dead. Seeing no escape, some jumped to their deaths from the city walls while others tried to cross the encircling lines of the Roman army and their allies. One escapee was caught picking through his own feces to extract the gold coins he had swallowed in order to hide them. "The rumor ran round the camp," recorded Josephus, "that the deserters were arriving stuffed with gold. The Arab Unit and the Syrians cut open the refugees and ransacked their bellies.... In a single night nearly two thousand were ripped open."[408] In the year 1014 A.D., at the Battle of Balathista, the Byzantine Emperor Basil II defeated the Bulgarian army of Tsar Samuel. The Emperor ordered his troops to put out the eyes of all 15,000 prisoners, "leaving one man in every hundred with one good eye so that he could lead his comrades home." Samuel, it is said, died of sorrow.[409] Early in the thirteenth century A.D., the barons of northern France began to organize a Crusade against the Albigensians, or Cathars of southern France, a splinter group who interpreted Christianity in a different way and challenged the Church's authority. The Pope blessed the fighting and set up the Inquisition to assist in rooting out heretics.[410] The accused were tortured on the rack until they confessed whatever they were told to say, and then burnt at the stake. When the crusaders captured the city of Béziers, legend has it that the Papal Legate accompanying the soldiers told them not to bother to separate the Catholic citizens from the Albigensian heretics, but to "kill them all and let God sort them out."

This type of cruel act, carried out by authorities across cultures and through time, seems stunning to us now. But what is truly astounding is the high degree of public support they often enjoyed. In the fifteenth century, for example, a chronicler records that the people of Bruges (in what is now Belgium),

> cannot get their fill of seeing tortures inflicted, on a high platform in the middle of the marketplace, on the magistrates suspected of treason. The unfortunates are refused the death blow which they implore, that the people may feast on their torments.[411]

When Christopher Columbus and his men first went ashore in the Caribbean on October 12, 1492, the Europeans were intrigued by the naked, seemingly peaceful agriculturists they met. But within a tragically short length of time they came to regard them as a less-than-human outgroup.

Bartolomé de Las Casas, who immigrated to the islands in 1502, records how the Spanish colonists,

> made bets as to who would slit a man in two, or cut off his head at one blow; or they opened up the bowels. They tore the babes from their mother's breasts by their feet, and dashed their heads against the rocks....They spitted the bodies of other babes, together with their mothers and all who were before them, on their swords.[412]

On May 26, 1637, during the war with the Pequot Indians in the New England Connecticut Colony, a Puritan army commanded by John Mason surrounded a small wooden fort in which "six or seven Hundred" Pequot Indians were sheltering. Mason ordered the wooden palisade surrounding the fort set on fire. Only seven Indians escaped alive. In an early use of the word "holocaust," which simply means consuming by fire, Mason said the massacre was an act of God who "laughed at his Enemies and the Enemies of his People to scorn making [the Pequot] as a fiery Oven."[413]

There is a museum built on one of places where the Japanese Imperial Army killed thousands of Chinese after the capture of Nanking in December 1937. Now called Nanjing, the city was the Chinese capital at the time. Today it is a booming part of the Chinese economic growth. Walking down a long ramp to enter the museum, the visitor is framed by glass walls at either side holding back hundreds of skeletons, which somehow give the impression they are struggling to rise up and avenge their massacre. In six weeks, the Japanese may have killed as many people as died in the atomic bomb attack in Hiroshima eight years later.* The killings took place one by one and they were done by individual soldiers with swords, bullets, and bayonets. According to Tang Sheng-chih, the general appointed by Nationalist leader Chiang Kai-shek to defend Nanking, competitions were held to see who could kill fastest. "In each team [of Japanese] one soldier beheaded prisoners with a sword while the other picked up the heads and tossed them in a pile. After a while they cut the victims' throats because it was quicker. There were no signs of remorse and the soldiers were laughing."[414] Some of their victims were buried to their waists in pits and then dogs were set on them. One Buddhist monk who refused to rape a woman for the voyeuristic pleasure of his captors

---

* In such massacres estimates of deaths vary greatly, in the case of Nanking from 40,000 to 360,000.

was castrated with a sword and bled to death. The attacks on women were so brutal, some died from internal injuries following rape.

In the eighteenth century, when the Hawaiian chief Kamalalawalu defeated Kanaloakuaana he had his rival's eyes put out before he had him executed.[415] When the Maori chief Tamaiharanui was captured by his enemy Ngai Tahu in 1830, he was tied to a tree. The widow of a man he had killed cut the chief's throat and drank his blood. Her son then tore out Tamaiharanu's eyes and swallowed them.[416] In England, until the end of the eighteenth century, the punishment for treason was hanging until half-dead, followed by castration and disemboweling—something a chimpanzee might also understand. In Duluth, Minnesota, less than a century ago, a crowd of 10,000 people watched three young black men tortured and killed in just one of many of lynchings throughout the U.S during that time.

In 1994, Hutus in Rwanda killed an estimated 800,000 Tutsis. In the 1960s, in neighboring Burundi, the Tutsis killed 100,000 Hutus. The Hutus were mainly Bantu cultivators while the Tutsis were mainly Nilotic herdsman, but historically the distinctions were often blurred and intermarriage was common. In the 1930s, however, the Belgian colonial administration had reinforced and intensified outgroup boundaries by issuing identity cards that labeled anyone with ten cows or more a Tutsi. During the 1994 genocide the Hutus called Tutsis "cockroaches" and the Hutu president proclaimed that the Tutsis "have no right to live among us." Most of the killings were done with machetes and many by adolescent gangs. Taking into account the absolute population of the two countries, a slaughter on a similar scale in the United States would have killed almost thirty million people over the course of only a few weeks. This type of evil always has two characteristics: It is a slaughter which depends first on a blind hatred of an outgroup, and second, on strong loyalties to other members of the team doing the killing.

Evil ethnic and racial hatreds can lurk below the surface for years and then explode with lethal force when triggered by some external event. When I first went to Kenya in about 1970, the population of Nairobi was perhaps half a million. By 2008, the population had grown to 2.5 million with 700,000 people living in the Kibera slum—which in 1970 had been a grove of trees near the Nairobi National Park. I have worked in that slum, and I never asked whether my friends there belonged to the Kikuyu or Luo tribes—they were just brave people trying to remain healthy in a place without proper sewers and eke out a living in the narrow pot-holed muddy streets between long lines of one-story shacks. When I drove past Kibera just a few days before the presidential elections in December 2007, the place was perfectly peaceful. But when the Kikuyu candidate for president rigged the election, vicious fighting broke out between the dominant Kikuyus and other tribes. The election was the spark, but old ingroup/outgroup identities and the even more ancient predispositions of team aggression provided the fuel. Across the country, about 800 people were killed, including some forty who were burned to death in a church in Eldoret, a town north of Nairobi. Between 300,000 and 600,000 had to flee their homes. What is often overlooked in reports of these events is that this did not happen in a failed state—it happened in a previously stable country that just happens to have undergone rapid, dangerous population growth and faces increasing stresses as a result.

For all the examples of group atrocities, there are certainly cases where evil is the lone act of a deranged individual. A few people—and a few chimpanzees—are just frankly antisocial. Presumably, such cases are the result of something going grievously wrong in a brain that has been built by a particular combination of genes and then submitted to a particular set of environmental pressures, so that it places almost everyone in an outgroup. When such individuals act alone, they are antisocial. But when they gain control over groups or even whole nations, they join the ranks of history's greatest villains. Joseph Stalin certainly earned his place on that list. The Soviet dictator introduced the *gulags*, or prison camps, following the purges and sham trails of the 1930s. Historian Allan Bullock writes that Stalin had an "ever-present fear of conspiracy, overthrow and assassination, and satisfied that desire for revenge that remained as strong as ever in a nature in which there was not a trace of magnanimity or human feeling and calculation was reinforced by instinctive cruelty."[417] The numbers Stalin killed directly and indirectly are incredible. Five to seven million people died in the Ukraine and elsewhere in the 1932–1933 famine, when Communist Party agents confiscated the harvest.[418] Two million Russian prisoners of war were killed, or sent to Siberia, after they were "freed" by the victorious Red Army in 1944. Stalin executed thirteen of his fifteen army commanders, and 154 out of 186 divisional commanders. On a single day—December 12, 1937—Stalin approved the execution of 3,167 prisoners. "Pity was unknown to him," writes Bullock. "He continued his routine as before, dictating letters, receiving officials, holding meetings and attending the theater. He showed the same lack of compassion when men with whom he had worked closely for years were executed." Most of us reserve the ability to dehumanize others for distant groups, but Stalin's brain seems to have dehumanized even those most of us would call colleagues or friends. He apparently enjoyed having people killed. "To prepare one's plans minutely, to slake an implacable vengeance, and then go to bed," said Stalin, "there is nothing sweeter in the world."

"Virtuous Citizens Acting in the Name of Righteous Causes"

Still, most evil is done not by monstrous figures, but by people we would consider normal. Arthur Koestler, the author and philosopher, worked as a journalist in the Spanish Civil War and was captured and sentenced to death by General Franco's fascists. He escaped execution thanks to British intervention, and his words at the beginning of this chapter capture the dark heart at the center of all evil. Rare is the Stalin-like individual, who lacks empathy and who seems to take pleasure in inflicting pain or watching pain inflicted on others; common is the "virtuous citizen acting in the name of righteous causes." In trying to understand evil, it may in fact be easier to explain the truly monstrous—Hitler, Stalin, Pol Pot, Caligula—than it is to account for the rest of the apparently normal population who are in many ways kind and loving folk, but who end up accepting evil around them, and find it easier to participate than to stand up and oppose it.

On December 12, 1937, Stalin did not kill 3,167 people personally. Hitler and Himmler did not kill nearly six million Jews and several million more homosexuals, gypsies, and prisoners of war with their own hands. Where did the executioners and torturers come from? Were the USSR or Germany uniquely equipped with Stalin and Hitler clones incapable of empathy? It has been said that the Nazi holocaust against the Jews "defies explanation." Daniel Jonah Goldhagen in *Hitler's Willing Executioners* concludes the Germans were the victims of "an extraordinary, lethal political culture." "The camp system," he writes, "allowed for the expression of the new Nazi moral disposition, one which was in its essential features the antithesis of Christian morality and the Enlightenment humanism."[419] But such phrases are descriptive, not explanatory. It is tempting, but misleading, to try to make the rest of the world "normal" by demonizing Hitler. An evolutionary approach to understanding evil is at the same time both more humbling and more challenging. Hitler was a sociopath, but the holocaust had millions of active participants and passive observers. Nazi propaganda and administrative procedures placed the Jews in an outgroup and tapped into the human propensity to beat and kill individuals in outgroups. And there is no reason not to think that many of us today might also have become willing executioners had we lived in Germany in the 1940s. A few centuries earlier, it had been written of the Jews:

> They hold us captive in our country. They let us work in the sweat of our noses, to earn money and property for them, while they sit behind the oven, lazy, let off gas, bake pears, eat, drink, live softly and well from our wealth. They…mock us and spit on us, because we work and permit them to be lazy squires who own us and our realm.[420]

The author was Martin Luther, the founder of Protestantism.

At the very time the Nazis were perpetrating obscene cruelties against other human beings, they were also passing the most advanced animal protection laws in Europe. Soon after coming to power in 1933, they banned vivisection, and regulated the shoeing of horses and even the boiling of lobsters. In what is likely to remain the ultimate oxymoron and symbol of the insane ability to dehumanize our own species, Hermann Göring threatened to "commit to concentration camps those who still think they can treat animals as inanimate property."[421]

Nine out of ten men in the German battalions selected to kill Jews in World War II were over forty years old. Most were married and had families they loved. Only a minority actually belonged to the Nazi party. They were ordinary men. The 101st Police Battalion, which was posted to round up Jews east of Warsaw, consisted of about 500 men. Between June 1942 and November 1943 they killed or deported to death camps over 80,000 Jews. Their commander, Major Wilhelm Trapp, invited any men who found the task of hunting for Jews in their beds or hiding in cellars and then shooting them at short range, to be unacceptable to be assigned to other tasks. Only a few chose to transfer. One officer conceded that execution duty "was entirely not in his spirit, but he had received this order from higher authority." He reminded the men to "think of their own women and children in the homeland who had endured aerial bombardment." It was an illogical way to arouse feelings of revenge—but such is the case in many tit-for-tat strikes in war.

By and large, German soldiers in World War II settled down to killing Jews with all the de-individualization and dehumanization characteristic of a chimpanzee raid. Often the soldiers walked beside their victims on the way to some newly prepared mass grave before ordering them to lie prone so they could be shot more efficiently. Children and babes in arms were killed. The 101st's Police Battalion's physician, a Dr. Schoenfelder, lent his medical expertise to the atrocity by drawing an anatomical chart in the sand, indicating where to shoot a person most effectively in the back of the neck. Sometimes the soldiers doing the killing had their uniforms soiled with brains and splinters of bone. A few were upset, especially on their first day, just as a visitor to a slaughter house might be alarmed by the noise, the falling animals, and the blood on the floor. But in the end, most settled into a routine. They took snapshots of the killing camps. Captain Julius Wohlauf was joined in his work by his wife Vera on their honeymoon, and she witnessed some of the roundups holding an ox-hide whip in her hand. On one occasion, some German Red Cross nurses came to watch the slaughter.

The military guards killing Jews seem to have been driven by the same set of loyalties that bond the basic human fighting force—the fear of being branded a coward by their comrades and the ability to dehumanize members of an outgroup. As one soldier said, "It did not require much courage to step forward [and ask to do some other task]." But they did not want to be seen as cowards. Even though the idea may be deeply upsetting or even offensive, it is difficult to escape the conclusion that the same set of drives which in one situation led to actions rewarded with a Medal of Honor or the Victoria Cross, in another environment led to convictions for war crimes. Flip the behavioral coin and the obverse of evil is courage. Which way the coin falls depends on how the environment tosses it.

The evolutionary logic of team aggression is to kill as many members of the outgroup as possible, at as minimum risk to yourself as possible. Eviscerating a live victim on top of the Templo Mayor in Tenochtitlan, beheading the citizens of Nanking in 1937, wielding the slave-driver's whip in the United States in 1790, cheering a lynching in Minnesota in 1920, feeding people into a gas chamber in 1944, or flying commercial jet liners loaded with fuel into skyscrapers in 2001 are all ways of expressing this same predisposition.

Fortunately, in the modern world a critical mass of people has come to

condemn such acts. Increasingly, with education, travel, and overall greater exposure to people of different cultures and backgrounds, we recognize
that outgroup distinctions are largely artificial—the "other" is in fact not
so different from ourselves. The dehumanization essential to raiding and
warfare is beginning to break down. Most of the six billion-plus people
now alive on this planet can probably identify one or more outgroups that
they dislike, but most have not, will not—and perhaps could not—kill
and torture its members. Instead of just bemoaning the fact that evil still
occurs, the perspective of evolutionary biology asks us also to focus on
the progress that has occurred and on how it perhaps can continue.

A World Ruled by Houyhnhnms and Yahoos

When Jonathan Swift wrote *Travels into Several Remote Nations by Le-
muel Gulliver* in 1726, he did not intend to produce the children's book
it has become. In his words, Swift wanted to "vex the world rather than
divert it." In Lilliput, Gulliver finds tiny people whose quarrels about
trivial matters mirror eighteenth-century religious and political disputes. In Brobdingnag he faces a race of giants who cannot comprehend
the corrupt homeland Gulliver describes. On his last voyage, Gulliver
visits the land of the Houyhnhnms and Yahoos. The Houyhnhnms are
intelligent, rational horses and the Yahoos degenerate, cunning, bestial
human beings. Initially, the Houyhnhnms dismiss Gulliver as just another lowly Yahoo, until one master horse befriends him. And as Gulliver learns more about these noble, sensate horses, he begins to respect
their culture, which he sees is based on reason and truth much more so
than his own. It was a prescient idea that has become more intriguing
with the passage of time.

Suppose for a moment that the big-brained, technically competent social animal that came to dominate and change the planet had evolved
from an ancestral horse. What might society look like? No doubt aggressive acts would occur between individual Houyhnhnms as they competed over sex or status. No doubt they would have displayed selfish as well
as altruistic acts, and there might well be considerable indifference to
the pain of others. But the terrible predisposition to band together and
rip apart individuals from neighboring herds or clans might never have
evolved. The ability to see other members of our species as something
deserving only scorn and annihilation grows out of the behaviors that

evolved to make team aggression possible. Observation suggests that this same-species antagonism is limited to exceedingly few species of animals—ourselves, one species of chimpanzee, and possibly wolves. And it is only in humans and chimpanzees that the impulse has become organized into raiding and warfare.

We can look at human cruelty in three different ways, each of which has profound implications for how we see ourselves and our relationship to one another. We can simply assume some people are good and some bad. Under this model, the Nazi guards in the Treblinka concentration camp in the 1940s were not fundamentally different from the rest of us. They had some evil bent, an inherited characteristic perhaps, which accounted for their sadism. Such a telling of the Holocaust story allows us, the good, to assure ourselves we are incapable of such atrocities. As an alternative, we could assume that everyone comes into this world a neutral, blank sheet of paper on which our culture and early experiences write our fate. Under this explanation, the Romans who flocked to watch 5,000 gladiators kill one another at the Emperor Trajan's bidding, or who crucified 6,000 slaves after the Spartacus revolt in 71 B.C., did so because that is what their culture taught them.

The biological perspective, which we believe to be true, does not allow such easy escapes from shared culpability. A biological approach to warfare invites us to look at ourselves in a new way. We suggest that the predisposition to form aggressive coalitions is so deep-seated within us that all humanity is compelled to live by two profoundly contradictory moral systems. We have the morals of the troop, expressed by "Thou shalt not kill," and the morals of the aggressive male coalition, also explicitly spelled out in the Old Testament, "And when the Lord thy God has delivered [a city] into thine hands, thou shalt smite every male thereof with the edge of the sword."[422]

"Who Is Willing to Destroy a Piece of His Own Heart?"

In the last analysis, the survival of any animal boils down to two things: food and sex. From an evolutionary perspective it is easy to explain competition between individuals of the same species—it is the force driving survival and reproductive success in an environment where there are never enough resources to keep everyone alive. There is no reason why competition between individuals of the same species should have a kind face. But the long, terrible catalogue of vicious, gratuitous, deliberate cruelty we see in warfare, and in warfare's first cousin slavery, goes beyond competition or an animal's unthinking attack to deliberate acts of evil. It is here that understanding our own evolution may help provide a more comprehensive and compelling explanation.

War, we have argued, is built on an inherited set of predispositions to kill members of an outgroup. Biology reminds us that we all have the seeds of evil within us. Writing in *The Gulag Archipelago*, Alexander Solzhenitsyn, who had spent eleven years in one of Stalin's prison camps, captures the Darwinian perspective on evil in a powerful way:

> If only there were evil people somewhere insidiously committing evil deeds, and it were necessary only to separate them from the rest of us and destroy them. But the line dividing good and evil cuts through the heart of every human being. And who is willing to destroy a piece of his own heart? [423]

The Blind Watch-Maker of Evolution, to use the evocative phrase biologist Richard Dawkins coined to describe the random, haphazard process of evolution, happened to assemble chimpanzees and human beings as highly social, very intelligent creatures with excellent memories and considerable altruism for their kin, but great hostility to individuals from an outgroup. The balance of these opposing drives differs in males and females, and males are particularly easily seduced into acts of team aggression in which small groups seek out and kill in the most vicious manner possible individuals of a perceived outgroup. This propensity developed because originally it enhanced access to resources and ultimately improved the reproductive performance of the successful raiders. Under this interpretation, we all harbor inherited predispositions, which can be triggered with fearful ease, to de-individualize and hate members of other groups. There is still part of us that, like the citizens of Bruges, "cannot get their fill of seeing tortures inflicted." In Solzhenitsyn's words, "the line dividing good and evil cuts through the heart of every human being."

As Swift suggested, if during its evolution a large-brained horse had achieved self-recognition and the ability to understand the individuality of others, then it might well have behaved with more consistent altruism than human beings do. A social animal that had not inherited the predisposition to kill members of other troops would not have developed a mechanism to "de-equinize" others. There would have been no putting out of eyes or impaling of enemies, no crucifying criminals, burning heretics alive, or drawing and quartering traitors—indeed the concept of traitor might not have existed. And this is the sad conclusion to which an evolutionary understanding of human behavior leads us: In the final analysis, only *Homo sapiens* and *Pan troglodytes* can be truly evil.

Ethical commentators have struggled with definitions and manipulated meanings to try to find one overarching philosophy that encompasses all of human behavior. But human beings are condemned to live in two ethical and behavioral worlds at once—the morality of empathy for and reciprocity with ingroups, and cold-hearted team aggression aimed at outgroups. Looking back on the Stanford prison experiment thirty years later, Philip Zimbardo concluded with a truth none of us want to hear: "Thus any deed that any human being has ever done, however horrible, is possible for any of us to do—under the right or wrong situation pressures." He added, "That knowledge does not excuse evil; rather, it democratizes it, shares its blame among ordinary participants, rather than demonizes it."[424] Our brains have evolved to harbor two opposite, conflicting, irreconcilable behaviors, and the one inevitably condemns the other—when we observe it in other people.

And yet, our biological inheritance is not optional, but neither does it necessarily condemn us to a life of cruelty. Ever since the first cities arose on the plains of Mesopotamia at least, we have been learning to live within larger and larger ingroups. Now we are finding ways to extend ingroup morality beyond national boundaries to embrace all of humanity. And we are learning to channel our team aggression impulses into more positive directions. Predispositions are powerful, but we already have proof that they are not destiny. Civilization at its best can, and often has, overcome a mind adapted to live in the Stone Age. In the coming chapters, we'll take a closer look at just exactly what that means.

12

THE FUTURE OF WAR

Means of destruction hitherto unknown, against which there can be no adequate military defense, and in the employment of which no single nation can in fact have a monopoly.

—HARRY S. TRUMAN, CLEMENT ATTLEE, and W. L. MACKENZIE KING, November 15, 1945[425]

TODAY'S MOST BRUTAL WARS are also the most primal. They are fought with machetes in West Africa, with fire and rape and fear in Darfur, and with suicide bombs and improvised explosive devices in Israel, Iraq, and elsewhere. But as horrifying as these conflicts are, they are not the greatest threat to our survival as a species. We humans are a frightening animal. Throughout our species's existence, we have used each new technology we have developed to boost the destructive power of our ancient predisposition for killing members of our own species. From hands and teeth tearing at isolated individuals, to coordinated raids with clubs and bows and arrows, to pitched battles, prolonged sieges, and on into the age of firearms, the impulse has remained the same but as the efficiency of our weapons has increased, the consequences have grown ever more extreme.

The evidence of history is that no advance which can be applied to the killing of other human beings goes unused. As scientific knowledge continues to explode, it would be naïve to expect any different. As if we needed any more reasons to confront the role of warfare in our lives, the present supply and future potential of WMDs should convince us that the time has come once and for all to bring our long, violent history of warring against each other to an end.

The nineteenth century was dominated by discoveries in chemistry, from dyes to dynamite. The twentieth century belonged to physics, from subatomic particles and black holes to nuclear weapons. The twenty-first century is set to see great advances in biological knowledge, from our growing understanding of the genome and stem cells to, it's a shame to say, new and expanded forms of biological warfare. In the past, each iteration of the application of scientific discovery to warfare has produced more horrible and destructive weapons. Sometimes temporary restraint is exercised, as in the successful ban on poison gas in the Second World War, but such barriers burst easily, as the deliberate bombing of civilians in the same war attest. Human beings have always appropriated new ideas to build increasingly formidable weapons and there is no reason to think that competitive, creative impulse will disappear on its own. As weapons become ever more horrifying—and, with the rise of biological weapons, increasingly insidious—it is no longer enough just to limit the use of one killing technology or another. We need to limit the conditions that lead to war in the first place.

It has become almost a cliché to note that we live in an increasingly complex and interdependent society. But this point is crucially important as we consider the future of war. Our cities once were fortresses, the walled sanctums where our ancestors sought refuge from marauders. The firebombing of the Second World War revealed a new urban vulnerability, but even that insecurity is nothing by today's standards. We live in giant cities, supplied with piped water and electricity, with trains in tunnels and cars on elevated roadways, with fiber optics under the pavement and air-conditioning plants for buildings with windows that cannot be opened. Our new urban centers have the vulnerability to terrorism and attack built right into them. Any modern city can be held hostage by a single Unabomber, brought to a halt by nineteen fanatical men, or devastated by any small raiding party drawing on modern scientific knowledge, from malicious computer programming to radioactive "dirty bombs" to infectious bacteriology. To understand the dangerous future of these WMDs, we'll first take a quick look at their history.

Poison Gas

On April 22, 1915, near the Belgian town of Ypres, the German army mounted the first poison gas attack in history. Fritz Haber, who would later receive the Nobel Prize for his work producing nitrogen fertilizer, labored day and night to develop chlorine gas into a weapon and supervised its first release in person. The 168 tons of gas deployed that day ripped a four-mile gap of gasping, suffocating men in the British lines. (The German commanders—as is so often the case when new weapons are used—had insufficient resources to exploit their opportunity.) In a revealing example of the difference between the attitudes of men and women toward war, Haber's wife Clara, who was also a chemist, begged her husband to stop his work on poison gas. After a dinner held to celebrate her husband's appointment as a general, Frau Haber shot herself in the garden—and Haber left the funeral arrangements to others while he traveled to the Eastern front to supervise the first gas attack on the Russians.[426] Unprepared, the Russians suffered 25,000 casualties. In one of the grimmer ironies in the history of dehumanizing others, while Haber was dismissed from the directorship of the Kaiser Wilhelm Institut in Berlin in 1933 because he was a Jew (he later escaped Nazi Germany), his invention, Zyklon gas, was used in the gas chambers of Nazi concentration camps to kill other Jews.

Despite the obvious horrors of gas warfare, the British began their own chemical weapons research in 1916. They tested 150,000 compounds including dichlorethyl sulfide, which they rejected as insufficiently lethal. The Germans disagreed, and took up its development. On initial exposure, victims didn't notice much except for an oily or "mustard" smell, and so the first men exposed to this "mustard gas" did not even don their gasmasks. Only after a few hours did exposed skin began to blister, as the vocal cords became raw and the lungs filled with liquid. Affected soldiers died or were rendered medically unfit for months, and often succumbed years or decades later to lung disease. At first the British were outraged at its use, but later they sent supplies of poison gas to their own troops in British India, for use against Afghan tribesmen in the North-West Frontier.[427]

By 1918, one-third of all shells being used in World War I were filled with poison gas. In all, 125,000 British soldiers were gassed, along with 70,000 Americans. Three weeks before the end of the war, the British shelled the 16th Bavarian Reserve Infantry with mustard gas. A young corporal named Adolf Hitler was blinded in the attack—and would later claim that the recovery of his sight was a supernatural sign he should become a politician and save "Germany."

Nuclear Weapons

Between the ages of eleven and seventeen, I was lucky to attend the Perse School in Cambridge, only a mile from the Cavendish Laboratory where much of the early work on atomic physics was conducted. Today, I teach at the University of California, Berkeley, an important site for early work on nuclear physics, and still the managing institution for Los Alamos National Laboratory in New Mexico, where the atomic bomb was developed. The knowledge to create the most destructive weapons in history was developed by clever men in pleasant surroundings, pushing the analytical power of their Stone Age brains to the limit. In that task, deep-seated human emotions and brilliant science clashed in complex ways.

The main motivating factor behind America's Manhattan Project was fear—fear that Nazi Germany would develop the atomic bomb first.[428] In the 1930s, a Hungarian theoretical physicist living in London, Leo Szilárd, foresaw that a nuclear chain reaction might be possible, and in December 1938, Otto Hahn in Germany conducted the crucial experiment confirming Szilárd's hypothesis. As a young German officer, Hahn had helped release the first chlorine gas at Ypres in 1915, but when the possibility of a nuclear weapon arose he had serious reservations, saying, "if my work should lead to a nuclear weapon I would kill myself." (Lise Meitner, another physicist, was the first to understand the potential of nuclear fission. She worked with Hahn in Berlin before being expelled from Germany because she was Jewish, and she refused any part in the development of the American bomb.) But while virtually every physicist who saw the potential for nuclear weapons recoiled in horror, scientific genies which can be weaponized are always difficult to keep in their bottles, and impossible during wartime. By the time Hitler invaded Czechoslovakia in March of 1939, science had advanced to the point that the best physicists in both Europe and America could see how an atomic bomb was scientifically possible. Soon, many would come to consider it necessary as well.

A German effort to build the bomb was launched, and headed by Werner Heisenberg, famous for his "uncertainty principle" of quantum phys-

Credit: Yasuko Yamagata/Hiroshima Peace Memorial Museum

Weapons of mass destruction make the ancient behaviors of war much more dangerous. On August 6, 1945, one atomic bomb killed over 65,000 people in Hiroshima and injured many more. A day later and a kilometer from ground zero, Yasuko Yamagata saw "[t]he corpse of a woman cradling her baby, both charred completely black, while the woman remained in running position, one leg raised." The image was seared into the artist's mind and she set it down on paper twenty-nine years later.

ics. Germany failed to make an atomic bomb by a wide margin, and there is some evidence, controversial to be sure, that Heisenberg and other German physicists had intentionally dragged their heels.[429] Whether true or not, it hardly mattered—Szilárd was convinced the Nazis were making progress and that only the Americans could beat them to the nuclear finish line. He drafted a warning letter, and together with Albert Einstein sent it to President Roosevelt. The Manhattan Project soon followed.

The U.S. tested its first atomic weapon in the New Mexico desert on July 16, 1945, just two months after the Allies accepted Nazi Germany's unconditional surrender. But the war with Japan raged on, and the new U.S. President, Harry Truman, struggled with the power he now controlled. "Even if the Japs are savages, ruthless, merciless and fanatic, we as the leader of the world...cannot drop this terrible bomb on the old Capital [Kyoto]," he confided to his diary. "The target will be a purely military one and we will issue a warning statement asking the Japs to surrender." In fact, Japan was on the verge of surrender and it might well have capitulated had they been told the Emperor could remain on his throne.* The Allies, however, insisted on unconditional surrender, and the Japanese refused. At 8:16 A.M. on August 6, a uranium-235 device called Little Boy was dropped on Hiroshima; a plutonium bomb, "Fat Man," was dropped on Nagasaki two days later. On September 2, 1945, the Japanese formally surrendered. The genie was out of the bottle.

Within months of the end of the war, Edward Teller, a Hungarian who was part of the team that had developed the U.S. bomb, was working on the hydrogen bomb, an even more powerful weapon. In the Soviet Union, Stalin had authorized work on an atomic bomb as early as 1942, and the Russians were helped initially by lease-lend shipments of uranium and other material from the U.S., and by Manhattan Project secrets leaked by the left-wing physicist Klaus Fuchs. His betrayal is said to have advanced the Soviet work by perhaps eighteen months, and captured German scientists added an extra boost after the war. Russia exploded her first atomic bomb just four years after the Americans. The British had their atomic bomb by 1953, the French by 1960, and the Chinese in 1964. Israel has never confirmed its membership, but is thought to have joined the nuclear club by the late 1970s.

Germ Warfare

The Shoshone Indians of Nevada, before battle, killed a sheep, drained its blood into a length of intestine, buried the draught in the ground to ferment, and then smeared their war arrows with the microbial brew. This would have guaranteed severe infection and probably death following even a superficial arrow wound. A 3,400-year-old clay tablet found in modern Turkey carries a cuneiform inscription with the intriguing phrase, "The country that finds them shall take over this evil pestilence." Molecular biologist Siro Trevisanato from Ontario, Canada, suggests that this may be a reference to a disease called tularemia which

---

* At the end of World War II, the U.S. submarine blockade was extremely successful and the U.S. had total air supremacy. Without oil and other essential products Japan did not present a military threat. Emperor Hirohito had sent a relative to the Soviet Union to act as an intermediary in peace negotiations with the Allies, but the Western public knew nothing of these overtures. The threat Truman and Churchill saw was from Russia claiming a place in the postwar occupation of Japan.

infects sheep, donkeys, rabbits, and human beings, and that it is the first instance of biological warfare in recorded history.[430] Tularemia is a highly infectious disease leading to a painful death from fever, skin ulcers, and pneumonia. It was the cause of serious epidemics in early civilizations stretching from present-day Cyprus to Iraq, and the historical record suggests that infected sheep and donkeys were driven into enemy lines in order to spread infection. During the French and Indian Wars (1754–1763), the British very likely gave hostile Indian tribes blankets infected with smallpox, and certainly considered the idea. Once you have dehumanized your enemy, the evidence is that it matters little which way you kill him. But biological weapons represent a particularly insidious and dangerous form of WMDs. They may lack the immediate gruesome effects of chemical weapons or the sheer destructive power of the atomic bomb. But they are inherently stealthy, potentially lethal on a global scale, and when living infectious organisms are involved, all but uncontrollable.

Both Japan and the U.S. worked on biological weapons during World War II, and the Japanese used anthrax and plague bacteria against the Chinese. U.S. research continued after the war until 1969, when President Richard Nixon renounced "the use of lethal biological agents and weapons, and other methods of biological warfare." The U.S. unilaterally destroyed its stockpiled biological weapons, a bold step which led to the 1972 Biological Weapons Convention. But although the convention was ratified by 140 nations, it lacked policing capacity and within one year of its passage, the Soviet Union began the largest biological weapons program in history. Vladimir Pasechnick, who would defect to the U.S. in 1994, reported overseeing 400 research scientists working on the program in Leningrad, with another 6,000 professionals throughout the country involved in the manufacture of huge quantities of anthrax and smallpox. Iraq also ignored the 1972 convention and in 1990, just before the First Gulf War, a factory south of Baghdad manufactured 5,400 liters of botulinum toxin. The Coalition forces had insufficient vaccines to protect their soldiers, and U.S. Secretary of State James Baker used diplomatic channels to let Saddam Hussein know that the U.S. would launch a nuclear response if attacked with biological weapons. By the time of the Second Gulf War, Hussein's biological weapon program had disintegrated.

As a physician, I must say that I find germ warfare to be particularly loathsome. There are three possible levels on which it could be waged, each more distressing that the one before. First, a bacterium such as anthrax, which is very stable, could be sprayed or spread around a community. Anyone who inhaled it would come down with a non-specific fever and fatigue, which looks like the onset of flu but, left untreated, leads to fatal pneumonia. An anthrax victim, however, could not infect another person. Second, an infectious agent, such as smallpox, could be used to start an epidemic. Third, a new and terrible disease could be genetically engineered that not only infects, but also avoids detection and resists treatment with our current arsenal of vaccines and antibiotics. This final scenario is the most chilling of all.

If anything qualifies as a miracle of modern medicine, it is the World Health Organization's use of vaccination to eradicate smallpox in the 1960s and 1970s. The last case of this ancient killer of millions was identified in October 1977 in Somalia. Yet the very fact of our medical triumph over smallpox makes it a particularly devastating weapon. The virus is highly infectious; causes severe, painful disease with a high rate of mortality; and unlike HIV, for example, is quite robust, and can persist in the environment for months or years. Unlike most viral diseases, it is possible to halt smallpox infection by vaccination after exposure. However, the smallpox vaccination must be given within the first forty-eight hours after exposure, and large-scale smallpox vaccination was stopped thirty years ago. A smallpox-based attack now could devastate a large population. But even if an outbreak were quickly contained,[431] it would bring a nation to a halt and be exceedingly frightening and painful.

All smallpox samples were supposed to be destroyed following eradication, with the exception of two batches. One is stored at the U.S. Centers for Disease Control and Prevention in Atlanta, Georgia, and the other at the Russian State Research Institute of Virology and Biotechnology outside Novosibirsk, Siberia. It is possible, however, that clandestine stocks were kept by Russia, Iraq, Israel, or some other countries, and shortly after 9/11, the World Health Organization decided to postpone the destruction of the final Russian and U.S. samples in case they are needed to provide scientific information to counter a bioterrorism attack in the future.[432]

Many other pox viruses and other infectious agents provided by nature could potentially be used as weapons. But the Frankenstein-like creation of novel germs is perhaps an even greater fear. A lethal virus might be assembled accidentally, as happened in Australia in 2000 when an experiment to sterilize rodent pests turned sour. The unintentionally lethal virus killed all the experimental animals, despite attempts at vaccination.[433] And the deliberate quest to make germ warfare more effective by genetically modifying existing bacteria and viruses has already begun. Sergei Popov, a Russian molecular biologist who worked in the Soviet biological weapons program, developed a microbe with the potential to cause a slow death from multiple sclerosis. "We never doubted," he said after defecting to Britain in 1992, "that we did the right thing. We tried to defend our country." His words echo those spoken by Werner Heisenberg and other German nuclear scientists after the Second World War almost exactly.

Biological agents need not kill to be effective terror weapons. In the case of rodent pest control, thought has been given to using a modified virus that would cause infected female animals to make antibodies against the coat surrounding their own eggs. As a pest control strategy, it would produce a generation of sterile rats. If a similar virus were developed against human beings, it might be years before a slowly emerging epidemic of infertility was even recognized as a deliberate attack. As one scientist has remarked, "the main thing that stands between the human species and the creation of a supervirus is a sense of responsibility among individual biologists." With an ever-growing population of scientists with the skill to manipulate the genes of bacteria and viruses, "individual responsibility" may prove a gossamer defense indeed.

Manufacturing Destruction

The nuclear arms race between the United States and the Soviet Union in many ways defined the mid-twentieth century. But in some ways we can learn even more from the nuclear confrontation that has played out on the Indian subcontinent. In 1948, Indian Prime Minister Jawaharlal Nehru, despite being an advocate of non-aggression and ending atomic tests, admitted that, if threatened, "no pious statements will stop the nation from using it that way." Nehru was right and on May 11, 1974, India detonated a plutonium bomb the size of the Hiroshima weapon. As the Indian threat increased, Zulfikar Ali Bhutto, then Pakistan's Foreign Minister, declared that his country would sacrifice everything to make an atomic bomb, "even if we have to eat grass or leaves or to remain hungry." Many people in that impoverished nation did in fact remain hungry as Pakistan poured its meager resources into a weapons program, which finally resulted in a series of nuclear tests in March 1998.

The disturbing lesson is that the technical and economic barriers to WMD acquisition are steadily dropping. The Manhattan project cost two trillion dollars in the money of the time, and involved an industrial effort as large as the whole of the U.S. automobile industry. Pakistan managed the same feat as an unstable third-world country with a fraction of the resources. If Iran and North Korea soon join the nuclear club, it will be in part thanks to nuclear secrets purchased from A. Q. Khan, the "father" of the Pakistani bomb. Perhaps most disturbing of all, there are thousands of pounds of high-grade nuclear material still in the former Soviet Union, left over from the Cold War. Some is unaccounted for, and much of the rest is poorly secured, vulnerable to purchase or theft by terror groups.

In much the same way, Germany's World War I chemical weapons were produced by the most advanced chemical industry in the world at the time. The sarin gas released into the Tokyo subway by the Aum religious sect in 1995, which killed seven people and made 2,000 ill, was made by a single, poorly qualified biochemist, Seichi Endo. Also in 1995, an American survivalist purchased plague bacteria on the open market from the America Type Culture Collection for just $300. Whether used by nations against their enemies, or by small bands of terrorists bent on causing ever greater fear, there is simply too little we can do to stop WMDs and their effects once they have been constructed. Our best hope of security is to encourage and enforce control, while also redoubling our efforts to understand and counteract the conditions that might lead to their use in the first place.

The Battle for Resources

We have already stated several times that all team aggression, all raiding, and all wars are ultimately about resources, even if the combatants aren't consciously aware of it. All life, in fact, at its most fundamental level is about competition for resources. Evolution has been driven by this competition for billions of years, and today's animals, plants, bacteria, protozoa, and fungi all exist because they competed successfully with their rivals in the past. If we are to have any chance of avoiding the wars of tomorrow, as the destructive power of today's weapons tells us we must, then we have to address this most basic of biological problems:

The fact that as the population of any species grows, the pressure on its natural resources increases and competition becomes more severe.

Biology has invented a million ways for plants and animals to compete with each other. A tree may compete for light by growing taller; early mammals competed with dinosaurs by only coming out at night; humans and chimpanzees—especially the males—compete for food, space, and reproductive opportunities by fighting with each other. Human wars may come wrapped in a veneer of religion or political philosophy, but the battle for resources is usually just below the surface. When Pope Urban II exhorted the nobles of Europe to join the First Crusade, he contrasted the lands where they lived, which had "scarcely enough food for their cultivators," with Palestine, where the Crusaders would be able to appropriate land from the Infidels.[434] In World War II, the need for land and resources was expressed as Hitler's concept of *lebensraum*, or "living space." "The aim [of] the efforts and sacrifices of the German people in this war," he wrote, "must be to win territory in the East for the German people." The Japanese attacked Pearl Harbor because they knew they had to destroy the American Pacific fleet if they were to access the Indonesian oil they needed to supply their industries. As we saw earlier, while rapid population growth and massive unemployment in some settings, such as the Gaza Strip, do not cause wars or terrorist attacks by themselves, they certainly make them more likely.

The predisposition for team aggression may be an inherent part of chimpanzee and human makeup, but the degree of competition for resources varies with the situation. For example, it seems that team aggression among chimpanzees is less common in the Congo, where there are more forest resources, than in Tanzania, where human encroachment has driven the animals into a limited area of forest. The human migrants who crossed the Bering Strait into the Americas about 15,000 years ago found a continent filled with large, easy-to-hunt mammals, and among their limited human skeletal remains we find no evidence of violence. But by about 5000 B.C., as numbers and competition increased, some human skeletons from hunter-gatherer societies in North America show evidence of scalping, or have arrowheads embedded in them.[435] A thousand years ago, in the American Southwest, the Anasazi and Fremont peoples were foragers who also grew maize. Some built elaborate cliff dwellings. The study of tree rings demonstrates that the area was subject to some decade-long droughts, and during these times the region seems to have been beset by raids and warfare. The population retreated to high pinnacles on the edges of deep canyons.[436] They hid small caches of grain in hard to reach places and positioned boulders to roll down on enemy clans. Human skeletons show signs of malnutrition, decapitation, and cut marks on long-bones suggesting cannibalism.[437]

Some Rousseauean anthropologists protest that reports of cannibalism represent a racist desire to denigrate other cultures,[438] but the scientific evidence suggests otherwise. Excavating an Anasazi site in the American Southwest dating from 1150 A.D., Brian Billman of the University of North Carolina at Chapel Hill found cooking vessels and the butchered remains of four adults and an adolescent. Sensitive immunological tests revealed evidence of human muscle protein in the pots; even more convincing, the same tests found evidence of human meat in preserved human feces found at the site.[439] When food is scarce, competition becomes increasingly intense and cannibalism, like team aggression, aids survival.

Critics have argued that the archaeological evidence for endemic violence in drought-ridden areas is too scattered and circumstantial to draw strong conclusions. A recent study of environment and warfare in contemporary Africa helps put that criticism to rest. Edward Miguel of the University of California, Berkeley, and colleagues Shanker Satyanath and Ernest Sergenti of New York University compared rainfall levels and incidents of civil conflict across the African continent, and found that as one increased, the other declined, with a statistical certainty of 95 percent.[440] Interestingly, the effect was found across many different cultures and irrespective of whether the country was well or poorly governed.

Competition for resources has led to violence everywhere we look. When Polynesian seafarers reached Easter Island about 1,300 to 1,700 years ago, they landed on a forested island full of flightless birds. By about 500 years ago, the trees had been cut down, the animals had all been eaten, and the clans, who identified themselves with the curious stone statues that still dot the island, fell to fighting each other. The population plummeted from an estimated 20,000 to just 2,000 by the time Europeans arrived in the eighteenth century. Here too we find archeological evidence of cannibalism, which lives on in the oral tradition of the islanders. A local insult used on Easter Island even today is, "The flesh of your mother sticks between my teeth."[441]

The thought that rapid population growth could increase conflict is hardly new, and certainly Thomas Malthus accepted this relationship in his 1798 *Essay on the Principle of Population*. As with so many efforts to interpret human behavior, however, the link between resource depletion and conflict has been obscured by extreme arguments. As Shridath Ramphal and Steven Sinding, then of the UN Commission on Global Governance and the Rockefeller Foundation, write, "there has been considerably more heat than light in the international dialogue" and efforts have been made that "suit a political, as opposed to a scientific interest."[442] Those looking at the same landscape of facts but through different lenses end up sparring instead of seeking synthesis. Nancy Peluso and Michael Watts,[443] colleagues of ours at Berkeley, castigate writers such as Robert Kaplan, author of *The Coming Anarchy: How Scarcity, Crime, Overpopulation, and Disease Are Rapidly Destroying the Social Fabric of Our Planet*, for making too direct a link between resource scarcity and conflict.[444] They point out, citing Karl Marx (who did in fact get a few things right), that economic patterns also help determine who controls and who has access to resources. No doubt some conflicts could be avoided by a more equitable distribution of resources; there is nothing contradictory in arguing for greater social and economic equality while also recognizing that high birth rates can overwhelm the ability of a finite region to sustain its human population regardless of such equality.

John May, the World Bank's demographer for Africa, has drawn attention to the demographic pressure that had built up in Rwanda by the time of the 1994 genocide. The population of Rwanda was two million people in 1950, and on average each woman had almost 8 children. By 1994, average family size had fallen slightly to 6.2, but the population had quadrupled to almost eight million, resulting in a population density of 292 people per square kilometer, the highest in all of Africa.[445] James Fairhead, an anthropologist from the School of Oriental and African Studies in London, adds an economic dimension to the analysis. Preceding the Rwanda genocide, Fairhead points out, agricultural land prices had reached an astronomical $4,000 per hectare in a country where many people lived on less than $500 a year. "Land," Fairhead concludes, "is worth fighting for and defending."[446] Tragically, the fighting which took place in 1994 left between 500,000 and one million dead. It was cast as an ethnic conflict, and senseless. Once its roots in resource competition are laid bare, however, the violent extermination of an identifiable outgroup takes on the all-too familiar logic of team aggression.

Can all conflict be reduced beyond even team aggression and resource competition, down to the single factor of population growth? It's not quite that simple, but a deeper investigation of the role of population increase shows quite clearly that growth rate and population demographics function as significant triggers for raiding, wars, and even terrorism. If we hope to reduce the number and severity of these violent incidents in our world, this is a relationship we need to understand. Peter Turchin of the University of Connecticut and his Russian colleague Andrey Korotayev provide important quantitative insight into the dynamic connections between population growth and conflict. In a careful study of English, Chinese, and Roman history, they showed a statistical correlation between an increase in population density and warfare, although not surprisingly the impact of population growth was not immediate but took some time to develop.[447] It is not the infant playing at the hearth but the hungry landless peasant twenty years later who causes the conflict. Adjusting for this and other variables (such as the fact that wars themselves tend to reduce population), and using robust data on population growth from church records in England along with historical data on conflict, Turchin and Korotayev found that intervals of relative peace and rapid population growth were followed by periods of conflict and slower population growth. Their study suggests that population growth accounts for a powerful 80–90 percent* of the variation between periods of war and peace. Even if the influence of population is substantially less than that, it remains outstandingly important. But here is the crucial point: Rapid population growth is not just an important cause of violent conflicts. In the contemporary world, population growth is a cause that can be contained by purely voluntary means.

In the past fifty years the world has accommodated rapid population growth tolerably well, although as rising oil and food prices suggest, this may not be true in the future. The combination of the industrial revolution and science-based technology increased global wealth at an astonishing rate. We have been a little like those first people to cross into North America, or the Polynesians who first landed at Easter Island, in more ways than one, however. Presented with vast new supplies of food, energy, building materials, and luxury goods our forbears could never have imagined, we have gorged ourselves on consumption, and we have driven

---

* Turchin and Korotayev focus on population growth. If one were to add the factor of population structure, especially the ratio of young men to older men, it is quite possible that essentially all the difference between war and peace would be accounted for.

our global population from just one billion people in 1800 to six billion in 2000. We live in a globalized world now, and worldwide population is expected to increase to over eight billion by 2030. The evidence of that increase is now all around us, in our polluted environment, our warming climate, our disappearing rainforests, and our increasingly degraded farmland: We are, as a species, in the process of proving Malthus's proposition that population will always outstrip resources.

Has the age of rapid resource expansion really come to an end? Human ingenuity continues as unchecked as our population growth, and we will no doubt find ways to squeeze more food, water, and energy out of the existing supplies. But there are natural limits on how far efficiency and invention can take us. Thomas Homer-Dixon, Director of Peace and Conflict Studies at the University of Toronto,[448] and Ambassador Richard Benedick,[449] who was the chief U.S. negotiator for the 1987 Montreal Protocol on atmospheric ozone levels, argue that resource wars will become increasingly common in many parts of the world in the twenty-first century.* Water, for example, is becoming a key constraint on development and quality of life in many places. Thanks to dwindling supplies and burgeoning populations, the Middle East and much of North Africa now have one-third as much water per capita as in 1960. Israel has already exploited 95 percent of the available water supply in the country, and uses it efficiently; there is no new supply to tap. In the Gaza Strip, seawater is contaminating groundwater supplies as fresh water is pumped out to supply the growing population.[450]

Egypt has depended on the Nile for irrigation, drinking water, and flushing its waste for thousands of years. But even that vast stream of water is now reaching its limits. Martha and I have watched millions of gallons of clear water pour over the Blue Nile falls near Bahir Dar in Ethiopia, and we have sat beside the origin of the White Nile at Jinja on Lake Victoria in Uganda. The two branches join at Khartoum in the middle of the Sudanese desert to make a vast, life-giving flow that has sustained forests, wildlife, and human populations since time immemorial. But by the time the Nile reaches the Mediterranean Sea, it is a sadly depleted shadow of its former self. In the year 2000, there were 170 million people in Ethi-

---

* Some commentators, such as Nils Petter Gleditsch (1998) of the International Peace Research Institute in Oslo, maintain that the relationship between human population growth and competition over resources is simply too complex to analyze, but the fact that relationships are difficult to quantify need not make them any less important. Others attempt the analysis and make a compelling case.

Credit: Amit Dave—Reuters

War, ultimately, is about competition over resources. Climate change and population growth are likely to make that competition more severe. With climate change and growing numbers of people depending on limited water sources, groundwater levels in parts of India have fallen by 25 meters in a decade. People came from miles around to this huge well in Gujarat, India during a drought in 2003. By 2025, 3 billion people worldwide will face limitations or downright scarcity of water.

opia, Sudan, and Egypt, all dependent on the waters of the Nile. There is significant demand for family planning in these countries, but for cultural and political reasons, that demand remains largely unmet. The populations of these three countries will continue to expand rapidly from 190 million today to a UN-estimated 337 million people by 2050. Population will more than double, but there will be no new water supply—all 337 million will be dependent on a source that is already under strain. In a region with a volatile mix of cultures, religions, and ethnicities, the added stress of severe water shortages may well be the spark that sets the team aggression impulse ablaze on a vast and horrifying scale.

And yet our consumption continues to increase. In recent decades, a billion new consumers have arisen in China, India, South East Asia, India, Brazil, Mexico, and parts of the former Soviet bloc. When the incomes of these newly affluent people are adjusted to take into account local purchasing power, their potential to buy better quality food, more consumer goods, and more automobiles will equal that of the U.S. While we should welcome the improved living standards and decreased poverty in many parts of the world, finite resources also make it essential that everything possible is done in the West and among the newly affluent to prevent runaway population growth. Norman Myers of Oxford University has shown that if the newly wealthy Chinese were to eat fish at the Japanese per capita rate, they would empty the seas, and if they used cars at the U.S. rate, they alone would consume today's total global output of oil.[451] In fifteen years, Martha and I have seen Beijing's and Shanghai's roads go from two-lane streets filled with bicycles to six-lane super-highways bursting with cars. The price of oil around the world continues to rise with the increased demand, and it is not going to fall to the low levels that Americans expected almost as a natural right just a decade or two ago. As competition for oil and other resources increases, will nations solve their differences through diplomacy, or through war?

Optimists point out that some countries, such as the Netherlands, are densely populated but still maintain a high standard of living. The implication is that good government and modern technology can help prevent the worst problems of expanding populations. But such arguments overlook the fact that we all need space to grow the food we need, to collect the water we use, and to absorb the pollution we create. Calculated realistically, the Netherlands has an ecological footprint fourteen times its area on the map, because it imports food for people and fodder for cattle, consumes drinking water that fell as rain in Switzerland, and pumps carbon dioxide from its power stations into the global atmosphere.

For billions of years, evolution has been driven by competition caused by the simple fact that, left unchecked, all living things can reproduce faster than their environment can sustain. Our population growth today is largely unchecked by hunger, disease, or predators, and it is highly likely that our numbers and industrial demands have already exceeded the environment's capacity to support them. Mathias Wackernagel in California, Norman Myers in England, and others calculate that we may have exceeded Earth's carrying capacity as long ago as 1975. According to these calculations, we *already* need a planet 20 percent larger than the one we have. Such estimates are difficult to make and open to criticism. But it doesn't take much more than an open set of eyes to realize that current human population growth and economic expansion are going to be impossible to sustain in the long term. Competition for resources is about to increase markedly. [452]

Lessons

Human beings are animated by curiosity. This same impulse to investigate our surroundings which today drives the scientific enterprise originally adapted our ancestors to a harsh, competitive environment. But unfortunately, the mixture of curiosity, the tendency to overreact when threatened, and unquestioning loyalty to our ingroup has become a lethal combination in today's world. We can expand the envelope of empathy to include greater numbers of people, but in times of war, or perceived threats to our safety, it too often collapses again.

Power, patriotism, and curiosity can drive even the most intelligent and informed men—and it is virtually always men—to turn new scientific discoveries into weapons of mass destruction. The witness of history seems to be that the predisposition to fight and to defend ourselves against attack is so powerful that human beings, once they perceive themselves to be in a life or death struggle of any kind, will always justify research and development of new weapons, however horrendous their effects. It is sobering to note how many winners of Nobel Prizes for science contributed directly or indirectly to the development of weapons of mass destruction—and how many achievements honored with a Nobel Peace Prize fell apart soon after they were awarded. If the Nobel Prize for physics is awarded for accomplishment, the Peace Prize seems very often to reward only effort. But this does not mean that true peace is impossible—so long as we understand the biology of war.

We live in very different evolutionary times than any of our ancestors. After 3.5 billion years of competition, life on Earth has reached its carrying capacity. More competition at this point means fighting harder over a constantly dwindling pool of available resources. As we seek ways to solve our environmental crises, address the warming climate, and combat emerging diseases and global poverty, our very survival as a species requires finding more ways to cooperate rather than compete. And thanks especially to WMDs, the survival of our species now also means bringing an end to war as we know it. It is time to leave our history of team aggression behind.

These are daunting challenges, to say the least. Each will require the commitment and individual efforts of literally billions of our fellow humans, as well as many careful, specific programs put into effect by entire populations. But there is one action that we must take, individually and as a world, if any of the others are to be successful. It directly contradicts some of our deepest evolutionary programming, but if we are to survive as a species, we must stabilize or even reduce population size. As we'll see in the coming chapter, to a very large extent that means recognizing that the natural tendencies of men are not consistent with the survival and well-being of their sexual partners, their children, and future generations to come. The most aggressive and violent aspects of men's inherited behaviors—summarized in the predisposition to team aggression—too often overshadow the more benign aims of women, especially that to have surviving and healthy children. Fortunately, women's impulses and aims are also based on deep evolutionary programming. All we have to do is create the conditions that allow them to be expressed.

13

WOMEN AND PEACE

WITH MARTHA CAMPBELL

The man or woman [who avoids having children] merits contempt as any visited on the soldier who runs away in battle.

—THEODORE ROOSEVELT, 1905

No woman can call herself free until she can choose consciously whether she will or will not be a mother.

—MARGARET SANGER, 1920

S OMETIME IN THE 1930S, an illiterate peasant left Yemen to look for a laborer's job in Saudi Arabia. Success followed, and before long Muhammad bin Awad bin Laden began employing others to work on road projects, and then on more jobs, like building royal palaces, rebuilding mosques (including the holiest in Mecca), and developing hotels. The Saudi Binladen Group became a multi-billion-dollar company with powerful connections. In the 1990s, the Group helped build the King Abdul Aziz Air Base, from which U.S. forces operated during the first Gulf War. Its business partners in that project included the Carlyle Group, which boasted George H. W. Bush and retired British Prime Minister John Major as advisers.

The company's success may have derived from hard work, good busi-

303

ness sense, and powerful connections. But its wealth all came from the Western thirst for oil, and the technology it used was all imported. Muhammad bin Awad bin Laden's private life remained that of a Yemeni patriarch. Altogether he married eleven wives, though he never had more than four at any one time. He fathered fifty-four children and might have had more if he had not died in an airplane crash in 1967. Still, despite his early death, bin Laden's life was a smashing evolutionary success: vast accumulation of wealth, frequent sex with subservient partners, and prodigious reproductive output as the result.

The seventeenth son of Muhammad bin Awad bin Laden, born in 1957, was named Osama bin Muhammad bin Awad bin Laden.[453] Osama's generation of wealthy Arabs split their time between jet-setting around the world and watching their own culture stagnate at home. All of the bin Laden brothers except Osama were educated in the West, and some married Western women. The family gave money to Harvard, Tufts, and Oxford universities. And yet they lived in a world of painful contrasts. Without oil exports, the foreign trade from all the Muslim countries of the Middle East combined would be less than that of tiny Finland. Even though Saudi Arabia has more oil than any other country, since it lacks access to family planning, the population has grown so rapidly that individual incomes have begun to fall. No doubt Osama, who studied engineering at Jeddah University, sensed the failure of his country to keep pace with the rest of the world, and he opted for a strident reassertion of traditional values. In 1979, under the influence of fundamentalist clerics and conservative thinkers, he joined the jihad against the Soviet invasion of Afghanistan. He supplied money and construction equipment, and late in the conflict, he fought briefly on the front lines. His focus since that time has, as we know all too well, shifted to a new enemy.

If Osama's career path has deviated from the family business, his personal life has hewed even more closely to Yemeni tradition, especially in terms of gender roles, than his father did. Women in his home in Saudi Arabia, and those who accompanied him outside the country, covered their hair with a *hijab*. When they left the house, the women were hidden behind burqas, full-length, head-to-toe coverings with only a mesh opening for the eyes. At home, they brought food to male guests who sat on cushions on the floor, but they never mingled with the men or took part in their conversations. Left alone, the men discussed medieval religious leaders such as ibn Taymiyya, the interpretation of dreams, and contemporary events in Palestine, Algeria, Chechnya, and the Philippines. Women were for sex and service, not partnership.

While most Western men might not want fifty-four children, many could fantasize about having eleven compliant wives and the ready access to sex and domestic service that it would imply. Few Western women, however, would wish to see themselves as a wife of the elder bin Laden, no matter how wealthy he may have been. Women in many traditional societies, not just fundamentalist Islamic cultures, live lives of limited choice and little power, however. Indeed, it is sometimes too easy to forget that even in the United States, women earned the right to vote less than one hundred years ago. Women in most traditional societies lack the rights and influence that women in modern democracies have fought hard to attain for the past 150 years. It is not difficult to imagine that for many of them, and indeed for many of our own great-grandmothers, the life afforded by the senior bin Laden's wealth might well have seemed the best of available choices.

But what if those women, Osama bin Laden's mother included, had more choices? What if they could choose to marry (or not) whom they pleased, or control when they had children, and how many? Or even just play a role as a full member of society? Given what we know about human evolution, based on contemporary sociological and demographic research and building on our firsthand experience with family planning and public health in many countries around the world, we feel that the impact of empowered women on a traditional culture can scarcely be overstated. A plausible case could even be made that if the mothers and wives of Al Qaeda members had been able to adopt modern family planning technologies as rapidly as Osama bin Laden and his colleagues took to cell phones and Stinger missiles, Al Qaeda might never have developed its bitter and fanatical perspective on the Western world.

Men Controlling Women

As we saw in chapter 6, women often suffer disproportionately in wartime, and in warring societies. But it is also true that when women have relatively more power and influence in family and political affairs, the impacts can be profoundly positive. This is not just feminist propaganda—there are solid biological reasons and measurable outcomes to support our assertion that one of the keys to gaining control over the male predisposition for team aggression is to ensure that women have options in their own lives. First, however, we need to understand just how few options women have had through much of human history. Limits on the role of women in traditional societies have most often been linked to control of sex and sexuality, which in turn are very frequently cast in a religious context. All of this gets us very far away from basic biology, but as we've shown, the male desire to control women's sexual activity can be traced to efforts to ensure paternity. This impulse lurks behind even the most extreme or bizarre expressions of religious obsession, misogyny, and sexual repression.

The men who brought down the Twin Towers on 9/11 covered the pictures of women on their motel room wall with towels, but they also rented pornography. Mohammad Atta, the leader of the nineteen terrorists, left a will specifying that women should only touch his corpse if they wore gloves, and that no woman should touch his genitals. As bizarre as their behavior seems, the 9/11 terrorists' disparaging, controlling attitudes toward women in fact have a long history.[454]

We don't know a great deal about the social roles of men and women during the earliest phases of human development. But in the shift from a hunter-gatherer way of life to settled agriculture, women lost out. Remembering back to chapter 6, women and men have a substantially different evolutionary relationship to sex and reproduction. In the strictly selfish terms of evolutionary logic, no parent wants to invest time and energy in another individual's offspring, thinking it is their own. A woman who has a baby always knows that she is the mother, but a man cannot be so certain. As our hominid ancestors began to nurture children, and not just tolerate them as chimpanzees do, presumably males wanted more reproductive certainty. And especially as settled agriculture and urban living arose and wealth began to be transferred from generation to generation, new social mechanisms developed to ensure paternity through the social control of women. Covering the body and even the face when outside the home, concubinage, prostitution, the cruelty of bound feet, and the horror of female genital mutilation all arose for the first time in settled agricultural societies.

Many of the controls on women, both historical and contemporary, have deep roots in religion and are underscored by culture. In Africa today, women make up half the population, do two-thirds of the work,

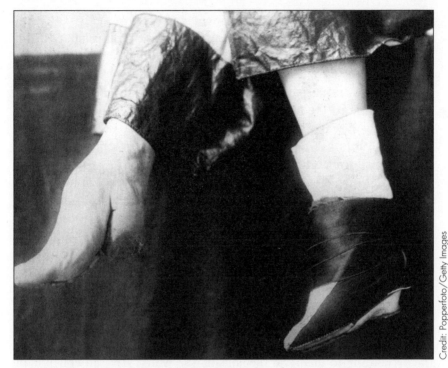

Credit: Popperfoto/Getty Images

For a thousand years, wealthy Chinese bound the feet of their young daughters, demonstrating the recurring male impulse to control women. (This photograph was taken circa 1880.) Not only could a woman with deformed feet not run away from her husband, foot binding also served as a symbol of a man's wealth, as he had to employ servants to do the work his crippled wife could not carry out. Small feet were perceived as erotic by men; for the girls, experiencing several years of pain for this purpose in early childhood no doubt reinforced the belief that pleasing a man was a woman's only purpose in life.

receive one-tenth of the gross income, and own one-hundredth of the property. There is an African saying, "Women and cows belong to men." In parts of Kenya, there is a tradition that when a woman's husband dies, or a young girl loses a member of her family, the "village cleanser" should have intercourse with the bereaved; otherwise she will bring misfortune to the village. No doubt it gives the man pleasure, but it is clearly demeaning and exposes the woman to yet another route of HIV/AIDS infection.

Much like team aggression, men's urge to control women and their reproduction shows up time and again throughout history and across cultures. In the Judeo-Christian tradition, it may be most eloquently expressed in the telling of the Biblical origin story. God made Eve as a companion for Adam, and Eve out-of-control brought great calamity down upon them both. Sex and sin have been linked ever since, with women cast as temptresses in need of control. The expression of that linkage has at times been bizarre, and deeply damaging. In the fifth century, Saint Augustine identified the concept of "original sin" with sexual intercourse and thought it was transmitted in the semen like some spiritual AIDS virus.[455] With lust or desire as the ultimate peril, Augustine argued that the only justification for sex was to conceive a child. As he put it, "That which cannot be done without lust must be done in such a way that it is not done for lust's sake." Contraception, obviously, would not sit well with that worldview, and Augustine preached that its use turned a wife into "her husband's whore."*

It is difficult to get into the minds of people living in the medieval world, but many of the practices and preaching which survive from the centuries following Augustine can be read as evidence of the male drive to control female sexuality and especially reproduction. Theologians seriously discussed whether it was more sinful to lust after a beautiful woman than an ugly one.[456] If a woman wished for intercourse on a holy day, then according to medieval theologians she would have an epileptic or congenitally malformed child, and "her impudence [should be quelled] with fasting and beatings." Medieval penitentials prescribed seven years of penance for oral intercourse.[457] Suffice to say, superstition, religiosity, and misogyny quite often went hand-in-hand throughout much of the history of Christendom. Again and again we see how cultures and religions frame women's sexuality as both powerful and dangerous, and as something that must be kept under control.

If we seem particularly passionate in our view of male control of women and its ties to religion, it is because of what we have seen. Our work has taken us to places such as the state of Bihar, India, where women are banished to the barn during their menstrual periods because they are considered unclean; we have been to countries like Pakistan where so-called "honor killings" still take place; and we have seen women in the Philippines have unsafe abortions because the Catholic Cardinal condemns the widespread distribution of modern contraception, having been taught

---

* Augustine records a number of youthful sexual encounters in his own writings (*Confessions*). On converting to Christianity he abandoned a long-term mistress and the son she had borne him. Augustine also praised his own mother for not complaining when his father abused her.

Augustine's tortured logic in seminary as a young man. Women, especially poor women, have often been poorly served by religious rules.

For all the wrong ideas and destructive traditions of medieval Christianity, its theologians did at least outlaw polygamy. And this has had profound and positive effects. Indeed, without the Christian teaching on monogamy, Western civilization might not have emerged in its current form.* As it happened, Christian culture gave rise to the scientific revolution, the industrial revolution, and the free market system, all of which have probably done more good than harm in the world. But that same Christian outlook outlawed family planning, punished abortion, and blocked scientific research on contraception—actions we would consider to have been profoundly harmful. In early modern Europe, the peasantry was so desperate to avoid excessive childbearing that marriage was postponed later and later, until in the early seventeenth century in Western England, the average age of first marriage for women reached almost thirty years. In the seventeenth and eighteenth centuries, women in big cities of Europe put the infants they could not support into foundling hospitals. With no safe source of infant food, these hospitals simply became socially acceptable ways of killing children.[458] Of the 14,934 babies admitted to the London foundling hospital in the 1770s, 10,204 died. In spite of people's obvious desire to limit their family size as the industrial revolution took off, the manifest and overt need for family planning was met by religious, legal, and social opposition.

Constraints on all aspects of human sexuality reached a high watermark in the nineteenth century. In the United States, the Reverend John Todd, a protestant minister, wrote a whole book condemning male masturbation—with the curiously appropriate title *A Student's Manual.* Sexual knowledge was so constrained in Victorian times that the English art critic John Ruskin, who had reviewed innumerable nude paintings of smooth-bodied hairless ladies, was unable to consummate his marriage on his wedding night when he discovered his wife had pubic hair. Marie

---

* It is a fair generalization that those contemporary nations which still endorse polygamy, such as Nigeria and Afghanistan, also seem furthest from developing into democratic societies. By contrast, Thailand has moved away from polygamy while becoming more democratic. King Chulalongkorn, Ram V, whom Western readers know mostly from the musical *The King and I,* abolished slavery and did a great deal to modernize Thailand, but he still had four queen consorts, numerous concubines, and fathered seventy-seven children. Today, the royal family is monogamous and his great-great grandson, the present King Bhumibol, has completed the modernization started by King Ram V. King Bhumibol is held in universal respect and he has helped stabilize the young Thai democracy, something that is unlikely to have happened if the royal family had not switched to monogamy.

Stopes, who was to become the pioneer of family planning in Britain, took several years to discover something was missing from her unconsummated marriage to Ruggles Gates—namely, that he was impotent.[459]

In the 1870s, Anthony Comstock persuaded the U.S. Congress to pass a series of laws that among other things, defined contraceptives as pornographic. (We'll return to these "Comstock Laws" in the next section; their impact on reproductive freedom and health are still being felt today.) During the nineteenth century the countries of Europe and every U.S. state passed laws against abortion.[460] Such laws did not stop women from trying to end unwanted pregnancies, but it did make their attempts more dangerous and painful, and subjected women to financial and sometimes sexual exploitation.* While the nineteenth-century European colonial empires brought basic public health measures to many parts of the world, they also exported their deep-seated hostility to any form of family planning. In their colonies to the south, cleaner water, better hygiene, better food, and vaccines all helped bring infant and child death rates down to levels never seen before, but these were not accompanied by the means to decide whether or when to have a child. As birth rates remained high, unprecedented survival rates brought about increasingly rapid population growth.

The Fight for Reproductive Freedom

Speaking very broadly, one of the themes through much of human history is men laboring to keep women under control, and benefiting from their success in achieving this. The full picture is infinitely more complex than that, of course, and strong and especially privileged women have always found ways to influence their own lives and the course of history, no matter how rigorous the societal controls on their official roles might have been. But something quite remarkable began to happen in European culture a couple of hundred years ago. Inspired by the Enlightenment, and spurred on by the social changes of the Industrial Revolution, women began to claim their right to participate in private and public life as full members of society. This struggle for equality remains a work in progress today, but the advances that have been made are dramatic on many levels, from the personal to the global. We'll take

---

* In a study conducted in Ethiopa, one of our students found that about 10 percent of women self-reported having to give sex as well as money to obtain an illegal abortion.

a closer look at two aspects that pertain especially to the place and nature of war and team aggression in the modern world.

First, there is the basic fact that empowered women tend to counterbalance the most chaotic and violent aspects of men's predisposition for brutal territoriality and team aggression. The predisposition itself is disproportionately male, and while women may also feel aggressive toward an outgroup, the decisions they make and the actions they take as a result are often quite different. Second, and on a still more fundamental level, when women have the choice to control when and how often they have children, they opt increasingly for smaller families, and often start childbearing later in life. This has important consequences for the size and age structure of human populations, with smaller families and later reproduction leading to stable population size, and the lower ratios of volatile young men which, as we saw earlier, can dramatically decrease the likelihood of violence, raiding, and war. Explicitly in the latter case and implicitly in the former, reproductive freedom for women is a crucial and necessary precondition for bringing an end to war as we know it.

For the two hundred thousand years of early human existence and up to the nineteenth century, the world's population growth was slow almost everywhere because of late puberty (a result of poor nutrition), prolonged breastfeeding (which limits women's fertility), and a tragically high infant death rate. That these natural restraints were removed is one of the great achievements of our species, but nothing comes without a cost. With the natural controls on our population growth gone, it became impossible to limit family size without some artificial technology.*[461] The history of modern family planning is one long, painful battle between women struggling to choose when to have a child, and legal and cultural constraints very often imposed by men determined to limit these choices. The barriers that have been raised between contraception and safe abortion and the women who need them are so universal that they are sometimes overlooked by demographers.[462] They seem to be invisible to many economists as well, some of whom still have not grasped that having a child is not a simple rational economic decision like buying a refrigerator. Rather, it is a continuous struggle to separate sexual intercourse, which

---

* The use of the word "natural" in family planning is interesting. St. Augustine explicitly condemned periodic abstinence (the rhythm method), which the Vatican now calls a "natural" method. This method is useful for a small number of women who enjoy equality and partnership in marriage, but it is of no help to the many women around the world who will be beaten by their partners or replaced by another wife if they refuse sex.

is frequent across all cultures, from pregnancy and childbearing. Many demographers seeking explanations for patterns of family size seem unaware of the fact that heterosexual couples have intercourse hundreds or thousands of times more frequently than is needed to conceive the number of children they would like.

The first articulate efforts anywhere in the world to promote contraception were made at the beginning of the nineteenth century. English and American Utopians and Free Thinkers, such as Francis Place and Robert Dale Owen, took the lead. In England in the 1870s, Charles Bradlaugh was tried for the criminal act of republishing a book on family planning written in 1832 in the U.S. by Charles Knowlton. The trial attracted great publicity, raising in Britain a new awareness of family planning. Primitive condoms and spermicides began to be sold, *coitus interruptus* became more widely used, and illegal abortion rates rose. In 1860, one quarter of English women had at least eight children, in a pattern of fertility similar to Afghanistan's today. After the publicity of the Bradlaugh trial, family size began to fall, and by 1925 only one in fifty married couples had eight children or more. In America, average family size fell from 7.0 in 1800 to 3.5 in 1900.

In Britain at the beginning of the twentieth century, the most vitriolic opposition to family planning often came from those groups that had the smallest families—the clergy and the medical profession. What little early support there was for contraception came from those who were in direct contact with the women who needed it most. In America, Margaret Sanger's life-long crusade for family planning began when, as a community nurse caring for a young mother after an illegal abortion, she found herself unable to respond to the woman's request for contraceptive advice and then saw her patient die from a second botched operation.[463] In 1917, Sanger and her sister opened the first family planning clinic in the U.S., in Brooklyn, New York. Women lined up around the block for advice, but using the anti-contraception Comstock laws the police moved in and closed the clinic down.[464]

Margaret Sanger became the first president of Planned Parenthood and she helped found the International Planned Parenthood Federation in 1952. I joined the IPPF headquarters staff as its first medical director in 1968, and found it inspiring to work with many feisty, independent, strong-willed, sometimes tiresome but never boring family planning pioneers from forty countries, many of whom had worked directly with

Sanger. Sanger was a woman of immense energy and determination, and she almost single-handedly found the money and the scientists to develop the oral contraceptive pill in the 1950s.[465] Sanger was also one of the first people to draw the link between family planning and peace. In 1920, with the memory of the Great War fresh in her readers' minds, she wrote,

> In every nation of militaristic tendencies, we find the reactionaries demanding a higher and still higher birthrate. Their plea is, first, that great armies are needed to defend the country; second, that a huge population is required to assure the country its proper place among the powers of the world....As soon as the country becomes overpopulated, these reactionaries proclaim loudly its moral right to expand...and to take by force such room as it needs.[466]

In the decades that followed the closing of Sanger's first clinic, ways were contrived to keep clinics open and offer at least some poor women the same privileges richer women had to manage their own reproduction. It took until 1965 for the U.S. Supreme Court finally to overrule the last of the Comstock laws—just one year before Sanger died at age eighty-six. In January 1973, in *Roe v Wade*, the Supreme Court went on to strike down all state laws outlawing abortion.

For some readers, even the mention of safe abortion as a necessary and important part of a woman's ability to manage if and when to have a child will be deeply offensive. Yet statistics show that, on average, every woman in the world now entering her fertile years will have an average of one abortion. The rates are surprisingly similar across developing and developed countries (39/1000 and 34/1000 women/year respectively).[467] The sad fact is that although contraception has improved beyond recognition in the past half century, misinformation about side effects remains common (for example, beliefs that oral contraceptives cause cancer, which they do not) and access is often limited (for example, the Pill remains a prescription item, even though it is safer than many over-the-counter drugs such as aspirin). Every country with an average of two or fewer children per family has widespread access to safe abortion, while no country with a high or even moderately high birth rate does. These latter countries tend to have large numbers of illegal and often unsafe abortions. In the Philippines, where abortion is prohibited by law, over 500,000 of the procedures are performed each year. Many are done in unsafe conditions by untrained providers, and maternal mortality is high.

Indeed, it is impossible to achieve low maternal mortality without access to safe abortion. Having said that, there is robust evidence, especially from some parts of the former Soviet Union, where abortion rates used to be very high, that good contraception can greatly reduce the need for abortion, although it will never eliminate it.

Most Western people think abortion should be available within certain limitations. At the extreme ends of the spectrum are those who believe every woman has an unfettered right to abortion, and those who think an embryo, even if smaller than a speck of dust, has an absolute right to life. The basic problem, it seems to us, is not agreeing or disagreeing with the belief that abortion is murder, but accommodating in a respectful way to a variety of sincerely held opinions on a complex topic. Religious assertions about when life begins are philosophically parallel to religious beliefs about life after death. They are both strongly held by different groups, but they are beyond the realm of science to prove or disprove. All big cities in Europe and North America have churches representing a variety of Christian denominations, as well as synagogues, mosques, and probably Hindu temples. Each teaches a different interpretation of life after death. As conflicts over abortion turn on differing interpretations of life *before* birth, and also of the value of that life, it should be no more surprising to have an abortion clinic in a city where many people are sincerely opposed to terminating a pregnancy than it is to have Buddhists and Baptists teaching different interpretations of eternal life. Discussions about religious tolerance are an appropriate way to frame abortion decisions, while arguments over the "rights" of the embryo versus the "rights" of the mother are assertions of absolutes that foreclose all reasoned argument. In a democratic, pluralistic Western society where church and state are separated, establishing a consensus on access to safe abortion should be possible.

In Western countries, a great deal of progress toward reproductive freedom has been made. Condoms and spermicides are no longer prescription items, unmarried women can obtain the Pill, and it is difficult to remember that even as recently as the 1970s, the majority of sexually active women spent a great deal of time worrying whether they would see their next menstrual period. But the controversies have not gone away. In the U.S., safe abortion remains a hot button issue. In Kenya, Catholic bishops made a bonfire out of burning condoms, in a country where 500–700 people are dying from AIDS every day. Martha and I both remember the shock and tears among our Catholic friends in 1968, when Pope Paul

VI issued the Encyclical *Humanae Vitae* (On Human Life), upholding the ancient ban on artificial birth control. In doing so, the Holy Father overturned the recommendations of some of his own bishops and Pontifical Commission, who had voted to accept some aspects of modern contraception. As one conservative theologian tried to explain, "If the church sent all those souls to hell, it must keep maintaining that that is where they are."[468]

The Central Question

Reproductive decisions are made one by one, by individuals. But the impact of those decisions in aggregate can be profound. A longstanding and important policy division in international development circles demonstrates this perfectly. It is generally agreed that the explosive human population growth of the last several generations is a real problem, a threat to human health, prosperity, and environmental security. But what is the relationship between family size and poverty? Many have argued that family size will fall automatically once socio-economic development takes place. It is true that wealthier, more empowered families tend to have fewer children, but this argument puts the cart before the horse. Studies show that the poorest families also have the greatest unmet need for contraception. Where family planning is easy to get, as in Thailand, there is little difference in family size between rich and poor, illiterate and well educated. But where it is difficult to get, and particularly where the focus has shifted away from family planning in the past ten to fifteen years, then wide gaps exist in family size between rich and poor, and between the educated and uneducated, and the disparities tend to be getting worse.[469]

Here is what many people have failed to understand about family planning: Rather than reduced family size being a beneficial side effect of development, access to family planning can help drive the development. The growth of human populations is actually a variable that is open to change just by giving women options—and when women choose to limit their family size, it actually helps to speed the transition from poverty and powerlessness to improved economic conditions. The truth is that no countries, except for a few oil-rich states, have escaped from poverty while still maintaining a high birth rate.[470] If the commonly held position that wealth and education are necessary precursors of smaller family size is true, then women in the many parts of Africa where people are getting poorer would be consigned to having large families indefinitely. If women's having control over their own reproduction truly helps them escape poverty, however, then making family planning realistically available becomes a win-win policy. It would help meet the widespread desire of many women (and most men) in developing countries to have healthier and more educated children, help them lift themselves out of poverty, and also serve the self-interest of their communities as well as all of us by reducing the likelihood of wars and terrorism.

In 1967, Professor Kingsley Davis, a well-known demographer at the University of California, Berkeley, insisted that family size would fall only when women were educated and incomes improved. He rejected the idea of making family planning available in poor countries as "quackery" and "wishful thinking."[471] Experience has proved him wrong, but we still hear people say that parents want many children to help with farm labor, or to care for them in old age because there is no social security system. And they insist that poor women won't use contraception until they are educated and until incomes improve. If it were true the world would indeed face a somber choice: If women won't have smaller families until they are richer, but they can't become wealthy until they have smaller families, then policy makers must either stand aside and watch the poor suffer, or find some way to force people to have fewer children. This is the box Kingsley Davis found himself in the 1970s. He supported Indian Prime Minister Indira Gandhi's coercive family planning policies, endorsing the need for a "totalitarian government" "ruling a docile mass of semi-educated but thoroughly indoctrinated urbanites...accepting passively what is provided for them."[472] The Indian policies are now universally condemned, but like the grin on the Cheshire cat, the idea that education and wealth are prerequisites for smaller families persists to this day.

Again and again, surveys conducted in poor countries from Argentina to Zimbabwe show that women denied the benefits of literacy are having more children than they want. Wherever women have had realistic access to family planning options and information, no matter how poor they may have been, family size has declined. In South Korea, when modern methods of contraception were made available and backed up by safe abortion, family size fell from 6.0 in 1960 to 1.8 children thirty years later. In Thailand, the Population and Community Development Association (PDA)[473] set a nation-wide example by involving shopkeepers, schoolteachers, village leaders, and others in selling contraceptives at a small profit. They offered voluntary sterilization, especially vasectomy,* and the law on abortion was broadly interpreted to allow safe early abortion. Family size fell also from 6.0 to 2.0 within thirty years.** It is nearly the same story in Mexico, Brazil, Colombia, and as we've mentioned, dramatically so in parts of Bangladesh. Remarkably, in one rural area of that country, where Dr. Zafrullah Chowdhury has built on local skills to make a range of health services and excellent family planning available, the reproduction rate has reached replacement level, or two children per family.***[474] In 1970, when women in both Bangladesh and Pakistan averaged over six children, anyone who dared to suggest that women anywhere in Bangladesh would be having just two children by the year 2000 would have been laughed out of the room by demographers and social scientists. Revealingly, in Pakistan, where family planning is not realistically available, the decline in family size has been slow; in 2007 Pakistani women were still averaging 4.1 children per woman. A difference in average family size of one or two children may seem small, but over time it makes a huge difference to the size of a country's population.****

Iran presents particularly a convincing and highly relevant example of the power of family planning in reducing family size. When a census was held near the end of the 1980–1988 Iran-Iraq War, informed decision-makers in the country could see that rapid population growth was outpacing growth in the GDP—a recipe for increased poverty. They were concerned also about protecting the environment, which inevitably suffers when pop-

---

* In the 1970s, PDA performed a record number of vasectomies on the King's Birthday—a public holiday in Thailand—and petitioned for a slot in the *Guinness Book of Records*. The unofficial guardian of world records unfortunately displayed a very Western shyness with all things sexual and did not reply.

** When the corresponding transition from large to small families occurred in the U.S., contraception and abortion were illegal. It took five times as long for the change to take place.

*** Some people die before they can reproduce, some choose not to, and some are sterile so the average number of children to achieve replacement fertility in an industrialized nation with a low death rate is actually about 2.1.

**** In 1950 the Philippines and Thailand had the same population (nineteen million). In 2000 the Philippines had an average family size of 3.2 and a population of seventy-five million, while Thailand had an average family size of 2.0 and a population of sixty-one million. The big difference, however, will come in the next fifty years, in which time the population of the Philippines is projected to reach between 106 and 150 million, while Thailand's will be only between sixty-two and eighty million.

ulation growth outstrips resources.* The religious leadership endorsed family planning for health reasons, and a High Council on Family Planning was established. The ministry of health allocated money to make contraceptives available. Iran started to manufacture its own oral contraceptives and condoms. Vasectomy and female sterilization were both offered as options at no cost, and they are now quite widely used. Before young couples can marry, they are required to receive family planning instruction. In the decade from 1986 to 1995, average family size fell from 5.6 to 3.3 children, and today family size stands at 2.0, with many provinces below replacement level.[475] The Koran and Islamic teaching can be friendlier to family planning than some Christian traditions. But as the example of Iran shows so well, lowering population growth does not require heavy-handed, top-down programs such as China's One Child policy. In most cases, when the means to control family size are provided, along with correct supporting information, women and most men are glad to use them.

Given all the inflammatory rhetoric we hear about and from the Islamic Republic of Iran, it would be easy to believe that this is a particularly militant society. The truth is much more interesting than that. As we saw in chapter 4, the young fundamentalists who took over the U.S. embassy in Tehran in 1979 have matured into liberal seekers of democracy. Even more importantly, the new generation of young men in Iran has totally rejected terrorism. The contrast between Iran of the 1970s, with its large families and restricted roles for women, and Iran in the opening decade of the twenty-first century, with its two-child families, is impressive. In the 1970s, Iran spawned what may have been the largest number of warrior suicides in the history of the world, when tens of thousands of boys and young men died attacking Iraqi positions or running out to blow up land mines with their bodies. As mentioned in chapter 5, some who died were as young as ten years old. They had been partially brainwashed into seeking a martyr's death, but even so the whole horrible episode could not have occurred without some degree of support from young radicals. In the twenty-first century, Iranians have not volunteered as suicide bombers, whereas each of their neighboring Muslim societies, in which women are overburdened with childbearing and young men cannot find employ-

---

* The impact of population growth on the environment is genuine but complex. Wealthy nations consume a disproportionate amount of timber and raw materials from around the world. Rapid growth in developing countries leads to slash and burn agriculture in forests and over-grazing on desert margins. Poverty drives people to hunt animals such as gorillas in Rwanda for meat.

ment, continues to generate fanatical warriors eager to become martyrs. Iran, like Pakistan, may acquire atomic weapons. We would consider any nuclear proliferation to be deeply unfortunate, but a nuclear Iran might actually be less worrisome than a nuclear Pakistan. It is true that Iranian President Mahmoud Ahmadinejad has shown the dangerous bluster, belligerence, and hatred that we associate with team aggression, and he may indeed be guilty of supporting militants in Iraq. But in reality his power in Iran is quite limited, and demographic changes mean that the emergence of similar demagogues is much less likely in Iran's future. In a generation's time, when today's two-child families are mature, Iran will be an intrinsically more stable and peaceful nation than Pakistan. Women will play a more equal role in civil society and there will be fewer of the cohorts of angry young men without employment or the opportunity to marry who are certain to fill the cities of Pakistan in twenty years' time.

When 9/11 occurred, some Islamic nations rejoiced openly, but Iran was one of the few to express sympathy for the United States, even holding symbolic funeral processions for the Americans killed. The religious leaders still hold power and solemnly discuss whether to permit satellite TV dishes—though every house in every town seems to have one already. When I was in Iran in 2005 I found people exchanging "Mullah jokes"—how many mullahs does it take to screw in a light bulb?* We must remember the country's respectful sympathy for the United States after 9/11, and its remarkable achievement of purposefully reducing family size in order to avoid increased poverty, preserve the environment, and maintain good health. Iran has been leagues ahead of the majority of countries facing similar challenges.

Family planning is wanted by hundreds of millions of people who do not have it. It is relatively easy to provide, and the costs are exceedingly low compared with other aspects of health and development, to say nothing of military expenditures. Nevertheless we are not investing nearly enough of our foreign aid in family planning in comparison with the need. Americans spend more on costumes and candies for children at Halloween than their government spends on family planning worldwide. The subsidy the U.S. paid to bring in one division of Polish soldiers to help with security in Iraq after the 2003 invasion was more than the U.S. spends on contraceptives and condoms for couples in developing countries who want them. The British government's budget for family planning

---

* Answer: Four. One to hold the light bulb and three to turn the ladder.

and reproductive health, which is relatively generous by world standards, is equivalent to the cost of half a pint of beer per citizen per year.

Unfortunately, barriers to family planning take a long time to decay. There is a chronic under-investment in international family planning and over-medicalization is pervasive. Oral contraceptives remain on prescription in Uganda, which means the rich can buy them in any pharmacy (in most developing countries, as foreign travellers often know, prescription laws are not enforced), but the poor cannot afford them. Regulations in Ethiopia limit the use of injectable contraceptives to doctors and nurses, and there are few or no doctors or nurses in the rural areas. In fact, village health workers can be taught to give shots perfectly safely. The battle between the sexes continues to be played out in the arena of contraception and safe abortion in ways that end up denying women the knowledge, devices, and drugs that they need to interrupt continuous childbearing. Even Japan, despite its economic and social sophistication, took forty years to approve the sale of oral contraceptives, but only six months to approve Viagra. Today, 120 million families in poor countries—including those that spawn or sponsor terrorism—cannot get access to the contraceptives and safe abortions they need. In sub-Saharan Africa, over 95 percent of families cannot afford modern contraceptives. Birth rates are rising as a result, not because families want more children, but because they lack realistic alternatives.

Family Planning, Violence, and National Security

Kenya provides a vivid example of the effect of too many unintended pregnancies. In 1970, you could take an evening stroll around Nairobi without concern. Today, the hotels warn guests to be careful when they step outside, and the shopkeepers close at dusk. Virtually every wealthy house is surrounded by a wire fence and has a guard to protect its residents from theft and violence by the unemployed, who have no honest way to earn a living. Kenya's population grew from six million in 1950 to over thirty-five million today. There is no doubt in our minds that the population structure due to this rapid growth—the high proportion of young males to older males—served as tinder for the eruption of anger and violence following the December 2007 presidential elections.

At the beginning of the twentieth century, Margaret Sanger was right to point out that military leaders often demand "a higher and still higher birth rate." Joseph Stalin, Adolf Hitler, Nicolae Ceausescu, and Idi Amin were among the vilest and most cruel dictators of the twentieth century and each one took specific steps to restrict access to family planning and safe abortion. When Stalin came to power in 1929, he began to reverse the previously liberal abortion law, and abortion remained illegal until he died in 1955. When Adolf Hitler was made German Chancellor in 1933, long before he invaded Czechoslovakia or Poland he set about invading German bedrooms by closing family planning clinics. He claimed contraceptives were a "violation of nature" and thought abortion should be "exterminated with a strong hand." Positive pregnancy tests had to be registered and countersigned by the local mayor. Within a few months of the defeat of France in 1940, Marshall Petain and the Nazi-dominated Vichy government ranked abortion with treason and sabotage, punishable by solitary confinement and hard labor for life. In 1943, Madam Marie Louise Giraud, a French laundress convicted of performing twenty-seven abortions, became the last person in history to be executed for providing abortion services. She was guillotined for the crime of helping poor women try to survive the brutal conditions of war and occupation.[476]

When Idi Amin came to power in Uganda in 1971, he first banned miniskirts and then enforced the law against family planning and abortion with new severity. In 1966, the Romanian dictator Ceausescu reversed a previously liberal abortion law, and contraception and voluntary sterilization were made illegal. The birth rate doubled nine months later, but then it began to decline as illegal abortion mills were set up. Soon, abortion-related deaths pushed the maternal mortality rate to the highest in Europe. Childless couples were taxed and doctors were paid only if the women workers in factories they oversaw had an appropriately large quota of babies. The day after Ceausescu was killed in December 1989, the very first act of the new, more democratic government was to take away the hated restrictions on family planning and abortion. The maternal death rate fell rapidly, while birth rate remained level.

For half a century, military analysts have pointed to rapid population growth in certain countries as threatening global security. This line of thinking is totally divorced from the humanitarian desire behind much of the international family planning effort, and quite separate from concern for the impact of population growth on the environment or on economic growth. There are obvious synergies to be had, but people working in the field of international family planning have for the most part either

322   SEX AND WAR

been unaware of the military thinking, or perceived it as potentially coercive—that rich countries might want to stop poor countries from having large families to limit the number of potential enemy recruits. For their part, the military analysts have not yet understood that rapid population growth is a variable open to change through purely voluntary means. The potential for cooperation and mutual benefit here is substantial. The use of contraception, and when appropriate, safe abortion, is basic to the autonomy of women; it contributes to the health of women and their children; it improves access to education and accelerates economic progress; and, as we argue, it helps make societies less violent and the world as a whole a safer place. Family planning provides a key that can open the door to better health, well-being, and security.

American Generals Dwight Eisenhower, Al Haig, and Colin Powell, along with the less well-known but highly influential General William Draper, have all been aware of the need to support international family planning. In 1957, then-President Eisenhower asked General Draper to lead a commission on U.S. international military assistance, and Draper soon became convinced of the importance of slowing rapid population growth.[477] He was trained as an economist, and after World War II he played a pivotal role in the reconstruction of Germany and Japan. As Chairman of the Washington-based Population Crisis Committee from 1965 until his death in 1974, Draper melded a deep humanitarian desire to alleviate human suffering with a clear understanding of the relation between rapid population growth and the inability to achieve economic development, particularly because of the inability to achieve higher per capita income. Draper lobbied the U.S. Congress to support the work of the International Planned Parenthood Federation, and he helped launch the United Nations Population Fund (UNFPA).* Interestingly, General Draper in 1972 was one of the first to articulate that "population *control*" was not the right term to use, because what mattered was slowing population growth on a voluntary basis.

As with everything involving war and terrorism, there is never a single explanation or a one-shot solution. There are simply relationships of varying strengths among many interacting variables. The world's richest countries with low birth rates have also spawned ter-

---

* One of Draper's achievements was to conceive the Berlin Airlift, which circumvented the USSR's blockade of that city after WWII. When I worked with him he was in his mid-seventies, and as a thirty-five–year-old I had trouble keeping up with him.

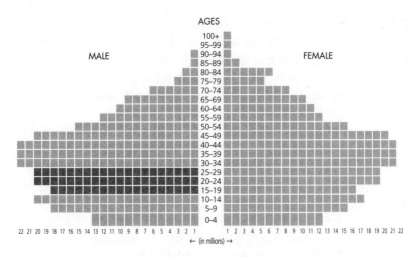

South Korea (2005)

The rapid fall in family size over the past thirty years can be seen in this population pyramid for the Republic of Korea. Currently the country has a large work force supporting relatively few children (the average family size is now 1.5), helping to drive economic growth. The low ratio of men aged fifteen to twenty-nine, in relation to older men and women, facilitates political stability. Today, less than one in ten of the population is over age sixty-five, but this will rise to one in three by 2050.

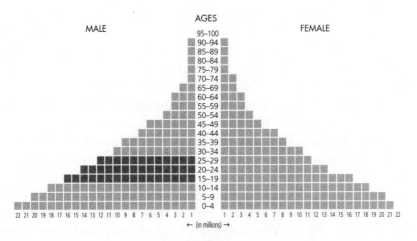

Iraq (2005)

The population of Iraq continues to grow rapidly and the average woman has 4.6 children. The adult work force must support the 40 percent of the population aged fourteen or under. The high ratio of men aged fifteen to twenty-nine in relation to the older population (many of whom are unemployed or under-employed) facilitates civil violence.

rorists. Undoubtedly crackpots and their sects will continue to arise. Every so often they will kill some people. But it is much more likely that the most frequent and destructive attacks will come from zealots outside Western society, from communities where women are generally deprived of family planning choices, the birth rate is high, and unemployment is higher still.

Many political leaders have grasped the importance of offering women family planning. President George W. Bush's grandfather, Prescott Bush, was treasurer of Planned Parenthood in 1947, and he is thought to have lost his first race for the U.S. Senate in 1950 because of this link. As U.S. ambassador to the UN, Prescott's son George H. W. Bush saw family planning as solving the "great questions of peace, prosperity, and individual rights that face the world." He was forced to renounce these earlier beliefs in order to run as Ronald Reagan's vice presidential nominee in 1980, although a shadow of his former views reemerged when he chaired the first U.S. government task force on terrorism in 1986. He noted that, "Fully 60 percent of the Third World population is under 20 years of age [creating] a volatile mixture of youthful aspirations that when coupled with economic and political frustrations help form a large pool of potential terrorists." Sixteen years later the National Academies, in their role as non-partisan scientific advisers to U.S. policy makers, published a report called *Discouraging Terrorism: Some Implications of 9/11.* Its authors used almost the same words to describe the relationship between terrorism and population to Bush's son, President George W. Bush: "With respect to economic and social conditions, many societies that foster terrorism are characterized by high population growth and large numbers of disadvantaged youth and by extreme poverty and inequality."[478]

Neither the task force nor *Discouraging Terrorism* specifically recommended investing in international family planning. But that important link was made as early as 1974, in a secret memorandum of the National Security Council (National Security Study Memorandum, NSSM 200). It concluded that rapid population growth "will weaken unstable governments, often only marginally effective in good times, and open the way to extremist regimes," and threatened "severe damage to world economic, political, and ecological systems and, as these systems begin to fail, to our humanitarian values." The Memorandum noted that high birth rates are generated by absence of "means of fertility control," and recommended a comprehensive effort including enhanced support to international family planning and improved educational opportunities and job creation for women. NSSM 200 was requested by President Nixon, it was signed by Henry Kissinger, and its conclusions were endorsed by President Ford.* There is strong evidence that the Roman Catholic Church lobbied effectively to block implementation, and the Memorandum was not declassified and made public until 1989.[479] Its recommendations were never implemented. If they had been, Osama bin Laden would have had a smaller population of angry young men to incite to violence, and the world today might well be a significantly different place.

The 9/11 Commission Report also is explicit about the link between population and terrorism:

> By the 1990s, high birthrates and declining rates of infant mortality had produced a common problem throughout the Muslim world: a large, steadily increasing population of young men without any reasonable expectation of suitable or steady employment—a sure prescription for social turbulence. Many of these young men, such as the enormous number trained only in religious schools, lack the skills needed by their societies. Far more acquired valuable skills but lived in stagnant economies that could not generate satisfying jobs.... Frustrated in their search for a decent living, unable to benefit from an education often obtained at the cost of great family sacrifice, and blocked from starting families of their own, some of these young men were easy targets for radicalization.[480]

And still we don't talk nearly enough or nearly as frankly as we must about sex and family planning. In part this is a cultural shyness, a leftover from our Puritan and Victorian roots. The political volatility of contraception, abortion, and helping others to limit their family size if they wish also help prevent the 9/11 Commissioners, the CIA, and other agencies and policy makers from taking the next logical step. Western nations have a golden opportunity to fight poverty, protect the environment, and reduce the likelihood of violent conflicts. The only cost is the relatively small one of helping women in countries with high birth rates reduce the burden of their repeated childbearing, but we and our children will pay a heavy price if we fail to act.

---

* President Ford wrote in 1993 that, "Mrs. Ford and I have consistently supported the Pro Choice point of view and will not change. I was very disappointed with the 1992 GOP Platform on this issue."

Church and State

In 2003, Jack Glaser and Frank Sulloway, faculty members of the Gold-man School of Public Policy, Berkeley, together with authors from Stanford and the University of Maryland, analyzed speeches and interviews from almost 23,000 people. They were looking to identify common themes in politically conservative speech and behaviors. In addition to a tendency to support the status quo, the researchers found that conservative leaders tended to accept inequalities, and were often intolerant of ambiguity and uncertainties, or as President Bush put it, "I know what I believe and what I believe is right." Unfortunately for women, poverty, and world peace, part of that belief system seems to include the idea that women must not have control over their own reproduction.

According to Glaser and Sulloway's analysis, the conservative male perspective tends to be territorial, trigger-happy, lacking in empathy for the less fortunate, and seeks to control women. In other words, it is the stance of team aggression, and this is just as true in Afghanistan as it is in America. Just after the fall of the Taliban in 2001, first lady Laura Bush said, "The fight against terrorism is also a fight for the rights and dignity of women." Unfortunately, as with NSSM 200 in the 1970s, very little has been done to turn her words into action. By and large, military hawks continue to oppose family planning and fundamentalist Christians and conservative Roman Catholics, who are theologically far apart on many issues, have united to oppose international family planning programs. The Vatican has used its Observer State status at the United Nations to pressure governments, and particularly the U.S. government, not to make family planning a part of foreign aid. In 2002 Martha attended the United Nations Asia and Pacific Population Conference in Bangkok, Thailand. Of the thirty countries taking part, only the U.S. government—participating on behalf of its Territory of Guam—sought to put constraints on access to family planning for the region's countries, which are home to 66 percent of the world's population. The consensus-driven UN rarely takes votes, but in this case the U.S. was defeated when every delegation voted against the American stance.

Population and family planning have been pushed off the political agenda not only by religious forces but by a complex of additional pressures. At the United Nations Conference on Environment and Development in Rio de Janeiro in 1992, known as the Earth Summit, a powerful bloc of delegates from developing countries, concerned with inequities in international economic power, deliberately weakened attention to population growth. Even more powerful today are the voices of some women's groups, especially in the United States. In 1994, women's groups were successful in drawing attention to the many needs of the forgotten half of the world's population at the United Nations International Conference on Population and Development at Cairo. But in the process, they inadvertently made it politically incorrect to even talk about population growth. For a decade, discussion of population growth was pushed off the international agenda, based on the misguided and unfortunate assumption that discussion of population is inherently detrimental to women because it may lead to coercion. The attention to women's needs was admirable, but the strategic sidelining of attention to population and family planning has, ironically and sadly, slowed efforts to give more women realistic choices about their own childbearing.[481]

Free-market economists make up yet another important force from the right, arguing that population growth is good because growing markets create prosperity. They tend to believe that as young people reach working age they will contribute to the economy, even though the empirical evidence, as we have seen, is that in countries with rapid population growth young people merely join the lines of the unemployed. Many economists also claim that natural resource scarcities can be compensated for by technologies and price adjustments, and they too have been influential in reducing U.S. political interest in international family planning assistance.

In the decade following the 1994 Cairo conference, as the development community's focus drifted away from family planning, international investment practically evaporated.[482] In many parts of the world, rich women continued to find ways to limit family size. But in Bangladesh, northern India, and parts of Sub-Saharan Africa, previous declines in average family size among the poorest women either stalled or actually began to rise again. Slowly, attention is returning to the population factor and, for example, in 2006 expert hearings by the British Parliament concluded that six out of eight of the well-publicized Millennium Development Goals relating to such issues as reducing poverty and increasing education are unlikely to be achieved without slowing rapid population growth through improved access to voluntary family planning. To hold class size constant in a rapidly growing population, the developing world needs to train and deploy 2,000,000 additional teachers each year.[483]

Women Waging Peace

Strategic thinkers who fail to understand how women, reproductive choice, and social stability are connected have overlooked a very important front in the long-term "war on terror." Giving women power over their own lives needs to be at the core of any broad, integrated plan to stop the next generation of terrorists.

We have suggested how and why the inherited predisposition toward team aggression is different in men and women. It is a simple observation, but it opens the door to a whole set of interlinked and promising strategies for reducing conflict, built around the empowerment of women. Two straightforward efforts need to be made:

- Everything possible should be done to ensure that women enjoy legal, economic, political, and social equality with men, including the ability to play an equal role in civil society. The more directly women can express their own, less violently competitive biological agenda, the better placed the next generation will be to make objective decisions about threats to their security. Although there will always be exceptions, on average, women politicians, leaders, and voters are less bellicose than men.
- Women must have access to contraception and safe abortion in order to control whether and when to have a baby. Women cannot be free and they cannot acquire and use education or employment, or take leading roles in civil society, unless they can decide when and when not to be pregnant.

Looking at the connection between demographics and security, three things happen when average family size falls in any country with high fertility today:

- Unemployment falls and economic growth has a chance to accelerate. Although competition for resources will still exist, politically stable countries are more likely to handle competition without conflict.
- The ratio of younger to older men in the population declines and the number and influence of volatile young males is reduced.
- As population growth rates fall, access to education improves, especially for girls.

The predisposition to war is an inherent part of human behavior, but so is the urge for peace. As women became more politically active in the twentieth century they sometimes organized non-violent demonstrations against male militarism. In 1982, 10,000 linked hands to form a human chain around the U.S. Air Force base at Greenham Common, England, protesting the deployment of 96 U.S. cruise missiles at the base.

Credit: Dave Caulkin—AP

All of these changes feed back into one another, setting up a virtuous spiral that reduces the risk of war and terrorism. Just as sex and violence are linked in the behavior of many mammalian males, so in the human species the empowerment of women and the possibility of peace and freedom from terrorism are united in important and genuine ways. Sadly, the political leaders who follow military affairs most closely are often also those who are least interested in the status of women in poor countries, and these same men are sometimes hostile to family planning and the need for safe abortion. They have yet to realize that in taking these positions, they are making the long-term challenge of global security more difficult, not less.

The Pill Is Mightier Than the Sword

Several million years ago, evolution thrust a Faustian bargain on female hominids. In order to cope with the huge asymmetry between the brief

pleasure of male fertilization and the female's far longer burden of pregnancy, breastfeeding, and childcare, evolution, through a series of blind
mutations, tailored a relationship between frequent sex and male participation in child rearing. Unlike in other apes, with their vivid vulval
swellings, the timing of ovulation in human females became concealed,
while at the same time females became both attractive and receptive to
male sexual advances throughout the menstrual cycle. Frequent, largely infertile sex generated a bond of love or necessity between sexual
partners, which helped ensure the survival of offspring. The male, thus
bonded, was more likely to provide resources, shelter, and protection
from aggression by other males.

And so it went until humanity made the transition from a hunter-gatherer way of life to settled living. That shift often ended with women in far
more subservient positions. Protection from aggression by other males often took the form of controlling and sequestering women. In some agricultural societies, women were kept inside the home, cut off from the rest
of society. Elsewhere they were still constrained by numerous pregnancies, years of lactation, and a lifetime of childcare. The more male team
aggression increased, the more women suffered. For millennia, there has
usually been no advantage in access to resources or security for women
buffeted by the tides and currents of male fighting.

Over the last two centuries the idea of legal equality for women has
gained ground. But only in the past half-century have women begun to
exercise genuine control over their reproductive lives, establishing genuine autonomy. This biological and political revolution still has not reached
many parts of the world. Between about 1960 and 1990, the developed
world in a low-key way tried to accelerate the diffusion of laws and attitudes that increase female autonomy, and began to offer women in poor
countries the information and the technologies they needed to manage
how many children to have and when to have them. This progress has
been halted or even reversed, largely as a result of pressure from conservative Christian groups in the U.S. Even though Europe, where religious
beliefs are more nuanced and less powerful, has not gone down the American road, the U.S. policy reversal has had a chilling global effect.

For all its political, cultural, and religious layers, the opposition to empowering women and offering family planning continues to be driven
by the conflict between the fundamental reproductive agendas of men
and women. The perspective of evolutionary psychology reminds us

how deep-seated is the male predisposition for violence and the ability to dehumanize an outgroup. It is an understanding that underscores the dangers of nuclear and biological warfare in a world of diminishing resources and booming populations. But evolutionary psychology also suggests new perspectives on waging peace. It would be an exaggeration to say that a global commitment to encourage female autonomy and make family planning choices available will by itself make the world a peaceful place. And yet, unless women can achieve greater equality and are also enabled to have control over their own childbearing, then efforts to address the many other sources of conflict and terrorism will be much less likely to succeed.

14

STONE AGE BEHAVIORS
IN THE TWENTY-FIRST CENTURY

Regime change in Iraq would bring about a number of benefits for the region. Extremists in the region would have to rethink their strategy of jihad. Moderates throughout the region would take heart, and our ability to advance the Israeli-Palestinian peace process would be enhanced.

—U.S. VICE PRESIDENT DICK CHENEY before the invasion of Iraq[484]

WHY DO NATIONS GO TO WAR? Why are intelligent leaders, commanding virtually unlimited resources, subject to such crass misjudgments as the above quotation reveals? A war that was supposed to pay for itself turned into the worst foreign policy disaster in U.S. history—how could we not ask why? It's easy enough to detect the vestiges of early human evolution in our history, but what about in the present? Unfortunately, as we look at recent examples of wars and acts of terror, we can see that our biological past is with us still. Calling these symptoms of our evolutionary heritage "Stone Age behaviors" is convenient shorthand for this bundle of human male predispositions that adapted us to a past environment but are often highly destructive in the modern world. The signs are all there: mindless hatred of outgroups reinforced by beliefs in the supernatural, loyalty among bands of young men, overreaction to danger, overconfidence during warfare, and petty but deep-seated rivalries and competition among members of the same troop. But it is important to remember that the term "Stone Age behaviors" isn't so much a criticism as it is the recognition of reality. A recurrent theme of this book is that *all* men are subject to Stone Age behaviors, and they are not all bad. These evolutionary hangovers frame our behavior when men and women select their sexual partners, when we rush to defend a child, or when siblings fight for their parent's attention. In a very real way, we could also simply call them "human behavior." Fortunately, men's natural territoriality and aggression are just one aspect of human behavior—but recognizing just how present and pervasive they are in our lives today is an important part of limiting their impact on our future.

History can be read in many ways. One is as an extensive catalogue of violent and often gratuitously cruel conflicts between clans, communities, religions, and political systems. Those who seek it soon find a never-ending sequence of rape and slaughter executed with flint arrows and bronze swords, wooden ballistae and iron cannons, men-of-war built of oak and dreadnaughts constructed of steel, canvas biplanes and supersonic bombers, and perhaps in the future atomic weapons and intercontinental ballistic missiles. There is a significant probability that we in the twenty-first century will see one or more cataclysmic wars, complete with nuclear bombs or some other weapon of mass destruction. But like someone who has just been diagnosed with a serious cancer, we also live with the hope that new insights will stave off catastrophe. As we start to understand why we fight, and what the consequences have been, perhaps we will find the motivation to take serious, informed action to rein in our Stone Age behaviors once and for all.

Science has spawned weapons of mass destruction, but it has also enabled us to examine our evolution and confront the darker side of our behavior, especially that of males. While there has been peace between industrialized nations for sixty years, the predisposition to fight still arises, hydra-headed, in new forms. Since the year 2000, mass killings in the Sudan and fighting in Lebanon, Somalia, Congo, and elsewhere have all underscored in blood the human willingness to hate an outgroup. The price tag on weapons of mass destruction continues to fall, and more nations are seeking to build atomic bombs and develop long-range rockets. For North America, Europe, and Australia, terrorism has emerged as the alternative face of war in the early twenty-first century, with 9/11 and the bombings in Madrid and London demonstrating that human males can always find new ways to express team aggression. Despite the overall decrease in the number of wars and deaths from fighting, the world seems to have become—and perhaps is—an increasingly dangerous place.

Evolutionary insights are commonly dismissed as overly deterministic, in this case that war is "in our genes," and we are thus condemned to perpetual warfare. Or they are seen as opening the door to the worst sort of nineteenth-century social Darwinism, suggesting that we should let the strongest win because that's what evolution "wants."*[485] But our inherited predispositions are not predestination, and our evolutionary past provides neither a moral compass nor a strategic plan for the future. As emphasized earlier, all evolutionary biologists would accept the assertion in the Seville Statement, cited in chapter 2, that, "War is biologically possible, but it is not inevitable." Recognizing the roots of warfare in the coalitionary raids common to male *Pan troglodytes* and *Homo sapiens* is not pessimistic; it is realistic. In order to secure peace we need to understand the nature of war, and if needed we must question accepted beliefs and sincerely held religious faiths. Civilization at its best tames our Stone Age behaviors and like Katharine Hepburn's character in the *African Queen*, quoted in chapter 1, we can be proud to rise above our inherited nature.

9/11

The terror attacks of September 11, 2001, turned a $100 billion U.S. government surplus into a trillion dollar deficit. It led to a new government bureaucracy—the Department of Homeland Security—spending $42 billion a year. It brought the airline industry to its knees and bankrupted two European airlines.[486] It led to a limited invasion of Afghanistan, the overthrow of the Iraqi regime, and increased chaos in both countries. At the time of this writing, the price of oil has tripled and the United States has pushed itself to an economic precipice. Not one of these effects happened because almost 3,000 people were killed and two spectacular buildings among many in one busy city came crashing down. They happened because the reactions of a great democracy

---

* The Nazis are sometimes framed as being driven by social Darwinism. In fact, Nazi mythology emphasized race but rejected evolution. Heinrich Himmler, the head of the SS, believed other races had descended from apes, but that the Aryans had descended directly from heaven and had been preserved in ice since the world began.

and its leaders were driven by Stone Age behaviors rather than rational analysis.

Luck was on Al Qaeda's side on 9/11, and from Osama bin Laden's perspective, the attacks were hugely successful.* The rubble from the World Trade Center Towers has long since been carted away, however, and there has not been a second attack on a similar scale in America. The ability to use aircraft as weapons has been almost eliminated by the simple strategy of strengthening the door leading to the flight deck of airliners.[487] In 2006 the British police thwarted an ambitious plot to blow up ten airplanes over the Atlantic, and 9/11 itself might have been stopped if introop rivalries had not prevented the CIA from sharing information about two of the hijackers with the FBI.**[488] Electronic monitoring is increasingly successful in tracking money transfers and other electronic communications. Ayman al-Zawahiri, Osama bin Laden's chief strategist, and Abu Musab al-Zarqawi, Al Qaeda's leader in Iraq, have both been killed. These effective interventions, it should be noted, did not require a military invasion.

As we saw earlier, our propensity to defend ourselves physically and aggressively, once threatened, is unusually powerful. As noted in chapter 6, when we lived in small clans that often raided one another, if we found one or two of our number dead our survival might depend on a forceful emotional drive to fight back immediately. In the fifth century before Christ, Sun Tzu wrote in The Art of War, "kill one, frighten ten thousand." The first killings in any conflict continue to be well-reported, while the thousands or millions that die later receive much less attention. Part of the impact of 9/11 came from the fact it was so totally unexpected by the public at large, and this probably helps explain why 6 percent of New Yorkers reported symptoms of post-traumatic stress after 9/11.[489] For every individual in Tokyo who was affected by sarin gas during the 1995 Aum Shinrikyo attack, fifty others feared that they were contaminated.

For tens of millions of Americans, the totally unexpected broad-daylight attacks on that clear September morning looked like the beginning of a war. A poll on the fifth anniversary of 9/11 found that only 14 percent

---

* A public opinion poll in Pakistan in 2007 found Osama bin Laden, with 46 percent approval, was more popular than President Musharraf at 38 percent. Both were more popular than President Bush, who met with the approval of just 7 percent of Pakistanis.

** One FBI officer in Minneapolis was berated for worrying over the fact that Islamic extremists were taking flying lessons in America. He retorted that he was only "trying to keep someone from taking a plane and crashing it into the World Trade Center." That was one month before 9/11.

of Americans thought the terror attacks were "not as serious" as the 1941 attack on Pearl Harbor.* In reality, while 9/11 was certainly horrifying, it was an isolated attack by a scattered band of fanatics, not a coordinated effort by a powerful nation to destroy the entire Pacific fleet, as a prelude to all-out war. But there is an important parallel between the two attacks. In December 1941, only days before the Japanese attack, a remarkable 88 percent of Americans asked if the United States should declare war on Germany and Japan said no.[490] After the attack on Pearl Harbor, there were few second thoughts about going to war.

With the American wars in Iraq and Afghanistan, it is easy to forget that in relation to current populations, fewer people are dying from all forms of team aggression than in the past. During five years of fighting in the American Civil War, one in fifty of the total population was killed. Five years of slaughter in World War I killed over one in fifty of the British population and one in every twenty-five Germans. Today, developers are competing to build houses on Civil War battlefields, and war cemeteries in Flanders and Verdun, with their long rows of white head stones, are quiet, sad places. Fewer and fewer visitors know personally the graves of any who were killed. The broken towns have been rebuilt, the trees have regrown, and the trenches have been filled in. In World War II, London, Coventry, Hamburg, Dresden, Hiroshima, and Nagasaki were all seriously damaged or destroyed by bombing. In a world filled with WMDs and Stone Age behaviors, we can't rule out a return to such wholesale slaughter. But it is important to remember that nothing so terrible has happened again in over six decades.

Afghanistan

After 9/11, the Taliban in Afghanistan became a perfect target for revenge and the perceived need to fight back. Nobody liked the Taliban, particularly for their unconscionable treatment of women, although in reality they were not totally unlike previous iterations of Afghan society. But in Osama bin Laden, they harbored a ready-made villain. The political decision to portray the American bombing of Afghanistan as part of a "war on terror" was reinforced by the media and significant public support. But when Al Qaeda members and other fighters were captured,

---

* Interestingly, while 9/11 was perceived as "war" by many, the insurance on the loss of property was not set aside as a war risk exclusion.

the U.S. was unwilling to give them the status of prisoners of war under the Geneva Convention, as this would have denied the U.S. the opportunity to interrogate them. The ill treatment of prisoners undermined much-needed support for the West among moderate Muslim groups.

The military action in Afghanistan was not a war, but a series of skirmishes, mainly fought by mercenaries from the pre-existing Northern Alliance of Afghan clans who had been fighting the Taliban since 1996, paid and supported by the U.S. Several hundred, or possibly several thousand, Taliban soldiers were killed in an intense American aerial bombardment. A battle with casualties of that small number would have been a minor, unrecorded skirmish during the Battle of the Somme or Stalingrad. Most people are surprised to learn that the number of Afghan women who die from pregnancy, childbirth, and unsafe abortion every year—16,000—greatly exceeds the number of Afghan men killed in the 2001 U.S. war on terrorism.

Some of the Al Qaeda leaders who had made Afghanistan their base were chased into the mountains near the Pakistan border, and bin Laden claims that he and 300 Al Qaeda confederates were in trenches spread over a square mile. The U.S. failed to kill or capture bin Laden, however, because of a series of elementary mistakes. First, Secretary of Defense Rumsfeld scrutinized every demand for larger forces in Afghanistan. He failed to understand the need for an adequate force on the ground, "nickel-and-diming," in the words of one commander, every request for troops.[491] Second, there was an overemphasis on high-tech warfare. The all-important air strikes in Afghanistan were controlled from a base in Saudi Arabia, where the Combined Air Operations Center rotated its staff every ninety days; even a chimpanzee could not establish a trusting relationship in so short a time. Third, like two apes competing for the alpha position, the key commanders of Operation Anaconda—Lieutenant General Michael "Buzz" Moseley controlling the air and Lieutenant General Paul Mikolashek on the ground—had an icy relationship. Lastly, the generals underestimated Al Qaeda's willingness and ability to fight. Bin Laden escaped and, as the experienced *Army Times* reporter Sean Naylor wrote, "As ever in combat, it was left to captains and sergeants to bear the consequences of mistakes by generals."[492]

Since the invasion, Afghanistan has had an election and a brief window of opportunity in which some elements of democracy and gender equality might have taken hold. But that hope was thrown to the winds when U.S. forces were redeployed to Iraq, leaving Afghanistan with few-

Much of Afghanistan is a feudal society dominated by competing warlords. Women are largely marginalized, with roles restricted to domestic service, sex, and childbearing. In order to leave home, this woman would have required her husband's permission, and then could only do so while wearing an all-concealing burqa.

The Northern Alliance fighters behind her helped overthrow the Taliban shortly after this photograph was taken in September 2001. One armored personnel carrier (APC) like this one costs over $1,000,000, while building a school in Afghanistan costs perhaps $20,000. More investment in education—especially for girls and women— might make it less necessary to patrol the streets with APCs in a generation's time.

er ground troops to maintain security and prevent the Taliban reemerging. In Kosovo and Bosnia in the 1990s, NATO deployed twenty armed peacekeepers for every 1,000 of the population. In Afghanistan in 2007, there was just one soldier for every 1,000 Afghans and the Taliban had regained control of parts of the country. Some 200,000 hectares of land were devoted to growing poppies to feed the international drug trade and fund the Taliban's resurgence.[493] In the same year, the U.S. spent $7 billion a month in Iraq. Had just a small proportion of that huge sum been available for Afghanistan, we might have helped make Afghanistan into a viable state with established rights and economic opportunities for all its citizens. Instead, Stone Age behaviors of revenge and overconfidence have left us with a broken country, an emboldened enemy, and abundant resentment.

Iraq

The invasion of Iraq in March 2003 was not a good plan poorly executed. It was a disastrous idea implemented in the worst possible ways. The
initial U.S. and British invasion was cleverly carried out by well-trained,
well-armed professionals who (despite some grievous lapses later) behaved with courage and discipline. However, the occupation rapidly
turned into chaos as bands of brothers came together spontaneously to
fight along religious and ethnic lines with weapons culled from Saddam
Hussein's former arsenals. By any reasonable measure, the sum total of
human suffering in Iraq is greater after the invasion than it was before,
even though the country had been ruled by a sadistic tyrant. U.S. soldiers were to end up fighting for longer than they did in World War II or
the Civil War. The younger front-line officers adapted to conditions on
the ground to survive but became increasingly critical of their generals,
who permitted the troops to be drawn into an endless guerilla war.[494]

The policy behind the Iraq invasion was developed by a small group of
would-be alpha males. The first Gulf War in 1991–92 had thrown Saddam Hussein's invasion forces out of Kuwait but had not reduced Hussein's potential to develop WMDs. In 1992 Paul Wolfowitz and Lewis
"Scooter" Libby wrote a Defense Planning Guidance arguing that the U.S.
could and should embark on preemptive strikes against any country, such
as Iraq, that might be developing weapons of mass destruction. Wiser
minds in that first Bush presidency, such as Secretary of State James Baker, rejected the idea. But Cheney, Rumsfeld, and others joined a growing
band of belligerent neoconservatives, or neocons, to formulate the 1997
Project for the New American Century (PNAC).[495] The PNAC's aims included promoting "American hegemony" and what it called "full-spec-
trum dominance"—which means the integration of military power on the
ground, at sea, in the air, and in space. Several neocons had close links
with hardliners in Israel. The PNAC saw a "unipolar" world in which
no country could challenge America's military might, and it advocated
forceful action against "rogue" states, especially Iraq. In typical primate
fashion, the alpha males did not foresee that a decade later their "full
spectrum dominance" would be defeated, and they would be humiliated
by holy warriors who espoused a medieval theology and planted improvised explosive devices in ditches.

Seen through the lens of evolutionary psychology, the war in Iraq is a
dramatic example of men's overconfidence. Dominic Johnson of Prince-

ton University has studied this type of Stone Age behavior experimentally by asking 200 men and women to take part in a computer game. The volunteers were put in charge of a fictitious country involved in a border dispute over a diamond field. Each was given a $100 million war chest and then allowed to negotiate, threaten their adversary, fight, or do nothing. Men were five times as likely as women to launch unprovoked attacks. Those who rated their own self-confidence highly were both more likely to fight and more likely to lose these computer war-games.[496] In Iraq, an overconfidence game was played out not on computer screens but with tanks, mortars, and human lives. The Bush administration grossly overestimated the scale and the immediacy of the threat posed by Iraq, while Saddam Hussein made a calamitous underestimation of America's willingness to use force. For Bush and Hussein both, the overconfidence that had helped our male ancestors conduct and defend against raids in the Stone Age backfired.

Saddam knew perfectly well that he had no weapons of mass destruction, but he kept up the illusion, partly because he thought Israel might attack if they knew the truth. In late 2002, he allowed UN arms inspectors full access "in order not to give President Bush any excuse to start a war," but it was too late and his action was interpreted as subterfuge. Like Hitler, Saddam believed he had "unique abilities." Earlier he had prevented the Iraqi Intelligence Service from analyzing U.S. attitudes because, as he said, "that is my specialty."[497] In 2003, he was convinced that the U.S. would not attack, and when the fighting did begin he honestly believed the U.S. would never reach Baghdad. He ordered his best aircraft literally buried in the sand so that they could be used to maintain the regional balance of power after the U.S. was defeated. Like French commanders in World War I, he believed the élan and fighting spirit of his troops would overcome modern weaponry. Meanwhile, Saddam looked at his elite troops, the Republican Guard, as a threat to his power. Every Republican Guard commander was spied upon and could not "even so much as start a tank without permission." Saddam did not, of course, plan the insurgency that consumed the country after his overthrow, but he did unwittingly set the scene: He ordered ammunition supplies to be dispersed across the country so his army could use them if he needed to confront future rebellions against his own regime—supplies which the Shiite and Sunni militias have been using.

While a moderate degree of overconfidence can be adaptive, the extreme shown by Saddam Hussein would not have survived in a hunter-gather clan where a leader is acknowledged by his fellows but depends on their continued support. In the modern world, Hitler, Stalin, and Hussein held power because they controlled the guns and men needed to kill anyone who challenged them. Bush's power relied on public support and many Americans found his confident swagger reassuring—a Stone Age leader to calm Stone Age fears. The American President's overconfidence was partly based on a born-again Christian's certainty that he was on the right path, guided by God. "I believe," said Bush in his 2004 acceptance speech to the Republican Convention, "freedom is America's gift to the world. It is the Almighty's gift to every man and woman."*[498] Like Hussein he disregarded intelligence reports, paying "little or no attention to prewar assessments by the Central Intelligence Agency that warned of major cultural and political obstacles to stability in postwar Iraq."[499] Just as Saddam sincerely believed the U.S. would not invade Iraq, Bush genuinely assumed American troops would be welcomed as liberators, that democracy would take hold after the war as it had in Germany and Japan after World War II, and that the U.S. would withdraw victorious after a brief occupation.

The Iraq war has gone on longer than World War II, but only one in 170 U.S. citizens had been engaged in active duty by 2007. In proportion to the population, twice as many men were *killed* in the Civil War as have served in Iraq. The troops who serve are all volunteers, and they come disproportionately from minorities and economically disadvantaged communities at a time when America has increasing disparities between the rich and poor. Nevertheless, the American troops are superbly trained and well-equipped, and they have performed admirably. But thanks to Stone Age thinking, they have been thrown into a situation where victory is all but impossible.

The evolved predisposition of a troop of chimpanzees, or our Stone Age ancestors, is to divide all individuals into two groups—ingroup or enemy. As George Kennan, the American diplomat in Moscow, saw in the struggle against communism after World War II, it is wise to exploit fissures in an enemy's ranks. That is why the U.S. built links with Yugoslavia's Marshall Tito when he stood up to Stalin. By contrast, the Bush strategy of "You're either with us or against us" gave America's enemies

---

* Bush used "I believe" twelve times in the speech and made many coded references to the Old Testament and to the Book of Revelation, which fundamentalist Christians clearly understood.

common cause and alienated many moderate Muslims who were truly sympathetic after the attack on the World Trade Center. It almost doesn't matter that Saddam Hussein had no WMDs, or that there was no connection between Iraq and 9/11 and Al Qaeda didn't show up there until well after the invasion. Many Americans, thanks to Stone Age behaviors, were itching for a fight after the 9/11 attacks, and Bush was going to scratch that itch, reality be damned.

Exactly 500 years before Bush plotted his regime change in Iraq, Niccolò Machiavelli was writing *The Prince*. The Renaissance Italian political theorist knew nothing about evolution, but he understood human behavior and aggression better than anyone in the Bush administration, writing,

> It must be realized that there is nothing more difficult to plan, more uncertain of success, or more dangerous to manage than the establishment of a new order of government; for he who introduces it makes enemies of all those who derived advantage from the old order and finds but lukewarm defenders among those who stand to gain from the new one.[500]

Israel and Lebanon

In the 1967 Six Days War, Israel soundly defeated Egypt, Syria, and Jordan. The Israelis occupied the Gaza Strip, the West Bank, and the parts of Jerusalem they had not previously held. One million Palestinians were either displaced to refugee camps in Lebanon, Jordan, and other countries, or they continued to live in the newly occupied territories. With a total fertility rate of between 5.5 to over 7, the number of Palestinians living in the occupied territories has mushroomed to nearly four million, with several million more refugees in surrounding countries. Over a quarter are under age fourteen and one-fifth are fifteen to twenty-three years old, and nearly all of the latter have no secure employment. The Palestinians in Lebanon are not allowed to own property or compete for jobs with the Lebanese. The Lebanese army controls entry to the refugee camps, but fundamentalist groups have free rein inside. From Beirut to Tripoli, the occupied and displaced Palestinian communities are perfect incubators for anger and radicalization. The situation can only deteriorate further as the Palestinian population doubles yet again in the next twenty years.

In 1982, Israel used overwhelming airpower and ground forces to invade Lebanon in an effort to crush the Palestine Liberation Organization. The invasion forced the PLO to move to Tunisia, but hostility toward Israel only grew. Shortly afterward, Hamas emerged in Palestine (partly with Israeli approval as a religious/social organization) and Hezbollah emerged as a radical force in Lebanon. In June 2006, Hamas killed two Israeli soldiers and captured another, and Hezbollah shot rockets from Lebanon into the Israeli city of Haifa. In retaliation, Israel launched hundreds of air strikes against Lebanon. But even the smartest bombs are useless against six-foot rockets launched from an iron frame no bigger than a bicycle by a tiny team of warriors. Of the 3.4 million people in Lebanon, less than one in 1,000 is a Hezbollah warrior, and so predictably, most of the casualties from Israeli bombing raids were civilians. A thirty-day air and ground campaign by Israel into Lebanon ended up uniting the enemies of Israel (and the U.S.) while dividing their allies. It was another case where the lessons of history were overridden by Stone Age behaviors.

There was ample evidence that the 2006 Israeli air campaign would be a tactical failure and backfire politically. Using U.S. built warplanes, and dropping a supply of bombs rushed from the U.S., Israel flew over 1,000 combat sorties in a few days, with the goal of creating "a rift between the Lebanese population and Hezbollah."[501] In fact, and predictably enough, the two became more united—just as a politically uncertain Britain united under German bombardment of London and other cities. The Bush administration explicitly delayed negotiating a cease-fire to give Israel time to achieve its goals, believing the illusion that more bombing would bring a "lasting peace" as opposed to a "quick fix." In fact, as a result of the bombing, peaceful solutions in the Middle East became more difficult.

Once we allow ourselves to dehumanize our enemies, we assume they will not behave like us: We are brave and being bombed unites us; they are cowards and bombing will force them to back down. But loyalties toward the ingroup and willingness to hate an outgroup are just as strong in an enemy as they are in us. The massive use of airpower to crush a militant group which enjoys significant support within a society—as Hezbollah does—is doomed to fail. It merely inflames the passions of those under attack. An opinion poll showed that three-quarters of the people of Gaza endorsed the abduction of the Israeli soldier even though half understood they might suffer from Israeli reprisal.[502] All this may sound self-evident, but our Stone Age behaviors run so deep that in our policies and in our tactics we often make the assumption that the other guys are somehow fundamentally different—that we will go on fighting but they will roll over and play dead. They do not.

Darfur

TV cameras cover violence in the Middle East, but a war of Old Testament ferocity has been going on in Darfur, western Sudan, since 2004, largely unnoticed. Here the goal of Sudanese Arab gun-wielding, horse-riding, *janjaweed* militias is to eliminate the non-Arab population. Up to 300,000 people had been shot, hacked, or starved to death by the end of 2006.[503] Thousands of women have been raped. Raids in Darfur by small bands of men with simple hand-held weapons—even machetes in some cases—are killing more people, both in absolute numbers and as rate-per-thousand of the population, than modern computer-guided weapons systems or supersonic fighters dropping fragmentation bombs.

The fighting in western Sudan is driven by two of the same basic principles as all team aggression: competition for resources, and rapid population growth. The Darfur region is bigger than Britain. The *janjaweed* militiamen are pastoral Arab tribes who have long despised the black African farmers who practice settled agriculture. In a drought-ridden semi-desert the best and most contested land is the 8,000 square mile Jebel Marra region, with its ample water and rich volcanic soils, farmed by the black African population. Musa Hilal, an Arab *janjaweed* tribal leader, says, "Africans have killed Arabs for years over grievances about land and water."[504] As we have seen, it is characteristic of team aggression for those who do the killing to perceive themselves as victims.

There is another conflict over resources playing out in this terrible slaughter. The region has oil reserves and China, which is helping exploit that oil, has supported the Sudan in the UN by blocking action to deploy a realistic number of peacekeepers. As was concluded in chapter 10, a fragile beginning has been made in restraining a few of the worst horrors of war, but there is still a long way to go. Competition over oil among rich nations, and over land and water in an increasing number of areas, can easily swamp the human capacity for empathy.

Upstream Strategies

We now know all too well what a hasty, impassioned, Stone Age response to a terror attack looks like. How would a more measured, effective approach be different? The first step would be to take a dispassionate look at our adversaries, and assess the real sources of danger.

Estimates of the size of Al Qaeda vary from a few thousand to perhaps 18,000. The latter figure is derived from an estimate of 20,000 jihadists thought to have been trained in Afghanistan to fight against the Russians.[505] Most of those trained in Afghan camps, however, had nothing to do with Osama bin Laden. Maybe 4,000 were recruited into Al Qaeda and probably only a few hundred swore an oath to bin Laden, some of whom have now been captured or killed. If we assume the Iraq war will cost at least a $1 trillion, then the U.S. will end up spending between $100 to $250 million per member of Al Qaeda in the "war on terror." Never in the history of human endeavor has so much been spent on fighting so few.

The overreaction to 9/11 and the Stone Age tendency to frame our enemies as the epitome of evil have created a mythical bin Laden of colossal proportions. It is a bias that obscures the much more serious threat of angry young Muslims who despise and hate the West. The invasion of Iraq and the overreaction of Israel in Lebanon have greatly multiplied this pool of men, some of whom are willing to sacrifice their lives in suicidal attacks. Sadly, some second-generation immigrants in Europe—and probably the U.S.—belong to this group.

A more reasoned response to terrorism, and to the ultimately more important threat of war with weapons of mass destruction, is to ask what achievable steps could be taken to reduce the risk of future wars and terrorist attacks. People who work in public health sometimes talk about upstream solutions, solutions that seek to prevent the cause of problems rather than simply treat its symptoms—preventing smoking rather than treating lung cancer, for example. We suggest that we look upstream for opportunities to prevent terrorism and wars. If the rich nations can develop a sensible control of the weapons they manufacture and sell, establish a statesman-like foreign policy, ensure a free press, empower women, and help reduce the conditions which trigger team aggression in young men, then the risk of warfare and incidence of terrorist attacks could both be reduced in a generation's time.

"The Means of Our Own Destruction"

Aesop, the Greek teller of moral fables, wrote, "The haft of the arrow has been feathered with one of the eagle's own plumes. We give our enemies the means of our own destruction." Two and half millennia later his insight remains sadly apt.

Immediately after the Taliban were routed in Afghanistan in 2001, one of my colleagues flew to Kabul to rent an office for an arm of the United Nations. He found unexploded Russian and American ordinance in every room of the house he chose. The modern world is filled with mind-boggling quantities of explosive materials just waiting for zealous terrorists to use. Hundreds of thousands of tons of new explosives are manufactured annually, both for the mining industry and for armaments. Ukraine alone has 2.5 million tons of munitions left over from the Cold War, including six million mines and numerous shoulder-launched missiles. Hundreds of tons of weaponry have been shipped from the former Soviet Union to fuel wars in Africa. It is possible to destroy high-explosive devices safely, but it costs money and at the present rate it will take the former Soviet Union fifty years to dispose of its surplus stocks—long enough to invite a great many people to buy or steal the stuff. The use of part of the $42 billion budget of the Department of Homeland Security to subsidize the destruction of weapons in Russia and Ukraine could be a good way to block the upstream risk of future terrorist attacks.

Like alpha chimps, political leaders often feel compelled to engage in displays of strength. Western nations seem likely to maintain their technical hegemony and to go on developing the best fighter aircraft and most ingenious pilotless robots, the most reliable rockets and the newest, most powerful weapons of mass destruction. Conversely, we can be reasonably certain that Islamic fundamentalists and other extreme groups, by the very nature of their thinking, will not blossom into creative scientists.

If we adopt a Hobbesian view of the world, we should assume that sooner or later every invention that can be, will be turned into a weapon. It follows that civil administrations should be much more vigilant in the control of many everyday materials. Ramzi Yousef, who made the bomb exploded in the first attack on the World Trade Center in 1993, purchased 1,500 pounds of the chemical urea and 130 gallons of nitric acid needed from City Chemicals, Jersey City, New Jersey. He added one hundred pounds of aluminum powder and mixed it all in a garbage can. As noted in chapter 12, biological weapons are also relatively easy to make. To test this, in the late nineties the U.S. Defense and Threat Reduction Agency set up an experimental factory entirely stocked with fermentation vats, Petri dishes, and other equipment purchased on the open market in the U.S. and overseas. They found to their alarm that this low-tech effort had the potential to make sufficient anthrax to fill a series of weapons. As India, Pakistan, and North Korea well know, nuclear fuel from civilian electric power plants can be reprocessed to make nuclear bombs; the lesson of history is not that such fuel *might* be made into weapons, but that it *will* be.

The spread of weapons of mass destruction could be slowed by a more careful, less cynical regulation of the export of Western commercial technology, as well as more thoughtful and prudent sale of actual armaments, from jet fighters to handguns. Unfortunately, Western democratic societies don't seem able to stop exporting their weaponry to less-stable countries, let alone able to control less obviously military materials. The "war" on terrorism is all too often undermined by short-term commercial profits. France, Russia, Britain, and the U.S. compete with one another in selling weapons that are often then used against them a few years later. Between 1989 and 1996 the U.S. sold $117 billion worth of arms.[506] Describing the Western companies that supplied Pakistan with the machinery needed to produce weapons-grade plutonium, A. Q. Kahn said, "They literally begged us to buy their equipment."[507] Israel shipped $50 to $100 million worth of arms to Iran in the first year of the Iran-Iraq War. One in five Jewish manufacturing workers was engaged in making weapons for Iran, the leaders of which in public called Israel "Little Satan."[508]

Even when a technology such as that used in biological warfare is ubiquitous, there is no reason to abandon a common sense level of vigilance. The Germans should have known that if they sold Saddam Hussein the equipment for building a pesticide plant it could also be used to make chemical weapons. The U.S.-based non-profit repository of bacteria and other biological cells, the American Type Culture Collection, should have guessed when they shipped lethal bacterial strains to the Iraqi Ministry of Higher Education that their bugs might be passed on to weapons manufacturers. A more focused upstream effort is needed to control the key technologies related to weapons development. It is a responsibility that all rich nations must accept, and none should be permitted to break rank. France has a particularly poor record here, although the U.S. and U.K. are not far behind.

Diplomacy

Diplomacy and treaties between competing groups arose with the use of language and were greatly reinforced by the invention of writing. They have no parallel in chimpanzee life. Diplomacy applies ingroup rules to potentially hostile outgroups, and it penetrates the barriers that originally evolved to keep competing groups apart. The human code of honor, which we looked at in chapter 3, can be so strong that national leaders will pay a heavy price to honor a treaty. Britain went to war in 1939 because it had signed a treaty with Poland. When Japan attacked Pearl Harbor and declared war on America, Hitler honored his treaty with Japan and declared war on America as well, even though he must have seen it made the possibility of German defeat more likely.

The Cuban missile crisis led to a ban on atmospheric nuclear tests and to the 1968 Treaty on the Non-Proliferation of Nuclear Weapons (NPT). Almost two hundred sovereign states have now ratified the treaty, but India, Pakistan, and Israel have refused to sign, and North Korea has withdrawn. Article IV of the treaty asserts every nation has an "inalienable right [to] develop, research, produce and use nuclear energy for peaceful purposes." The trouble is peaceful nuclear power generation overlaps with making bombs. At the same time, some progress has been made. In September 2005 at Warren Air Force Base, Wyoming, the last MX missile, with its ten nuclear warheads, was decommissioned.[509] These powerful missiles had been ordered by President Carter and deployed by President Reagan in the 1980s amidst considerable controversy. The U.S. had 9,680 nuclear bombs at the end of the Cold War, and it has 5,235 today. [510] There is no national security reason, other than missile envy, why that number should not be brought down to a thousand or less. For the past forty years the NPT has mandated good faith negotiations on "general and complete disarmament under strict and effective international control," but for a species which has evolved sophisticated ways of dissembling, it is probably whistling in the wind to imagine that further proliferation can be prevented.*

---

* The only nation to have developed nuclear weapons and then totally destroyed its supply of warheads has been South Africa; it did so before joining the Non-Proliferation Treaty in 1991.

The Media

When I was in a provincial city in Afghanistan in 2002, drinking water came from a leather bucket lowered into open wells along the main street. There were no sewers or central electric supply, but when our small group's visit was on the local TV news, we were greeted the next day by strangers speaking Farsi, "Saw you on television last night, Sir."

Today, people increasingly get their information not from their neighbors, their school, or church, but through broadcast media and the Internet. In the modern world, the way a war is reported can determine its outcome. The publication of grainy still images of prisoner abuse at Abu Ghraib taken by U.S. soldiers themselves marked a turning point in attitudes toward the Iraq war in America and across the Islamic world. Ideally, a free media should make us more thoughtful about wars and terrorism, but that is not always the case. At the height of the First World War, U.S. Senator Hiram Jackson said, "The first casualty when war comes is truth." The heavy guns on the Western Front could sometimes be heard across the Channel in England, but the horror of life in the trenches was kept from the public in an effort to maintain support for the war at home. In the Second World War, Germany and Russia ran huge propaganda machines. The Japanese government imprisoned anyone challenging the *kokutai*, or "Imperial System."* In Britain, the Emergency Powers (Defence) Act passed at the outbreak of World War II gave government censors almost unlimited powers.**[511] In June 1942, when America achieved a stunning victory at Midway by sinking four Japanese aircraft carriers with the loss of only one of their own, even this victorious news was inexplicably kept secret.

But widespread censorship, especially in the West, has become all but impossible. During the Vietnam War the U.S. military were required to transport journalists to the front, and for the first time, images of war came straight to the TV screen. Granted, those being killed were three-inch-high images, interrupted by soap and cigarette advertisements. Yet the steady beat of the nightly news told everyone that American soldiers

---

* Unlike Nazi Germany, where thousands died for criticizing the government, the Japanese authorities believed they could mold critics into cooperative citizens, and only one person was executed.

** For example, during the critical evacuation of over 300,000 British soldiers from Dunkirk, after their defeat by the Germans in 1940, there was no British war correspondent assigned to report on the evacuation, only journalists piecing together a story as the bedraggled troops landed back in England.

were dying and Vietnamese peasants in conical straw hats were being incinerated by napalm. Media coverage of the ground war in Vietnam forced President Nixon to turn to air attacks, especially in Cambodia, where there were no TV reporters to cover the falling bombs.*

In the 1991 First Gulf War, 147 Americans died and 10,000 to 20,000 Iraqis were killed. TV images of the "Highway of Death," with Coalition air power gunning down Iraqis retreating from Kuwait, had an immediate effect on policy, including then-President Bush's declaration of cease-fire just one hundred hours after the war began. The 2003 invasion of Iraq received even more intensive coverage, including by journalists "embedded" in forward units. Video cameras caught tiny slices of the fighting in color and TV audiences were regaled with real-time images of vehicles in clouds of dust driving past dead camels rather than thoughtful commentary on the goals and progress of the war. TV cameras avoided close-up images of bloody and dismembered corpses and, as in most wars, euphemisms were coined for killing, such as "attrition" or "degrading" forces.

War-reporting in the future is likely to become yet more instantaneous and difficult for governments to censor. Not only has public demand for unfettered, reality-based reporting increased, but communications technology is making potential journalists of us all. Video cameras as small as golf balls can transmit real-time images, and amateur blogs can spread them to millions of viewers. The new generation of cell phones, PDAs, and satellite phones can send and receive pictures and information practically anywhere. A TV crew in Afghanistan in 2001 heard a U.S. marine phoning his wife telling her to record the TV news that day, as his unit would be on it. During the Iraq insurgency the one booming business has been the sale of mobile phones. The limitation of the media is that, however extensive the coverage, it will continue to be interpreted by brains evolved in the Stone Age. For example, President Bush is sincere when he speaks of a "war on terror" and the "decisive ideological struggle of the twenty-first century," or dubs terrorists the "successors to Fascists, to Nazis, to Communists and other totalitarians of the twentieth century."[512] Such phrases receive support not because they represent a valid analysis of the world since the year 2000, but because they play well with an audience viewing

---

* Even though the U.S. professed to recognize Cambodian neutrality, one million tons of bombs were dropped on that country. For comparison, during World War II the Germans dropped 80,000 tons of bombs on Britain.

aggression through brains evolved in a time when human beings lived in clans of a few hundred people and team aggression was endemic. But as we have seen in the examples above, clear, unbiased information remains a much-needed antidote to these maladaptive interpretations, whether intended to deceive or arising merely as a byproduct of hominid evolution.

Education

A free media should be able to help tame the human predisposition for violence, but only in the context of a reasonably educated body of viewers, readers, listeners, and text-messagers. Human behavior is a rich mixture of nature and nurture. We have focused a great deal on the "nature" part of that equation, the inherited predispositions honed by millennia of competitive evolution. But these impulses interact constantly with a changing set of cultural standards transmitted from generation to generation. Chimpanzees in different parts of Africa show incipient "cultural" differences, but humanity has uniquely triumphed over the mortality of each generation through the persistence and perpetual growth of cultural information accumulated over time. In sexual reproduction, we share our DNA with one other person. But the thread of knowledge and beliefs that we help transmit from the past into the future is woven by all members of our species.

Possibly the most important cultural influence in our lives is the level of emotional investment our parents, and most especially our mothers, make as we grow up. This investment in each child is rarely as great in a large family as it is in a small one. What a child learns at its mother's knee, in school, and in daily life can set a child's tolerance of and propensity for violence high or low, and adverse childhood experiences can have life-long effects. William the Conqueror was an illegitimate child, orphaned at eight, who saw one of his guardians murdered in his own bedchamber when he was ten years old. Is this part of why he laid waste the north of England, even destroying farm implements so those who were not massacred starved to death? Did Hitler lack empathy for his victims because his father literally beat him senseless when he was a child?

The first generations of universal, free, compulsory education in Europe produced the patriotic young men who fought to the death in 1914–1918. In the contemporary Western world, young people are more likely to question the use of force in settling international disputes. In the 1990s researchers at the University of Texas sent a self-administered question-naire to several hundred adolescents in the U.S., Russia, Finland, Esto-nia, and Romania. They found that American students were about twice as likely as the European students to agree with statements such as "War is necessary to settle differences between countries" and "A person has the right to kill to defend his or her property." Are American youths more trigger-happy because, unlike their European counterparts, their parents and grandparents have not known war on their soil? In the above survey the girls' opinions tracked in the same direction as the boys', but their re-sponses were consistently and sometimes markedly less aggressive. Only 10 percent of European girls agreed it is right to kill to defend property. In trying to develop an educational strategy to reduce the possibility of war, perhaps we can build on sex differences with some assurance, and expect that the growing political voice of women is likely to help deter militaris-tic adventures.

A child brought up in a kinder, gentler world may mature into a less warlike adult, but what then happens if his nation is threatened by a so-ciety that glorifies war? One reason that the Aztecs produced such fear-less warriors is that the boys were educated to tolerate pain and support violence. Children were whipped with thorny canes. Those who were re-calcitrant were held over the vapors from hot peppers, or tied naked to stakes. Adolescent pupils were expected to mutilate themselves by driving agave thorns into their legs or ears. While not so extreme, the 10,000 *ma-drassa* religious schools in Pakistan teach a strident fundamentalism. At a minimum, they take away a young boy's ability to learn other things as he spends several years doing nothing but memorizing the Koran.* In other settings, the *madrassas* are explicit recruiting grounds for extremists. As government schools fail to keep pace with rapid population growth, *ma-drassas* financed by oil-rich Islamic countries are providing an alternative education for the poor in several Islamic countries. An example of an up-stream intervention likely to make the world a safer place would be for the West to subsidize elementary education in countries such as Pakistan. They could do so for a fraction of the military budget, and thereby reduce the chance of terrorist attacks in ten to fifteen years. Given a choice, many parents would probably prefer not to send their sons to religious schools

---

* It is held that the true message of the Koran can be understood only in Arabic, so in a country like Bangladesh a child will memorize something he does not even understand, just a set of strange noises.

and, equally important, girls could begin to reap the benefits of education and participate more fully in their societies as adults.

The Haves and Have-Nots

The world is very clearly divided into "haves" and "have-nots." About one billion people live on less than one dollar a day, while the richest 20 percent of the global population earn three-quarters of the global income.[513] In 2007, the two wealthiest individuals in the world had more money than the combined GDP of the 45 poorest countries.[514] The terrorists who have attacked the West, and who no doubt will do so again, act largely because they live in a world of frustrated dreams. Their fundamentalist interpretations of religion and radical political philosophies offer an illusion of hope. Any of us might be attracted to such a perspective if we lived in the same environment.

It is difficult, or perhaps even impossible, for a country to lift its people out of poverty when the population grows at 2 or 3 percent per annum. Prior to the 1994 genocide in Rwanda, the average woman there had 8 children and the population had grown from two to seven million in forty years; Rwanda was the most densely populated country in Africa, with well over 800 people per square mile.* Philip Gourevitch, in his moving account of the Rwandan genocide, points out that the killers were young men, "motivated by the opportunity to drink, live to murder, and enjoy higher living standards than they were previously accustomed to."[515]

The nightmare for the high-fertility countries of Africa, Pakistan, northern India, Indonesia, the Philippines, Afghanistan, Bangladesh, and parts of the Middle East is the waste of human potential. Initiative, hard work, and even genius are rarely rewarded with advancement, it becomes more difficult to educate your children as the years go by, and life seems to get worse rather than better. Mohammad Atta, who led the attack on the World Trade Center, witnessed his family's way of life slipping backwards because of Egypt's overpopulation, unemployment, and poor government. The unemployed and underemployed men—the "men who lean against walls"—are in the majority in Egypt and throughout the Middle East. The average man in formal employment in the Gaza Strip has eigh-

* There is still no realistic emphasis on family planning in Rwanda, and the estimate for population in 2050 is between eighteen and twenty-one million.

Credit: David Guttenfelder—AP

How many people live in a city? It depends where you draw the boundary and if any census is accurate. There may be 10 or 15 million people in Lagos, Nigeria. The crowds and the traffic jams there are notorious, and the city's most densely populated areas have over 50,000 residents per square kilometer. The city grows by 600,000 more people each year and by 2015 it will be among the most populous urban centers in the world.

teen family members and other individuals dependent on him. It is a recipe for frustration, extremism, and violence.

Possibly, we "haves" will once again fight one another, but we have learned that modern weaponry could lead to our own destruction. In a sense, we "have" too much to want to risk losing it in Stone Age battles. The global poor have less to lose, and are likely to keep fighting other "have-nots" in a desperate struggle to gain a little more. An important group of have-nots already hate the haves, and justify their contempt by judging the wealthy as godless hedonists, deserving both divine and earthly punishment. Implementing fair trade practices, encouraging sustainable local development, and helping women to access the family planning they want and need would all help to lower this resentment, by narrowing the gap between wealthy and poor.

Sense of Scale

Our Stone Age brains are not evolved to comprehend the true dimension of the dangers now besetting us. Primates have lived in small social groups for forty million years. Human beings evolved in small, tightly knit social groups, and we remain intensely interested in intimate human stories while finding it difficult to assess events that play out on a global scale involving millions or billions of people. We need to defend ourselves when we are attacked, but also we need to be able to assess the risks and make appropriate investments to secure our safety for the future. We tend to overestimate risks we don't easily understand, such as peaceful nuclear energy generation, while underestimating the risks of those things we control, such as driving a car.

The modern media fulfills the primate need to know about other individuals, but it lacks a sense of scale. TV and newspapers feed Stone Age perspectives by, for example, headlining individual courtroom dramas of little significance to anyone not personally involved, while bypassing other stories that are more likely to affect our lives. Since 2001, deaths from terrorist attacks in all regions of the world[516] have been a vanishingly small part of all deaths across the globe. According to the Rand Corporation, 16,075 people died in terrorist attacks worldwide between 9/11 and 2006, with the largest number (5,408) in Iraq. By contrast, over 1,000 deaths occur each day attributable to smoking in the U.S. alone.[517]

We perceive large changes, taking place over decades, as less important than a few people being killed at the same spot and time. For example, while the newspapers will headline the unexpected death of twenty people, global population increases by a staggering 200,000 more births than deaths every day. This fact, and its fundamental role in everything from environmental destruction to the 2007 election violence in Kenya, never even makes the newspapers. In the case of warfare and terrorism, this lack of scale is paired with an overreaction to perceived threats and a tendency to paint any enemy as universally bad. TV reporting can make a skirmish appear as if it is a genuine battle. During the fighting in Afghanistan in 2001, the UN flew in 800 journalists to Kabul. In some cases TV cameras from a dozen nations repeatedly transmitted images of the same tank, greatly amplifying the apparent scale of the conflict. Possibly there were almost as many journalists as there were Al Qaeda fighters. The Afghan invasion looked like a real war but wasn't.

Our lack of a sense of scale and willingness to invest disproportionately when we perceive ourselves to be threatened are so strong that we fail to see how American concern with "defense" throws the entire world out of kilter. Dwight Eisenhower, who knew the full horror of large-scale war, ended his presidency in 1961 by warning of the military industrial complex, "We must not fail to comprehend its grave implications." Military expenditures are more often driven by Stone Age analyses than calm reflection. Eisenhower is said to have muttered in the Oval Office as he ended his presidency, "God help this country when someone sits at this desk who doesn't know as much about the military as I do."*[518]

"My God Was Bigger Than His"

Lt. General William "Jerry" Boykin, Deputy Undersecretary of Defense for intelligence in the second Bush administration, is an evangelical Christian. He was involved in the failed effort to rescue U.S. hostages in Iran in 1980, commander in Somalia during the 1993 failed U.S. attack on Somali warlords, in charge of the failed hunt for Osama bin Laden in Afghanistan in 2002, and played a role in the Abu Ghraib prison debacle. Reflecting on the fighting in Somalia, Boykin told a religious au-

---

* Defense Secretary Rumsfeld flew as a naval pilot 1954 to 1957. President Bush served briefly in the Air National Guard. Vice President Cheney was of age to be drafted for Vietnam but sought multiple deferments as a student, married, and had a child exactly nine months after the draft extended to childless married men. In his own words, "I had other priorities in the '60s than military service."

dience, "I knew my God was bigger than his. I knew that my God was a real God and his was an idol." [519]

In a successful struggle to overcome alcoholism and with the help of American evangelist Billy Graham, George W. Bush became a "born-again" Christian, or as he puts it, "Our priority is our faith."[520] The younger Bush helped his father George H. W. Bush win the 1988 election by mobilizing the religious right, and in 2000 said he believed "God wanted him to run for president." Bush is reported to have told religious leaders, "I trust God speaks through me." When asked if he discussed issues with his father, the younger Bush remarked, "You know it is the wrong father to appeal to in terms of strength. There is a higher father I appeal to." During the run up to the Second Gulf War, he expressed his belief that "Events are not moved by blind change and chance [but] by the hand of a just and faithful God."* After 9/11, Bush called the war on terror, "This crusade." He comforted a parent whose child was killed in Iraq with the words, "If you truly believe the Scriptures, you will see your son again."[521]

The increasing influence of religion on political affairs in the new millennium is a vivid symbol of the triumph of Stone Age behaviors over scientific rationalism. As I said earlier, I went to an English public school where every day began with a Christian service—the several Jews in the school got on with their homework and joined the rest of us later. I still appreciate the ritual of the Anglican denomination and its flexible beliefs, which in my experience means you don't have to leave your brain at the church door. I value my Jewish, Buddhist, Parsee, agnostic, Hindu, and Muslim friends. When 9/11 occurred I was moved by the many e-mails I received from Muslim friends and acquaintances around the world expressing sorrow and sympathy. I understand the organization and appeal of evangelical groups and I have read every word of the Bible—in fact, twice. I couldn't help noting the numerous inconsistencies, for example in relation to the biological or miraculous conception of Jesus,** and I couldn't feel more strongly that a literal interpretation of the Bible is not just wrong, but dangerous in a real and pressing way.

I have quoted the harshness of the Old Testament several times in this

---

* The Israeli newspaper *Haaretz* reported that Bush told the Palestinian leader Mahmoud Abbas that God had told him to go to war in Afghanistan and Iraq, although the White House later denied this account.

** Two of the gospels tell of Jesus' virgin birth but Saint Paul is explicit that he was, in the beautiful words in the King James Bible, "born of the seed of David according to the flesh" (Romans 1:3).

book and I will add one more text. Unless one is prepared to assert that black is white, then the Old Testament is perfectly explicit in commanding the Israelites to commit genocide:

> Of these people which the Lord your God gives you as an inheritance, you shall let nothing that breathes remain alive, but you shall utterly destroy them; the Hittites and the Amorites and the Canaanites and the Perizzites and Hivites and the Jebusites, as the Lord has commanded you.[522]

This is the command of a warlike god indeed, and one that may well have helped the early Israelites vanquish their neighbors.

To a greater or lesser extent, all of the major religions are patriarchal, and all have inspired aggression, war, and terror. It was men, after all, who wrote the holy books. At their most extreme, religions condone the torture and execution of those who interpret a faith in a different way, or they inspire martyrdom to counter threats to true believers. The violence, of course, is always done to protect the immortal souls of those who have seen the truth. The problem is there is no objective way to determine whose "truth" should prevail.*

The antidote to religious fundamentalism is religious tolerance. In the sixteenth century, Europe tore itself apart in conflicts that turned on the assertions that Protestants and Catholics made about the path to salvation. Eventually, in 1648, the Treaty of Westphalia ended religious despotism and the terror of the Thirty Years' War by accepting that one nation could not impose its faith on another by force. It helped usher in the modern era of separation of church and state and tolerance for various beliefs. Modern science and technology blossomed most in the least theocratic countries, where parliamentary democracy, universal education, freedom of expression, equal opportunity for women, and access to family planning were to take root also.

Some nineteenth-century English thinkers saw the world as a ladder of increasingly sophisticated cultures, ascending step by step from savagery to a world of Victorian rectitude and high civilization. Some twentieth-century anthropologists tried to dismantle the ladder and proclaim each rung to be of equal validity. Certainly, nineteenth-century European imperialism had an inflated opinion of itself, and the evil cruelties perpetrat-

---

* Even the Aryan Nation's Church of Jesus Christ Christian claims to be "divinely ordained" and its members perceive they belong to a loving family even while preaching racial hatred and claiming Hitler should be regarded as an Old Testament prophet.

ed in its name, for example in King Leopold's Belgian Congo, are among the most terrible in history. Still, one could argue that societies built on religious tolerance and scientific honesty are in fact better than others, not only in a material sense but in the "pursuit of happiness" of its members. Modern secular society has been demonstrably more successful than any preceding civilization in bringing satisfaction to unprecedented numbers of people. Science, with its intrinsic honesty and its rejection of the supernatural as an answer to real world events, has proved the only medium in history capable of linking women and men of all cultures and all races in a common understanding of the real world.

To a large degree, the world has squared off into a handful of major organized religions. In 1900, half of the 2.5 billion people in the world were Christian, Muslim, or Hindu. As a result of differential population growth, religious evangelism, and cultural assimilation, by the end of the twentieth century a full 64 percent of six billion belonged to these same religious groups.[523] By 2025 it could be seven out of ten of a projected eight billion people. These numbers omit Buddhists, Confucians, Jews, Parsees, and myriad other beliefs, of course, but the point is that religions are particularly strong delineators of ingroup versus outgroups, and the lines seem to be hardening. As Timothy Shah and Monica Toft have written in *Foreign Policy*, "when people are given a choice between the sacred and the secular, faith prevails." As we suggested in chapter 8, religions probably arose and persist because they unite large groups of people, and when it comes to fighting, the larger group usually wins.

The one thing all religions agree on is that there is no objective way of unraveling religious truth. For the true believer, faith is a virtue, even when he or she must abandon reason and accept uncritically a religious story dating, in the case of the world's most popular religions, from a time when people thought the world was flat, diseases were a punishment for sin, and angels but not people could fly.* Sam Harris, in *The End of Faith*, asks his readers to "Imagine a future in which millions of our descendants murder each other over rival interpretations of Star Wars or Windows 98. Could anything—*anything*—be more ridiculous? And yet, this would be no more ridiculous than the world we are living in."[524] Contemplating the fact that more Americans believe in extraterrestrial visitors than in biological evolution, it is hard not to agree with his point.

---

* The Danish Protestant philosopher Søren Kierkegaard (1813–1855) argued that genuine faith was strengthened by believing in the absurd.

The world's great religions all teach love for our neighbors, but they also preach different routes to paradise, and promote the chosen over non-believers. Since the year 2000, 43 percent of lethal conflicts inside countries have been inspired or intensified by religion.[525] The list is seemingly endless and deeply depressing. Jews and Muslims in the Near East fire rockets at one another; Muslims from Azerbaijan and Christians from Armenia shoot each other; Singhalese Buddhists and Tamil Hindus kill one another in Sri Lanka; Orthodox Russians and Muslims in Chechnya blow up whole cities; Muslims and Christians in Ethiopia and Eritrea spent years in trench warfare; Hindus and Muslims in Kashmir have fought since 1949; Orthodox Christians and Catholics in the former Yugoslavia slaughtered one another from 1992 to 1995; and in 2006 Sunni and Shiite Muslims in Iraq were killing a thousand people a week.

And yet, there is hope. As E. O. Wilson predicted thirty years ago, evolutionary psychology is beginning to provide important new insights into ethics and religion.[526] There is increasing scientific recognition that the building blocks of human morality are emotional as well as cognitive. We are born with a moral intuition about fairness and not harming others. It is an instinctual response that is rapid and automatic, and which we seem to share with other primates. People do not search for confirmatory evidence for such intuitive responses, but they do draw support from any source that backs up what they already believe to be true. "Moral reasoning," it has been suggested, "is like the press secretary for a secretive administration—constantly generating the most persuasive arguments it can muster for policies whose true origins and goals are unknown."[527] It is not always easy for our more deliberative, cognitive responses to override these deep intuitive responses.

There is, however, an additional insight emerging, which is directly relevant to the origins of warfare. As we have seen, group solidarity, commonly reinforced by religious beliefs, makes for success on the battlefield. Human morality (and male chimpanzee behavior) involves shared behavioral norms which reinforce cooperation within a group and punish violations of these social norms. Such behavior is an extension of reciprocal altruism, sometimes called "indirect reciprocity," in which individuals read and respond to one another's reputations.[528] Free-riders are shunned while trustworthy individuals are rewarded by being incorporated in joint endeavors. Cultural practices eliciting cooperation and genetic traits encouraging indirect reciprocity can co-evolve. As psychologist Jonathan Haidt of the University of Virginia has written, "pre-agricultural human groups may have engaged in warfare often enough that group selection altered the gene frequencies as well as cultural practices. Modified genes for extreme group solidarity during conflict may have evolved in tandem with cultural practices that led to greater success in war."[529] The computer modeling of altruism and hostility carried out by Choi and Bowles, which we mentioned in chapter 11, demonstrates how such behavior could have evolved.

It is a somber but not hopeless diagnosis: Violence and aggression, empathy and love are all products of the human brain. In fact with the new technology of functional magnetic resonance, my colleagues in the Helen Wills Neuroscience Institute at Berkeley can ask someone a question or show them a picture and then see in what areas of their brain the blood flow increases as the subject thinks about a problem. For example, when people think about "fairness," specific frontal brain structures light up in the images.[530] Most people, however, still think of moral sentiments and religious convictions as transcendental things that come from outside of us, either reflecting some eternal truth, emanating from a supernatural power, or as instructions from a God who created us and who will reward or punish us according to how we restrain aggression or enhance empathy. History shows that this understanding of morality has not worked terribly well as a means to ending war; taking a biological perspective has a much better chance of success. Our survival as a species will not depend on divine intervention but on understanding our Stone Age behaviors. Once we do that, controlling them should become an achievable goal.

Connecting the Dots

At one level, the terrorist attacks, wars, and violence that have scarred the opening decade of the new millennium are sad and disillusioning. At another level, they are par for the historical course. What is easily overlooked is that more of the world is at peace now than at any other time in history. For billions of people, the chance of dying violently has never been lower, a higher proportion of children have access to secondary education than ever before, and an unprecedented rise in global wealth is lifting millions out of poverty.*[531]

---

* Between 1990 and 2005 GDP per person doubled in Thailand, tripled in Vietnam, and increased by a factor of five in China, while the percentage of people living on a dollar a day plummeted. Enrollment in secondary schools went from 30 to 50 percent in 1991 to 70 to 90 percent today.

Unfortunately, a number of our outdated Stone Age behaviors are making it difficult to build a safer, more secure, and more equitable world. In the U.S., the media's intuitive ability to portray news as a series of "human" stories divorced from a sense of scale came together with an administration with an unusually high proportion of overconfident alpha males. Add to this mix a tendency to overreact to the first killings in any conflict, and policy mistakes become not only possible, but likely. The decision to frame the response to a lucky (from the perpetrators' perspective) attack by a miniscule group of terrorists as a "war on terror" was just such a mistake, and a costly one. Instead of focusing on upstream factors that might prevent militant young men from hating the West, the decision to invade Iraq created an incubator for a vast new generation of terrorists. It has undermined much needed support from moderate Muslims and our global allies, and made the limitations of U.S. strategic power patently obvious to our global enemies. The invasion of Iraq has made Iran, which clearly intends to develop nuclear weapons, much more influential in the Middle East and left America unable to focus on or believably intimidate North Korea, a country which, by any standard, is an ugly dictatorship.

In an increasingly dangerous world we need to analyze future scenarios as objectively as possible, using the best available evidence. But we are becoming more religious, not less so. In democracies, religious fundamentalists are influencing policy and in the Muslim world fundamentalists are filling shortfalls in education. Fundamentalist teachings, whether Christian, Muslim, or any other religion, end up restricting and controlling women, which in turn makes wars and terrorism more likely. The twenty-first century is seeing a clash of cultures, but that clash is not between Islam and Christendom. Rather, it is between fundamentalism and reason. Reason's child, science, may entice us with new ways of killing one another, but it also teaches us to rein in our most aggressive tendencies.

15

Civilization at Its Best

We must be ready for peace when it comes.

—JANE GOODALL, 2000[532]

JANE GOODALL NEARLY DIED in a light airplane crash in Tanzania, suffered the trauma of having her students kidnapped and ransomed by terrorists from the Congo, and happened to be in New York on 9/11. She was also the first scientist to witness and understand the nature of team aggression in chimpanzees, and it is the time of these initial discoveries that she describes as "some of the most intellectually and emotionally challenging years of my life." Goodall's early belief that chimpanzees were "rather nicer than human beings" was replaced by the realization "that they, like us, had a dark side to their nature."[533] Perhaps that understanding makes her 2002 appointment as a United Nations Messenger of Peace all the more appropriate.

Whether we believe in God or in some spiritual essence, or are agnostic or resolutely atheist, all humans know how to walk the road of peace. The desire to bring an end to combat and strife has been particularly powerful during times of great belligerency. World War I was envisioned as "the war to end all wars," and in 1940, as Britain faced the real possibility of defeat, Churchill spoke poetically of a future where "the life of the world may move forward into broad, sunlit uplands."[534] Unfortunately, when we reach the crest of the hill we all too often face yet another valley of destruction.

The problem, as we have seen, is not that we do not know how to end wars or create a peaceful world where we enjoy those broad sunlit uplands. That—to use a military metaphor—is as plain as a pikestaff. Each year since the turn of the millennium the world has spent $800 billion on the military and $80 billion on educating the next generation. Surely if we reversed these expenditures, it would be a major first step along the path of peace. The question is not whether such a policy would make the world more peaceful, but rather, why won't we do it? Why is peace so universally desired and yet so agonizingly difficult to achieve?

If the arguments set out earlier are valid, then the male predisposition for team aggression has been with us for at least five to seven million years of hominid evolution. Team aggression is at the very marrow of our Stone Age behaviors, and for a range of explicable reasons, it is not going to go away just as a result of wishful thinking. Team aggression and killing members of an outgroup was a relatively low-risk way for the males who evolved the behavior to increase their access to territory and resources, and those who exhibited this behavior were more likely to pass on their genes to succeeding generations than those who did not. It may well have been an essential behavior for the rise of the first complex civilizations and city-states. But the world has changed profoundly, and wars and terrorism have become hugely costly, without providing any justifying benefit.

Territoriality became a defining characteristic of human behavior deep in prehistory, and the basic impulses to expand territory and power through team aggression have not diminished over the last 5,000 years of population growth and increasing division of labor. In fact, the spectacular improvements in science, technology, and communications that have taken place in the past two centuries have concentrated more and more of the power of the modern state in the hands of just a few alpha males.

It appears that given a particular environment, nearly all men are capable of attacking others, and some women will encourage men to kill those they identify as enemies. Most men, it seems, whether university students involved in the Stanford University prison experiment, or the upright family men who led victims into Nazi gas chambers, have the potential to do terrible things to a perceived outgroup. In Mozambique, a survey of 504 children who grew up during the three decades of war which convulsed that country from 1964 until 1992 found three-quarters had seen people killed, eight out of ten had witnessed abuse or torture, and 63 percent had seen rape or sexual abuse.[535] On the one hand, we are a uniquely violent animal. On the other, even behaviors that evolved long ago are open to change under a broad range of influences. Whatever the mechanism that translates our DNA code into daily behavior, the mental frameworks that evolution has given us are flexible guidelines, not rigid rules. In fact, as we saw in chapter 2, the expression of genes themselves can be altered by the environment in which a child grows up. All big-brained animals are evolved to be highly adaptable, and *Homo sapiens* more so than any other. Our genes have been selected for in a way that gives us a big brain able to make choices and override impulses and emotions when confronted by a complex, changing environment. Civilization at its best can rise above the pain and horrors of war.

"If You Want Peace"

In war, as in medicine, the correct diagnosis is pivotal. Basil Liddell Hart was the son of a Church of England clergyman and a company commander on the Western Front in World War I. He became a leading military theorist and coined the aphorism, "If you want peace, understand war."[536] The more fully we understand the nature that leads to war, the more likely we are to succeed in rising above that nature. An evolutionary filter lets through not only the gore-red flash of biological competition but also the ability to accept a rainbow of cultural influences on our behavior. Understanding the biological roots of our behavior invites us to ask a new set of questions about how to contain wars and terrorism.

The standard social science model, which dominates most of our discussions of war and peace, assumes that culture and the environment determine aggression and set the stage for conflict. Margaret Mead's mistaken belief that a newborn infant is a blank slate waiting to be marked with culture's message has infected the Western mind[537]—and a wide preference for this belief has sustained it for half a century. As long as we believe the 1986 Seville Statement's assertion that "It is scientifically incorrect to say that war or any other violent behavior is genetically programmed into our human nature,"[538] our thinking about the origins of war and terrorism will not advance. In a crude, shorthand form, this standard view suggests that if we stop giving little boys toy guns they will not

How to Make Peace Break Out
• Empower women with education and opportunities
• Increase the number of women in parliaments and legislatures
• Enable women to have the means to manage whether and when to have a child
• Help people prevent unintended pregnancies, thereby slowing population growth
○ leads to fewer volatile young men age 15–30 (as a ratio to older men)
○ reduces competition for resources
• Ensure universal, secular, scientific education
• Encourage knowledge of history and an understanding of our evolution from other animals
• Develop and maintain a free media
• Avoid supplying weapons to potential enemies

grow up to be soldiers. It expresses a naïveté that has seen its hopes shattered time and time again.

We all share the desire to live in a world without wars and terrorism. But ignoring reality will get us no closer to that goal. If we truly want to attain peace, we first have to make peace with the fact that the behavioral building blocks of war are indeed to be found in our nature. The evidence presented earlier suggests that practically any young man can be turned into a soldier, as the history of conscription and military training demonstrates. Given a particular set of circumstances, most men might also become terrorists. Solzhenitsyn's insight that "the line dividing good and evil cuts through the heart of every human being" comes much closer to capturing human nature than anything in the Seville Statement. Raids and wars are not deviant activities; they are the logical expression of deep-seated behavioral predispositions. The impulses underlying warfare are universal, but happily, they need not be universally expressed.

Slavery

The Anglican cathedral in Zanzibar, East Africa, is built on the site of a former slave market. The crypt once contained cells for holding slaves and the altar sits at the point where they were flogged. If we accept that slavery, as suggested in chapter 8, began as a spin-off of early warfare and depends on the same innate ability to dehumanize other groups of human beings, then it is impressive how thoroughly and broadly social attitudes toward slavery have changed. No Roman citizen could have imagined civilized living without slaves. Slavery was recognized by statute in the North American colonies[539] even before it became a legal institution in the Southern states. Jefferson and the other founding fathers of the independent United States struggled with the injustice of slavery but accepted its universality, rather as we continue to accept the universality of using military force as a continuation of diplomatic strategies.

By the end of the nineteenth century, one terrible part of the human predisposition to dehumanize others had come close to being abolished. Individuals still demean other groups and no doubt always will, and women, children, and men are still traded and exploited, beaten and made to work with little or no liberty or remuneration. But all nations, at least in their public pronouncements, now explicitly condemn slavery. The pen and the printed word broke the iron shackles and whips of slavery. Intellectually tenable and politically realistic objections to slavery are historically recent, but they have been wonderfully effective. In the mid-eighteenth century John Wesley and George Whitfield, the founders of Methodism, challenged slavery. The Quaker John Woolman, in his two-part work *Considerations on the Keeping of Negroes*[540] (1754 and 1762), categorized slavery as counter to Christ's exhortations to heal the sick, take in the naked, and feed the hungry. The crucifix on the altar in the Zanzibar cathedral is made from the tree under which David Livingston, an eloquent, life-long opponent of slavery, died. Eventually, in an increasingly literate world, religious leaders, novelists, and abolitionists replaced the aggression toward outgroups which we call slavery with the common bonds of humanity. In the twenty-first century, any nation engaging in an open, systematic trade in human slaves would find itself subject to economic sanctions and probably to forceful intervention.

Women

Peace needs strong allies in order to persist, and the ally that has been most consistently overlooked is the one that makes up slightly over half the human race—women. If we have focused in the preceding chapters on the role of men and their predispositions in creating war, we hopefully have not overlooked the crucial role of women in building peace. In Islam, a woman who dies in childbirth is accorded the honors given to a man who dies on the battlefield,* and the parallel celebration of "man the taker-of-life" with "woman the giver-of-life" contains an important kernel of truth.

The trend in twentieth-century academic sociology was to lessen and obscure the behavioral differences between men and women. "Sex roles in general, and aggressive and violent behavior in particular," writes Betty Reardon in *Women and Peace: Feministic Vision of Global Security*, "are determined by learning not biology."[541] Rather than improving the prospects for global security, arguments like these undercut them by disregarding the possibility of making women allies of peace. Evolutionary psychology both asserts and explains why men and women have always manifested often mutually antagonistic sex differences in behavior.[542] Men are evolved to be territorial and competitive, and to engage in team aggression. Women usually lived in territories men carved out, and benefited more through ingroup cooperation and social stability than through outgroup hostility and aggression. If evolution provides the poison root of warfare, it has also supplied an important antidote. We overlook women's powerful evolutionary heritage at our collective peril.

From Aristophanes' fifth century B.C. play *Lysistrata*, in which Greek women plotted to bring peace by withholding sex from their warrior husbands, to the Greenham Common camp, where women protested the deployment of American nuclear missiles in Britain in 1982, women have been consistently more pacific than men. Overall, women are less easily seduced than men by the raw power of weapons, or by blind loyalty to their country, right or wrong. It appears that women go to church more often than men, but they are less likely to embrace the strictest rules of their religion and they tend to be more tolerant of outgroups.** Whether a single head of state is male or female may be less important to a country's policies than the proportion of women with political power in the government and in the society at large. Countries with a high percentage of women in an elected legislature appear to be more peaceful. In Sweden, 45 percent of the parliamentary seats are held by women, while in Nigeria

---

* The Taj Mahal in Agra, India, is a monument to Mumtaz, one of the wives of the Moghul Emperor Shah Jahan who died in childbirth in 1630.

** Across forty-three countries, 26 percent of men and 36 percent of women rated their religious beliefs as "very important." Numbers were considerably higher in the U.S. (men = 59 percent, women = 61 percent) than in Great Britain (men = 21 percent, women = 42 percent), but the proportions were consistent. Across the same forty-three countries, 51 percent of men and 46 percent of women said they would not like to have a homosexual as a neighbor (U.S. men = 45 percent, women = 32 percent; Britain men = 36 percent, women = 26 percent).

women hold just over 3 percent. It is perhaps encouraging that since the 1994 genocide in Rwanda, the number of women elected as local representatives has risen to 26 percent.

As women are often more likely to seek consensus than men, they may prove better at reducing the tensions that lead to war. In 1966 in the struggle between Catholic and Protestant factions in Northern Ireland a small group of women from each side, many of whom had sons killed, founded the Northern Ireland Women's Coalition (NIWC). The women in the NIWC found it easier to cross the Catholic-Protestant divide in the search for peace than the men, and quite rapidly the NIWC became a significant political force for peace.[543] During the Sierra Leone civil war in the 1990s, market women were clever enough to sell soft drinks to both sides while the men continued to fight one another. Women Waging Peace is the appropriate name for a non-profit organization founded in 1999 in Washington, D.C., which seeks solutions to seemingly intractable conflicts.[544]

Reproductive Autonomy

As we have seen, some analysts and strategic thinkers have recognized that rapid population growth, with its swollen cohorts of unemployed, angry young men, increases the risk of conflict and terrorism. As noted earlier, the bipartisan 9/11 Commission described a rapidly rising population as "a sure prescription for social turbulence."

Unfortunately, policy makers and social commentators have not connected the dots, so they commonly overlook the importance of population growth. Even more rarely do they understand that the rate of population growth is a variable open to change. In 2002, I helped prepare the budgets to be submitted to the foreign aid donor community for the reconstruction of Afghanistan following the fall of the Taliban. My role was to focus on maternal health and family planning, while my colleagues looked at paying the health staff, buying medicines, and all the other things needed to put together a health system devastated by over two decades of fighting. Our instructions were to write a detailed three-year budget and make estimates for expenditures five and ten years down the road. As the team leader was closing his laptop, I remarked, "You do realize that in ten years' time there will be 50 percent more people in Afghanistan." Following an expletive, he tapped some keys and the budget jumped up by a few billion dollars: Humanitarians, development specialists, international bankers, military strategists, and the general public all seem to find it difficult to grasp the implications of rapid population growth.

As detailed in chapter 13, rapid population growth is largely the result of women lacking access to the family planning options they need. Women's inability to choose when to have a child also deepens the division between the "haves" and the "have-nots." High birth rates undermine economic progress, even when sound economic policies exist and there is little or no corruption. Social scientists studying conflict find a robust relationship between per capita income and the possibility of civil war—among the most destructive of all types of conflict.[545] A country with a per capita income of $250 and an average of over six children per woman, such as Malawi, has a 15 percent chance of experiencing a civil war in the next five years, while in a country such as Mexico, with per capita earnings of about $5,000, that chance drops to just 1 percent. Revealingly, while the relationship between income and civil conflict is strong, similar analyses fail to find a relationship between political repression and conflict. The struggle against injustice, it seems, is most likely to turn violent when people also perceive themselves as competing for resources—as happened in the 2008 ethnic fighting in Kenya following a corrupt election.

As women are given the means to control their own fertility, family size begins to fall. What has been called a "demographic dividend" then begins to kick in. Savings rise as the cost of educating and caring for the next generation falls, and women begin to achieve social and legal equality as they are able to contribute more to the formal labor force. All the Asian "tiger" economies of the 1990s were built upon highly successful national family planning programs in the 1960s. Hard work, clever planning, and open markets may all have contributed to the economic booms, but family planning was an essential element in the escape from poverty.[546] In addition, each of these countries had either *de jure* or *de facto* access to safe abortion. Most of the nations that have supported or spawned terrorism, in contrast, make it difficult for couples to have small families. Efforts to give women power over their own lives, enabling them to manage whether and when to have a child, represents an achievable, broad, and integrated plan to help shrink the next generation of terrorists and to reduce the probability of conflict between nations. In a generation's time it is likely Iran will have nuclear weapons, but because family planning is now available and women, on average, are having only two children and are moving toward greater equality, Iran will be much less likely to use these bombs than Pakistan, where families will be twice as large and unemployment rampant.

When I first visited Afghanistan in 1969 there were 12.5 million people. When I helped write the budgets in 2002 there were almost 30 million. By 2050 the population of this harsh, arid, largely desert land could explode to between 90 and 110 million. These extraordinary numbers hold huge implications for the welfare of Afghans, the structure of their society, the peace of the region, and the possibility of world terrorism. By 2050, there will be more men aged fifteen to twenty-nine than there were men and women of all ages in 1971. If we are correct in our arguments in this book, then unrestrained population growth of this type is a sure recipe for internal strife and regional instability. Afghanistan will become an incubator for a virtually limitless supply of angry male jihadists willing to die in America or Europe using whatever horrible, user-friendly weapons are available. Abu Bakr Naji, considered a chief strategist for Al Qaeda, sets out that group's approach in a document aptly named *The Management of Savagery: The Most Critical Stage Through Which the Umma Will Pass*. The *Umma* refers to the "community of believers," and Al Qaeda specifically targets for recruitment and chaos "particularly those [states] that are over-crowded."[547]

Policies

Offering women the option of family planning would be relatively simple if it were not for the evolved male drive to control female reproduction. Wherever my wife and I go in the world, we find that it is the barriers to family planning, rather than lack of demand for family planning, that keeps family size high.[548] We also recognize that there are always some women for whom it is easier to adopt the male reproductive agenda than to challenge it, but on the whole it has been male theologians, male legislators, and conservative male doctors who build and maintain barriers to family planning.

One memory from my 1969 visit to Afghanistan illustrates this last point and sticks in my mind. It was December and I was in a small, exceedingly cold clinic outside Kabul. The young Afghan doctor, with a puzzled look on his face, handed me something like a school exercise

Credit: Anjum Naveed—AP

Rape has often been institutionalized in societies where women have few rights. When she was thirty years old, Mukhtar Mai was gang-raped on orders from tribal elders in Pakistan because her younger brother was seeing a girl from a rival tribe. Normally, such "honor" punishments go unreported, and the raped women may commit suicide or be killed by their own families. In a remarkable display of courage, Mukhtar Mai took her case to court in 2002. Her assailants were imprisoned, but several have since been released, and her life has been threatened.

book. It was a sheaf of about fifteen pages of clinical and demographic information that an American "expert" required him to fill out for every woman fitted with an intrauterine contraceptive device (IUD). I looked it over and in fury tore it in half and threw it to the floor. In a country where girls are often married in their early teens, where a wife cannot leave the house without her husband's permission (and then only when fully obscured by a full-length *burqa*), and where to ask a man his wife's name can be read as an insult, it was asinine to even think of writing down, let alone asking about, a woman's periods, her family history, or how often she had intercourse. It is exactly this type of inappropriate, culturally insensitive, and irrelevant medicalization of family planning that prevents women from getting the options they need and deserve. Despite the obvious unmet need for family planning in Afghanistan, only a miniscule number of educated and partially emancipated women would even think of visiting a clinic known to ask so many unnecessary and embarrassing questions. The result has been predictable enough. In 1969 the average Afghan woman had between 7 and 8 children. Today, that average has not budged.

Could it have been otherwise? Certainly, Afghanistan is in many ways a terrible place to be a woman. Afghan proverbs include "One's own mother and sister are disgusting," and "Women belong in the house or the grave."[549] But even in such a setting, family planning is not impossible. In Bangladesh, another conservative Islamic culture where women also live in *purdah,* forbidden to leave the home, average family size has fallen from almost 7 in the 1960s to 2.7 today. When asked in surveys how many children they want, most young couples today say "two." Crucially, this includes most young men as well as their wives. What was the reason for this success in Bangladesh? A range of contraceptives has been offered and in many cases carried literally to women's doorsteps. There were no fifteen-page forms to fill in; contraception succeeded for the simple reason that contraception was made available. A similar strategy in Afghanistan in the 1970s would also have worked but was not tried. In a sensible world, the second Coalition airplane into Kabul after the fall of the Taliban in 2002 would have carried millions of packets of Pills and IUDs for women who wanted them. Life and death opportunities have been missed—and continue to be missed.

The Sword's Edge

We have entered the opening decades of the most dangerous century in human history loaded down with Stone Age predispositions and behaviors. We find ourselves balancing on a sword's edge: We could step one way and enter the "broad, sunlit uplands" of peace, or stumble the other way and find ourselves face to face with an authentic apocalypse far more terrifying than the child-like images of Armageddon propagated by contemporary Christian fundamentalists in America.

As I try to unite what I believe we know about our inherited predisposition for team aggression against an outgroup and male aggression and competition within an ingroup together with the history of technology and warfare, I am awed and dismayed by the choices we face. The ever-increasing capacity for destruction combined with inevitable human foibles and mendacity present a somber landscape. I first began writing this book before 9/11. Since that time, the profusion of terrorist attacks, foreign policy fiascos, and misapplication of military power by the U.S. and U.K., along with the unfolding cruelty of resource wars in Africa, have provided a truly dispiriting stream of evidence for the presence of Stone Age behaviors in our modern world. I would be happier by far if recent history had not supported the primary thesis of this book, but it has, all too clearly.

Winston Churchill once said, writing a book is an adventure.* It always takes more effort than you bargain for, but it can also reward you by leading down unexpected pathways. I began this book because I was a biologist and a physician who had also been co-teaching a course on the history of warfare. I wanted to discover if I could trace the transition from chimpanzee aggression to human raids and warfare in more detail. In the process, I saw more and more clearly that men and women might play very different roles in relation to aggression. I also saw that team aggression could manifest itself in ways other than overt war, including in football and soccer, mountain climbing and other adventure sports, and in street gangs. I realized how adaptive it might have been for our hominid ancestors to react strongly to signs of attack, such as finding a corpse. And I became convinced that predispositions that evolved long, long ago still influence our behaviors today.

---

* He went on to point out, quite correctly, that "To begin with, it is a toy and an amusement; then it becomes a mistress, and then it becomes a master, and then a tyrant. The last phase is that just as you are about to be reconciled to your servitude, you kill the monster, and fling him out to the public."

As we come to the conclusion of this book, I appreciate how difficult it is to communicate the depth and pervasiveness of the male predisposition for competition and team aggression in contemporary human affairs. For literally millions of years our male ancestors teamed up and went out to kill their neighbors. For many of our mothers, over many hundreds of thousands of generations, the least costly strategy was to go along with the male agenda of team aggression. It is a predisposition that leaves the contemporary world spending 3.3 billion dollars a day on the military,[550] while almost three billion people live on two dollars a day or less. For many, this level of military spending represents a good investment in a dangerous world—until we realize that the world itself need not be so dangerous.

Taking up an idea touched on in chapter 11, what if the big-brained, technically competent mammal that had taken over the world had been an intelligent, talking horse rather than an ape? An equine Mozart might have found it difficult to play the piano, let alone make one. But setting that fantasy aside, a species with no evolutionary history of team aggression would evaluate its competitors objectively, not de-individualize or "de-equinize" them and assume they were different and inferior. Unlike territorial primates, a horse society, however violent, would have no concept of patriotism or of being a traitor. There would be no 9/11 attacks, and no disproportional responses to threats. There would be no overconfident leaders making huge military blunders. An intelligent horse might gallop away from a fellow animal that died a violet death, but it would not assume that a mass attack was imminent and feel impelled to invest all its wealth in great armies. Intelligent stallions might contrive to pass legislation limiting the freedom of mares, but when horse Einsteins understood how to initiate nuclear fission, they would make a power station to keep their stables warm, not a weapon to vaporize everyone in sight. But we poor Yahoos, the degenerate human beings that Jonathan Swift imagined living beside a noble society of horses, are condemned by our biological history and our current choices to live in the shadow of weapons of mass destruction.

As emphasized in chapter 12, we live with a triad of weapons of immense potential destructiveness. Each of the more than thirty U.S. Poseidon submarines can carry enough nuclear warheads to destroy every city with a population of over 150,000 in a country as large as Russia. Nuclear, biological, and chemical weapons are becoming more accessible by the month, and there is a real possibility of wars or terrorist attacks that would turn 9/11 into only a minor footnote on the last page of our history as a species. We do indeed live on a sword's edge. If you are young, or have children—or as I do, grandchildren—then think about what the young of today can expect to see during their lives. Statistically, a child born today should have a good chance of living until the twenty-second century. But will our misshapen behaviors let it happen?

Children born today could live out their lives in an increasingly healthy and prosperous world, or they could die violent deaths at the hand of someone they never knew and who never knew them. The future is rife with dangers of immense proportions, and the predisposition for team aggression will be with us always. At the same time, measurable progress is taking place in curbing some types of warfare. As recently as 1939 the possibility of abolishing large-scale war between European states was unthinkable; today, war between France and Germany or America and Japan is all but inconceivable—we have become part of each others' ingroups. We are now too well-connected with our friends across these borders. But the greater threat of terrorist groups, for whom national borders have less meaning, remains. The most immediate danger is not that they will mix explosive liquids in an airplane lavatory, but that they will graduate to stealing nuclear weapons or developing biological warfare.

If political and military leaders ignore the lessons of past experience, as has most certainly been the case in the opening years of this new millennium, then the prognosis lies between the merely somber and the cataclysmic. If religious fanaticism remains constant or continues to grow in influence; if economic disparities increase; if rapid population growth is not abated in less developed and volatile parts of the world; and if, as seems certain, weapons of mass destruction spread inexorably, then the chances that an infant born today could die in a nuclear explosion or biological attack are frighteningly high.

In the past half-century, India and Pakistan have fought four wars with conventional weapons. Now they both have nuclear weapons. Will they go another fifty years without using them? Israel has nuclear weapons,[*551] and Iran surely would like to. Will they be used? If China, India, and other emerging economies continue to grow at the rate they have for the past decade, global competition for oil and other crucial resources could reach

---

* Experts agree that Israel has developed although never tested an A-bomb. The number of warheads is uncertain but a dismissed Israeli nuclear technician Mordechai Vanunu suggested in the 1980s that they might have over one hundred, and former president Carter said they had 150 warheads by 2008.

the boiling point within a generation. Other potential triggers for war abound, from the status of Taiwan to political wrangling over North Korea and Darfur. China probably holds a few hundred nuclear warheads. It has liquid-fueled rockets that could reach all of the Eurasian land mass, and it has a nuclear submarine that could carry ballistic missiles. Politically, China is as unlikely to launch an unprovoked nuclear attack as the U.S. is on China. But we cannot discount the possibility of human error or just one psychotic person who finds a way to gain access to the nuclear trigger. Every time the nuclear club gains a new member, the risk of unintended, as well as intended, use multiplies.*

The threat posed by biological weapons is less immediate but just as terrifying. Consider the following scenarios. What if, as the population of Niger explodes from 14 million today to 50 or 80 million in the second half the twenty-first century, angry, frustrated, hungry Muslim groups in that country join with their brothers in northern Nigeria to attack Christian groups farther south? Might some military dictator in the region then use his oil wealth to develop and deliver biological weapons? What if a small band of terrorists exploded a "dirty" bomb in San Francisco and a fundamentalist Christian U.S. president overreacted, aiming bunker-busting atomic weapons at whoever seemed a likely target? So many politicians are scientifically illiterate that such a president might believe against all evidence that a recently installed, trillion-dollar missile shield would actually work, only to discover that the computer codes contained errors and a full scale nuclear exchange had been triggered.

If we fail to tame our Stone Age behaviors, then these become technically possible and humanly plausible scenarios. Those who happened to survive a war fought with weapons of mass destruction might come to their senses and finally take steps to control our Stone Age behaviors. It is more likely, however, that many would become even more radicalized, and continue to fight in a downward spiral of destruction.

Unnatural Honesty

Human beings are social animals, yet we are also competitive, duplicitous creatures. We evolved to dissemble with, manipulate, and outwit

---

* It has been reported that an American manning a rocket silo once demonstrated that with a piece of string and a spoon he could override the carefully designed system requiring two men to initiate firing a missile.

one another. We all tell white lies. We all hold back information. Where we value truth and condemn serious deceit, as in our troops, we do so in order to counteract the perpetual temptation for free-riders to cheat. In a profound way, this makes the scientific process "unnatural," as it depends on sharing information among groups in the most transparent way possible. Ultimately, true scientific advancement depends on individuals and teams providing so much information that any critic can repeat the experiments and confirm or refute their conclusions.

Science is a uniquely successful way of thinking which has provided food, energy, global communications, and good health on an unprecedented scale. The scientific method has expanded our picture of the universe beyond all imagination, but it also gives us the knowledge to manufacture weapons of unimaginable destructiveness. Weapons of mass destruction are the deformed child of modern science. If they are used in terror attacks by some group of religious extremists, it will be because they have been copied or stolen, not because some fanatic thought them up—religious zealotry and good science rarely if ever go together. If it is true we have inherited a propensity to kill those who belong to an outgroup, then we are doomed to live in a world of contradictions. Our ingroup behavior is built on friendship, trust symbolized by a handshake or a kiss, unalloyed laughter, and peaceful secure sleep. Our outgroup behavior is to kill others of our species without motivation and to hate without reason—to dehumanize those who are exactly like us and to deindividualize those who just as easily could be our friends. Our inherited ability to dehumanize and deindividualize the enemy overrides what Winston Churchill once called the "silly conventions of the mind."

As this book has argued, we must expect human beings to continue to create new weapons as scientific knowledge expands, and assume that groups hostile to our society will exploit its intrinsic vulnerabilities. The standard social science model looks for the causes of war; the evolutionary model seeks ways to make peace break out, while always expecting the worst.

Until the nineteenth century, it was argued that the chemicals found in living organisms, such as urea, could be made only in the presence of some sort of "vital essence." In 1828 the German chemist Freidrich Wohler made urea in a test tube, rendering such vitalism obsolete. Recently, George Church and Xiaolian Gao made an artificial sequence of DNA in a Harvard University laboratory, demonstrating that no divine hand will be needed one day to create a biological organism.[552] Science has given our species the power of both the creator and the destroyer, which we once imagined gods alone held. If we are to separate the creator from the destroyer, then we need to begin by understanding our evolutionary inheritance and rising above our Stone Age emotions and ways of thinking.

In addition to full transparency and total honesty, science depends on asking the right questions, and questioning everything. If we are to find ways of controlling team aggression, then we will have to ask questions close to the soul of human experience. Many people believe that our noblest impulses concerning love, altruism, self-sacrifice, honesty, and compassion come from without—that they are bestowed by divine grace, and independent of the huffing and puffing of mere mortals. Moral truths, it is argued, are absolutes to be accepted without testing, especially when asserted with sincerity by charismatic individuals. Others believe, as the authors of this book all do, that ethics as well as evil are the legacies of evolution. We hold with conviction nearly all of the same moral precepts that others attribute to a transcendental origin. We need to ask, where do such standards come from?

If we stand back and look honestly at one another, we find that whether humanist or religious seeker, we all end up asking the same questions about the natural world and how the human mind uplifts, modifies, or undermines the ethical principles most people hold to be "true." A Buddhist or follower of Confucius might not believe that God gave Moses stone tablets inscribed with, "Thou shalt not commit adultery, neither shalt thou steal, neither shalt thou bear false witness." But they still adhere to the same moral codes. We believe that the evidence suggests this is because we are social primates. As such, we have evolved to behave in ways that we call moral, and we have evolved social and psychological mechanisms to reinforce these behaviors. Whether an evolutionary explanation or a Word of God assertion makes such a code more or less profound, or and more or less universal, is for the individual to decide.

E. O. Wilson has written, "The essence of humanity's spiritual dilemma is that we evolved genetically to accept one truth and discovered another."[553] Given our inherited predispositions, the majority of people even in this postindustrial society may be slow to make the switch from accepted belief to a worldview based on observation and testable ideas. Many people will continue to pray to their god to bring peace, while passionately asserting that another person's religion ferments war. As Abraham Lincoln wrote of the two sides in the American Civil War shortly before he was assassinated, "Both read the same Bible, and pray to the same God; and each invokes His aid against the other."[554]

Civilization as we know it is only a few thousand years old. Given our powers of destruction, can we expect it to last another few thousand years? Can we be certain that no Stalin-like man—and once again it is likely to be a man—will arise to control and use stockpiles of nuclear weapons, or release a bio-engineered, incurable, lethal disease? Seven years before World War II, Adolf Hitler said, "We will never capitulate— no, never. We may be destroyed, but if we are, we shall drag a world with us—in flames." In the final months of World War II he came close to fulfilling his prophecy. If Werner Heisenberg had made him his atomic bomb, he would have come closer still.

If we look at the record of human ingenuity, resilience, empathy, and love, we see much that is inspiring, and many last-minute escapes from potential catastrophe. The alternative to destruction is that nations become more rational and better educated, women play an equal role in society, family size shrinks, and states solve conflicts diplomatically. This is not a fairy tale; it describes the progress of relations between Canada and the U.S., or France and Germany. Much work remains to be done, obviously, and most especially where conflict is endemic and also likely to spill over and involve more stable countries. Today, the most dangerous parts of the planet are Africa, Pakistan, and the Middle East. Given time, slower population growth, and increasing autonomy for women, Iran, Turkey, Syria, Jordan, Egypt, and other Islamic states could begin the transition to less fundamentalist, more stable societies. Sub-Saharan Africa needs a variety of assistance provided in ways that actually reach those living on a dollar or two a day rather than ending up in the Swiss Bank accounts of their leaders. In order to escape from poverty, educate its children, and break the cycle of violence, Sub-Saharan Africa has an urgent need to slow population growth. Happily, as we have shown in chapter 13, there is already a large, unmet need for family planning and safe abortion; meeting that desire should be among our top priorities.

If we don't destroy one another with war first, then a thoughtful application of scientific knowledge could solve most of our other problems, serious as they are. Over the course of several decades, scientific and technical advances could develop sustainable sources of energy that do not add to global warming. Medical science could continue to improve health and longevity. Perhaps with some breakthroughs in sustainable agriculture, the world could feed a stable population indefinitely. Given good management, the oceans might someday recover from their current over-exploitation. It might take centuries for the cut and degraded forests of Asia, Africa, and South America to recover, but perhaps if we exercise wise, science-based stewardship of the land then some fragments of tropical forest could survive to flourish once more.

Perhaps in two or three fragments of African forest, a handful of chimpanzees would escape extinction. The females would continue to nurture their young, while males continued to compete for status and occasionally sally forth and kill their neighbors. They would provide a fascinating reminder to our descendents of just how close we had come, based on those same impulses, to causing our own destruction, had we not understood and tamed our Stone Age behaviors at the very last moment.

To the Reader

MY EARLY WORK IN OBSTETRICS and family planning took me into most of the world's countries, including a number of nations newly ravaged by war or in the throes of violent strife, where I saw the consequences of wartime rape and the terrible conditions for women and their families during and after all the bloodshed. My fascination with Charles Darwin and biological evolution has similarly deep roots, starting some thirty years ago. But it was only fifteen years later, after I had moved to the University of California, Berkeley, that I started to link together evolutionary behavior and warfare. I read voraciously everything I could find on humans' and other primates' actions in conflict settings, and I started writing what would become pieces of this book about the violent damage that we *Homo sapiens* inflict on others of our own species. When I started writing this book, my wife Martha Campbell would look at something I had drafted about human violence and say, "But that's something only men do." She suggested that the book I was working on wasn't so much about *human* violence, it was really about the behavior of men. Martha's insights have been instrumental throughout this book, and she is a co-author of the two chapters explicitly on women (chapters 6 and 13).

I have been lucky to work in a great many countries and perhaps even more fortunate to find pleasure and satisfaction in a variety of disparate disciplines. I tell my students that I always knew exactly what I was going to do in my life, I just kept ending up doing something entirely different. As a child my elder brother Bill, who became professor of zoology at Lancaster University, taught me a great deal of science. He helped me bleach and mount the skeletons of the dead animals I found and once, the chicken we had on Christmas Day during World War II in Britain, when a chicken was a rare treat. As a schoolboy, I wanted to be an archaeologist. I won a history scholarship to Cambridge, but then I switched to medicine. As a medical student I wanted to be a neurologist, but then I became fascinated by obstetrics. During my obstetrics education, I took time out to become a laboratory scientist studying embryology.

The first of the two passions that drive me arose when I was in my twenties. As a young doctor, most nights when I was on duty at the North Middlesex Hospital, I was confronted with the sad effects of a botched abortion, and had to take care of these women in the operating room. At a time when even contraception was illegal in some American states and when many family doctors in Britain refused to prescribe oral contraceptives to unmarried women, I became committed to giving women family planning choices. In 1968, I was appointed the first medical director of the International Planned Parenthood Federation, in London. Growing up during the Second World War, I had never been more than a few hours' bus journey from home and even France seemed an impossibly distant place. Now my work took me to every corner of the world. Ten years later I immigrated to the United States to head up a large humanitarian organization called Family Health International that worked on family planning in forty countries. There I reconnected with a friend from my laboratory days, Roger Short—one of the world's foremost experts on reproduction in elephants, kangaroos, primates, and many other animals. Together we took the organization I headed into AIDS prevention as well as family planning. But Roger also inspired my passion for the insights Darwinian evolution provides. This second animating passion matured slowly, but it changed the way I see the world just as surely as the first. Now, the blossoming science of evolutionary psychology has given me a new and better way to grapple with the conundrum of human violence and sexual behavior. Both Roger and I had separately met Jane Goodall, and we found the discoveries that she and other scientists, such as Richard Wrangham, were making about chimpanzee behavior spellbinding.

My interest in history, archaeology, and zoology and the things I had seen traveling the world and working with poor people in remote villages, as well as with highly skilled professionals in big cities, started coming together. I began to focus on one question: *Why do we humans, remarkably social animals with extremely large brains, spend so much energy on one thing—deliberately and systematically killing other members of our own species?* This book is an extended effort to answer that question, and I believe the answers are as profoundly important as they are fascinating.

Some years after beginning this book, I had the good fortune to run into a stunning article on the biological roots and social development of war by Thomas Hayden in *U.S. News & World Report*. Tom had trained as a biologist and metamorphosed into a professional journalist and writer, and we seemed to share a great deal of thinking on the nature of war. I invited him to join me in telling this story, and we both hope you will enjoy the results. Martha, a lecturer at Berkeley whom I have fondly called my "editor from hell" since we started publishing together in 1993, has joined us in writing parts of this book too, focusing mainly on the two chapters that deal specifically with women.

Many other people played important roles in the creation of this book. My understanding of Catholic theology relating to contraception was greatly expanded by the late Father Francis Xavier Murphy, an expert advisor to the Second Vatican Council. My knowledge of military history exploded—if that is the appropriate word—in 1998, when I had the privilege of teaching a course at Berkeley with Tom Barnes, a lawyer, military expert, and professor in Berkeley's History Department. Being at Berkeley, we enjoyed calling the course "Make War Not Love." My sincere thanks go to Dr. Assefaw Tekeste, who shared his experiences during his fifteen years as Eritrea's Minister of Health, using both an AK-47 and a stethoscope on the frontlines of the terrible war with Ethiopia. Thank you also to Dr. Zafrullah Chowdhury, who left a medical post in England to return home to Bangladesh to treat casualties in the War of Liberation, and Dr. Hamid Taravati of Mashad, Iran, who served as a doctor in the equally horrible Iran-Iraq war. Then there are those now dead, who had fought in World War II: General Andrews of the U.S. Army Air Forces—how I wish I had asked you more about flying a B-17 over Germany—and Dr. Bob Greenblatt—I should have listened more carefully to your description of surviving a *kamikaze* attack. And my own father and elder brother—Dad, I should have asked you more sensible questions about serving in the Royal Air Force in 1939 and 1940, and Tom, I should have asked you more searching questions about protecting refugees during the 1949 Partition of India and fighting in the Korean War.

I have been helped by graduate students who challenged my ideas and helped me track down references I couldn't find. Thank you Mike Musante, Angela Chen, Ben Bellows, Maura Graff, Brandon Swansfeger and Joseph Colangelo, now serving in the U.S. Navy. Friends and colleagues helped me by commenting on various drafts. Thank you Drs.

Steve Dell in Oakland, Tom Hall in San Francisco, and Ed Stim in Tokyo, Professor George Wolfe at Berkeley, Dr. Bradley Thayer in Washington, and my remaining brother, Professor William Potts in Lancaster, England, master at reconstructing the dead chicken, who since then has also reconstructed—after wading through in tall boots—large whales.

My co-author Tom would like to thank his father, historian Michael Hayden of the University of Saskatchewan, who read early drafts of several chapters and has spent a lifetime teaching Tom the importance of getting the facts right; and also his mother, the late Joan Hayden, who taught him that people matter most of all. His deepest gratitude goes to his wife and fellow writer, Erika Check Hayden, for reasons that go far beyond her help with this book. Tom and I both thank his agent, Michaela Curran, without whose good work our collaboration would not have been possible, and his colleagues Lauren Stockbower and Kelly Krause for their help finding images that we feel capture the essence of this book.

I have written or edited a dozen previous books, for the most part published by university or academic presses. This book is different—its broad perspective and broader relevance simply don't fit into what have become increasingly restrictive intellectual silos. It has been a pleasure finding Glenn Yeffeth and BenBella Books (no, not the Algerian guerilla, but Glenn's children's names run together) who focus on publishing quality over quantity and aren't afraid of introducing readers to ideas that might challenge as well as entertain.

Books ultimately are written by thousands of different people spread across time and geography, who influence us by their writings, their examples, and their discoveries. (Neither these past eminences nor our more immediate inspirations bear any responsibility for our mistakes or oversights, of course.) As we finish *Sex and War,* I realize how interests going back to my childhood, knowledge I have gained from teachers, books, and friends and colleagues across the globe, and experiences—sad, comic, and tragic—have come together in new and, for me, exciting ways. I am left with a sense of urgency about the state of the world, but also powerful optimism that understanding our biological heritage can help us to shape a future as individuals and as a species that is brighter than our past.

Malcolm Potts
Berkeley, California
March, 2008

References

1. Davis, G. *Interception of Pregnancy: Post-Conceptive Fertility Control*. Sydney: Angus and Robertson. 1974.
2. Eder, J. R. *Let's Go Where the Action Is! The Wartime Experiences of Douglas Campbell*. Knightstown, IN: JaaRE Publishing Inc. 1984.
3. Thornhill, N. W., Thornhill, R. An Evolutionary Analysis of Psychological Pain Following Rape. *Ethology and Sociobiology* 11:155–76, 177–93. 1990.
4. Zerjal, T., Xue, Y., Bertorelle, G. The Genetic Legacy of the Mongols. *Annals of Human Genetics* 72:717–21. 2003.
5. Moore, L. T., McEvoy, B., Cape, E., Simms, K., Bradley, G. D. A Y-chromosome signature of hegemony in Gaelic Ireland. *American Journal of Human Genetics* 78:334–8. 2006.
6. Betzig, L. L. *Despotism and Differential Reproduction: A Darwinian View of History*. New York: Aldine Publishing Company. 1986.
7. Potts, D. M., Potts, W. T. W. *Queen Victoria's Gene: Haemophilia and the Royal Family*. Stroud, Gloucestershire: Alan Sutton Publishing Limited. 1995.
8. Hagood, W. O. *Presidential Sex: From the Founding Fathers to Bill Clinton*. New Jersey: Citadel Publishing Group. 1996.
9. Alexander, G., Hines, M. Sex differences in response to children's toys in nonhuman primates. *Evolution & Human Behavior* 23:467–79. 2002.
10. Stillwell, Cinnamon. Honor killings: When the ancient and the modern collide. *San Francisco Chronicle*. January 23, 2008.
11. Rhodes, Richard H. *The Making of the Atomic Bomb*. New York: Simon & Schuster. 1986.
12. Rousseau, Jean Jacques. *Discours sur l'Origine et le Fondement de l'Inégalité parmi les Hommes*. 1754.
13. Montagu, Ashley. *Culture and the Evolution of Man*. New York: Oxford University Press. 1962.
14. Huizinga, J. (1938) *Homo Ludens: A Study of the Play Element in Culture*. Boston: Beacon Press. 1972.
15. Chapman, J. The Origins of Warfare in the Prehistory of Central and Eastern Europe. In *Ancient Warfare: Archaeological Perspectives*. Ed. Anthony Harding. Stroud, Gloucestershire: Sutton Publishing. 1999.
    Kokkinidou, Dimitra, and Nikolaidou, Marianna. Neolithic Enclosures in Greek Macedonia: Violent and Non-violent Aspects of Territorial Demarcation. In *Ancient Warfare: Archaeological Perspectives*. Ed. Anthony Harding. Stroud, Gloucestershire: Sutton Publishing. 1999.
    Whittle, A. *Neolithic Europe*. Cambridge: Cambridge University Press. 1985.
16. Keeley, L. H. *War Before Civilization: The Myth of the Peaceful Savage*. Oxford: Oxford University Press. 1996.
17. Rousseau, Jean Jacques. *Discourse on the Moral Effects of the Arts and Sciences*. 1750.
18. Wilson, E. O. *Consilience: The Unity of Knowledge*. New York: Alfred A. Knopf. 1998.
19. Trotter, Wilfred. *Instincts of the Herd in Peace and War*. London: T. Fisher Unwin Ltd. 1916.
20. Ardrey, Robert. *The Territorial Imperative: A Personal Inquiry into the Animal Origins of Property and Nations*. New York: Atheneum Publishers. 1966.
21. Lorenz, Konrad. *On Aggression*. New York: Harcourt, Brace & World. 1966.

22. Wrangham, Richard W. Evolution of Coalitionary Killing. *Yearbook of Physical Anthropology* 42:1–30. 1996.
23. Wrangham, Richard, and Peterson, Dale. *Demonic Males: Apes and Origin of Human Violence.* Boston: Houghton Mifflin. 1996.
24. Grant, P. R. *Ecology and Evolution of Darwin's Finches.* Princeton: Princeton University Press. 1986.
    Weiner, Jonathan. *The Beak of the Finch.* New York: Vintage Books. 1994.
25. Bodmer, Walter, and McKie, Robin. *The Book of Man: The Human Genome Project and the Quest to Discover Our Genetic Heritage.* Oxford: Oxford University Press. 1997. In *Talking Science.* 1994.
    Ridley, Matt. *Genome: An Autobiography of a Species in 23 Chapters.* London: Fourth Estate. 1999.
26. Watson, J. D., Crick, F. H. C. Molecular structure of nucleic acids: A structure for deoxyribose nucleic acid. *Nature* 171:737–8. 1953.
27. Balter, M. Speech gene tied to modern humans. *Science* 297:1105–6. 2002.
28. Goodall, Jane. *Chimpanzees of Gombe: Patterns of Behavior.* Cambridge, MA, The Belknap Press of Harvard University.
    Goodall, Jane. *In the Shadow of Man.* Boston: Houghton Mifflin. 1988.
29. Sulloway, F. H. *Born to Rebel: Birth Order, Family Dynamics and Creative Lives.* New York: Vintage Books. 1997.
    See also Townsend, F. Birth order and rebelliousness; reconstructing the research. In *Born to Rebel. Politics and the Life Sciences* 19:135–244. 2000.
30. Huebner, D. K., Lentz, J. L, Wooley, M. J., King, J. E. 1979. Responses to snakes by surrogate and mother-reared squirrel monkeys. *Bulletin of the Psychonomic Society* 14:33–6. 1979.
31. Fossey, Dian. *Gorillas in the Mist.* Boston: Houghton Mifflin. 1983. (Dian Fossey was murdered in her house in the gorilla reserve probably as a result of her impassioned defense of the animals from poachers and the threats of tourism.)
32. Goodall, Jane. *In the Shadow of Man.* Boston: Houghton Mifflin. 1988. Page 270.
33. Archer, J. Testosterone and Aggression. *Journal of Offender Rehabilitation* 21, 3–4:3–39.
    Archer, John. The Influence of Testosterone on Human Aggression. *British Journal of Psychology* 82:1–28. 1991.
34. Udry, Richard J., and Talbert, Luther M. Sex hormone effects on personality at puberty. *Journal of Personality and Social Psychology* 54:291–5. 1988.
35. Sapolsky, Robert. *The Trouble with Testosterone: And Other Essays on the Biology of the Human Predicament.* New York: Scribner. Pages 149–52.
    Menaghan, Edward O., and Glickman, Stephen E. Hormones and aggressive behavior. In *Behavioral Endocrinology.* Jill B. Becker, Marie Breedlove, and David Crews, Eds. Cambridge, MA: MIT Press. 1992. Pages 261–86.
    Archer, John. *Male Violence.* New York: Routledge. 1994.
36. Brooks, J. H., and Reddan, J. R. Serum testosterone in violent and nonviolent offenders. *Journal of Clinical Psychology* 52:475–83. 1996.
    Mazur, Allan, and Michalek, J. Marriage, divorce and male testosterone. *Social Focus* 77:315–30. 1998.
37. Mazur, Allan, Booth, L., Dabbs, James M. Testosterone and chess competition. *Social Psychology Quarterly* 55:70–77. 1992.
38. Schultheiss, O., Campbell, K., McMelland, D. Implicit power motivation moderates men's testosterone: Response to imagined and real dominance success. *Hormones and Behavior* 36:234–41. 1999.
39. Goldstein, J. S. *War and Gender: How Gender Shapes the War System and Vice Versa.* Cambridge: Cambridge University Press. 2001.
40. Dabbs, James M. Testy fellows. *Science.* April 26, 1991. Page 513.
41. Sapolsky, Robert M. Neuroendocrinology of the stress response. In *Behavioral Endocrinology.* Jill B. Becker, Marie Breedlove, and David Crews, Eds. Cambridge, MA: MIT Press. 1992. Pages 287–324.

42. Allman, J. A. *Evolving Brains*. New York: Scientific American Library. 1999. (Prozac, the widely prescribed anti-depressant, is a serotonin uptake inhibitor.)

43. Gesch, C. B. et al., Influence of supplementary vitamins, minerals and essential fatty acids on the antisocial behaviour of young adult prisoners. Randomised, placebo-controlled trial. *British Journal of Psychiatry* 181:22–28. 2002.

44. Knaden, Markus and Wehner, Rüdiger. Path Integration in Desert Ants Controls Aggressiveness. *Science* 305:60. 2004.

45. Caspi, A. et al. Role of genotype in the cycle of violence in maltreated children. *Science* 297:851–4. 2002.
    Lewis, D.O., et al. Violent juvenile delinquents: psychiatric, neurological, psychological and abuse factors. *Journal of American Academy Child Adolescent Psychiatry* 18:307–19. 1979.

46. Teicher, M. The neurobiology of child abuse: Maltreatment at an early age can have enduring negative effects on a child's brain development and function. *Scientific American*. March 2, 2002. Pages 68–75.
    Teicher, M. Wounds that time won't heal: the neurobiology of child abuse. *Cerebrum* 2:50–67. 2000.

47. de Waal, Frans. *Peacemaking Among Primates*. Cambridge, MA: Harvard University Press. 1989. Page 65.

48. Zuckerman, Solly. *The Social Life of Monkeys and Apes*. New York: Harcourt, Brace. 1932.

49. Kummer, H. From laboratory to desert and back: Social Systems in Hamadryas Baboons. *Animal Behavior* 32:965–71. 1984.

50. de Waal, F. B. M., and Johanowicz, D. L. Modification of reconciliation behavior through social experience: an experiment with two macaque species. *Child Development* 64: 897–908. 1979.

51. Francis, D. D., Diorio, J., Liu, D., and Meaney, M. J. Variations in maternal care form the basis for a non-genomic mechanism of inter-generational transmission of individual differences in behavioral and endocrine responses to stress. *Science* 286:1155–8. 1999.
    Francis, D. D., Szegda, K., Campbell, G., Martin, W. D., and Insel, T. R. Epigenetic sources of behavioral differences in mice. *Nature Neuroscience* 6:445–6. 2003.

52. Adam, D. War Is Not in Our Biology: A Decade of the Seville Statement. In *Violence: From Biology to Society*. James S. Grisolia, et al. Eds. Amsterdam, New York: Elsevier Science BV. 1997. (1699 of the postal ballots sent out by the American Anthropological Association supported the Declaration and 230 opposed it—J Benthall in *Anthropology Today*, 1989.)

53. Wilson, E. O. *Consilience: The Unity of Knowledge*. New York: Alfred A. Knopf. 1998.

54. Goodall's definitive summary of chimpanzee behavior is her *Chimpanzees of Gombe: Patterns of Behavior*. Cambridge, MA: The Belknap Press of Harvard University. 1986. She has also written widely for more general audiences (e.g. Goodall 1967), and photographs in the *National Geographic* magazine and television sequences illustrating chimpanzee behavior in Gombe have become familiar to many people. Many key primatologists, such as Wrangham and Pusey have worked in Gombe, and today Tanzanian observers maintain daily records of chimp behavior more than 40 years after Jane Goodall first set foot in the Gombe Stream National Park. Other books on wild and captive chimpanzee societies include:
    de Waal, Frans. *Chimpanzee Politics*. Cambridge, MA: Harvard University Press. 1982.
    de Waal, Frans. *Peacemaking Among Primates*. Cambridge, MA: Harvard University Press. 1989.
    de Waal, Frans. *Good Natured: The Origins of Right and Wrong in Humans and Other Animals*. Cambridge, MA: Harvard University Press. 1996.
    de Waal, Frans. *The Ape and the Sushi Master: Cultural Reflections of a Primatologist*. New York: Basic Books. 2001.

55. Wrangham, Richard W., McGrewe, W.C., de Waal, F. H., Heltne, P.G. (Eds) *Chimpanzee Cultures*. Cambridge, MA: Harvard University Press. 1994.
    Boesch, C. and Boesch-Achermann, H. *The Chimpanzees of the Tai Forest: Behavioral Ecology and Evolution*. Oxford: Oxford University Press. 2000.

56. Orangutans also demonstrate cultural differences and in the Suag swamp forest use a variety of tools. Carel von Schaik. *Among orangutans: Red Apes and the Rise of Human Culture*. Cambridge, MA: The Belknap Press of Harvard University. 2004.

Goodall, Jane. *Chimpanzees of Gombe: Patterns of Behavior*. Cambridge, MA: The Belknap Press of Harvard University. 1986.

57. Gagneux, P., Woodruff, D., Boesch, C. Furtive mating in female chimpanzees. *Nature* 387:358–9. 1997.

58. de Waal, Frans. *Peacemaking Among Primates*. Cambridge, MA: Harvard University Press. 1989.

59. Goodall, J. *Chimpanzees of Gombe: Patterns of Behavior*. Cambridge, MA: The Belknap Press of Harvard University. 1986. Page 241.

60. Boesch, C. and Boesch-Achermann, H. *The Chimpanzees of Tai Forest: Behavioral Ecology and Evolution*. New York: Oxford University Press. 2000. Page 137.

Boesch, C. Hunting strategies among Gombe and Tai chimpanzees. In *Chimpanzee Cultures*. R. W. Wrangham, W. C. McGrewe, Frans de Waal, and P. G. Heltne, Eds. Cambridge, MA: Harvard University Press. 1994.

61. Boesch, C. Cooperative hunting roles among Tai chimpanzees. *Human Nature* 13:27–46. 2002.

62. Goodall, J. *Chimpanzees of Gombe: Patterns of Behavior*. Cambridge, MA: The Belknap Press of Harvard University. 1986. Page 513.

63. Homeland defense in the wild. *Science* 295:957–58. 2002.

64. Hamilton, W. D. The Genetical Evolution of Social Behavior. *Journal of Theoretical Biology* 7:1–52. 1964.

65. In artificial conditions, for example when different strains of wild mice are confined in a laboratory cage, then hostility toward strangers of the same species may end in killing. See also:

Hall, K. R. L. Aggression in monkey and ape societies. In *The Natural History of Aggression*. J. D. Carthy and F. J. Ebling, Eds. New York: The Academic Press. 1964.

Southwick, C. H. Peromyscus leucopus: and interesting subjects for studies of socially induced stress. *Science* 143:55–6. 1964.

Huntingford, F. and Turner, A. *Animal Conflict*. London: Chapman & Hall. 1987.

66. Mech, L. D. Buffer Zones of Territories of gray wolves as regions of interspecific strife. *Journal Mammal* 58:559–74. 1994.

67. Orangutans, like other apes, have been driven out of the best habitats and now necessarily live at a low density which makes social contact unusual—Connie Rogers. Revealing behavior in orangutan heaven and human hell. *The New York Times*. November 15, 2005.

68. Pryor, K. and Norris, K. S. *Dolphin Societies: Discoveries and Puzzles*. Berkeley: University of California Press. 1991.

Reynolds, J. E., Wells, R. S., et al. *The Bottlenose Dolphin: Biology and Conservation*. Gainsville: University of Florida Press. 2000.

69. Sumner, W. G. *Folkways: a study of the sociological importance of usages, manners, customs, mores and morals*. Boston: Ginn Publishing. 1906.

70. Quoted in Sanderson, S. K. (2001). *The Evolution of Human Sociality: A Darwinian Conflict Perspective*. Lanham, MD: Rowman and Littlefield Publishers. 2001. Page 324.

71. *Morning Post*, June 22, 1915 as quoted in Knightley, P. *The First Casualty: From the Crimea to the Falklands: The War Correspondent as Hero, Propagandist and Myth Maker*. London: Pan Books. 1989. Page 84.

72. Sherif, M., Harvey, O. J., et al. *Intergroup Conflict and Cooperation: The Robber's Cave Experiment*. Norman: University Oklahoma Press. 1961.

73. Zimbardo, P. Pathology of imprisonment. *Trans-Action* 9:4–8. 1972.

Haney, C., Banks, C., Zimbardo, P. Interpersonal dynamics in a simulated prison. *International Journal of Criminology and Penology* 69. 1973.

Meeus, W. H. J., Quinten, A. W., et al. Administrative obedience: carrying out orders to use psychological-administrative violence. *European Journal Social Psychology* 311. 1986.

74. Zimbardo, P. G. The SPE: What it was, where it came from, and what came out of it. *Obedience to*

Authority: Current Perspectives on the Milgram Paradigm. Ed. T Bliss. Mahwah, NJ: Lawrence Erlbaum Associates. 2000. Pages 198–210.

75. Asch, S. E. Effects of group pressure upon the modification of distortions of judgment. *Groups, Leadership and Men. Research in Human Relations* 177. 1951.

76. Milgram, S. Behavioral study of obedience. *Journal of Abnormal and Social Psychology* 67:371–8. 1963.
   Milgram, S. *Obedience to Authority: An Experimental View.* Pinter & Martin. 1977.

77. Bandura, A., Underwood, B., et al. Disinhibition of aggression through diffusion of responsibility and dehumanization of victims. *Journal of Research in Personality* 9:253–69. 1975.

78. Boesch, C., and Boesch-Achermann, H. *The Chimpanzees of the Tai Forest: Behavioral Ecology and Evolution.* Oxford: Oxford University Press. 2000.

79. Goodall, Jane. *Chimpanzees of Gombe: Patterns of Behavior.* Cambridge, MA: The Belknap Press of Harvard University. 1986. Page 200.

80. Boesch, C. Hunting strategies of Gombe and Tai Chimpanzees. In *Chimpanzee Cultures.* R. W. Wrangham, W. C. McGrewe, Frans de Waal, and P. G. Heltne, Eds. Cambridge, MA: Harvard University Press. 1994. Pages 77–92.

81. Boesch, C. Cooperative hunting roles among Tai chimpanzees. *Human Nature* 13:27–46. 2002.

82. Campbell, A. *A Mind of Her Own: The Evolutionary Psychology of Women.* Oxford: Oxford University Press. 2002.

83. Daly, Martin and Wilson, Margo. *Homicide.* New York: Aldine de Gruyter. 1988.

84. Smuts, B. Male Aggression Against Women. *Human Nature* 3:1–44. 1992.

85. Daly, Martin and Wilson, Margo. *Homicide.* New York: Aldine de Gruyter. 1988. Page 210.

86. Tinbergen, N. *The Herring Gull's World.* London: Collins. 1953.
   Clutton-Brock , T. H., Guiness, F. E., et al. *Red Deer: Behavior and Ecology of the Two Sexes.* Chicago: University of Chicago Press. 1982.

87. Meijering, P. H. *Signed With Their Honor: Air Chivalry During the Two World Wars.* New York: Paragon House. 1988. Page 43, quoting Udet, E. (1935). Mein Fliegerleben.
   Berry, H. *Make the Kaiser Dance: Living memories of a forgotten war. The American experience in World War I.* New York: Doubleday. 1987.

88. Gray, G. *The Warriors: Reflections on Men in Battle.* Lincoln, NE; London: University of Nebraska Press. 1959. Page.144.

89. Perkins, I. Interview. M. Potts. Berkeley, CA.

90. Darwin wrote (Darwin, C. *The Descent of Man, and Selection in Relation to Sex.* Princeton University Press. 1981. Vol. 2, Page 405.) ". . . we are not concerned here with hopes or fears, only with the truth as far as reason permits us to discover it." Quoted Thayer, Bradley A. *Darwin and International Relations: On the Evolutionary Origins of War and Ethnic Conflict.* Lexington: The University of Kentucky Press. 2004. Page 102.

91. Band of Brothers has been used as the title of several books: Stephen Ambrose's (1992) account of the 101st Airborne Division in WWII, Squadron Leader William Grierson's (1997) account of the bombing campaign over Europe, and a 2001 TV miniseries in the USA.

92. Flick, Nathaniel. *One Bullet Away: The Making of a Marine Officer.* Boston: Houghton Mifflin Company. 2005.

93. Crutchfield, James A. *George Washington: First in War, First in Peace.* New York: Macmillan. 2005. Page 40.

94. Churchill,Winston. Malakind Field Force. 1898. quoted Bartlett J. *Familiar Quotations.* Boston: Little, Brown & Company. 1968.

95. PBS *Our Century.*

96. Bourke, Joanna. *An Intimate History of Killing: Face-to-Face Killing in the Twentieth Century.* London: Granta Books. 1999. Page 57.

97. Gray, G. *The Warriors: Reflections on Men in Battle.* Lincoln, NE; London: University of Nebraska Press. 1959. Page 31.

98. Bourke, Joanna. *An Intimate History of Killing: Face-to-Face Killing in the Twentieth Century.* London: Granta Books. 1999. Page 32.

99.   Ibid. Page 31, quoting de Mann, Henry. *The Remaking of a Mind: A Soldier's Thoughts on War and Reconstruction.* London. 1920.

100.   de Hartog, Leo. *Genghis Khan: Conqueror of the World.* New York: Barnes & Noble. 1989.

101.   Montague, C.E. *Disenchantment.* London: Chatto and Windus. 1922.

102.   Gray, G. *The Warriors: Reflections on Men in Battle.* Lincoln, NE; London: University of Nebraska Press. 1959.

103.   Manchester, William. *Goodbye Darkness: A Memoir of the Pacific War.* Boston: Little, Brown. 1980. Page 391.

104.   Grossman, Dave. *On Killing: The Psychological Cost of Learning to Kill in War and Society.* Boston: Little, Brown. 1995.

105.   Marshall, S. L. A. *Men Against Fire.* Gloucester, MA: Peter Smith. 1978.

106.   Gabriel, R. A. *Military Psychiatry: A Comparative Perspective.* New York: Greenport Press. 1986.

107.   Weintraub. Stanley. *Silent Night: the Remarkable 1914 Christmas Truce.* New York: Free Press. 2001.

108.   Lord, F. A. *Civil War Collector's Encyclopedia.* Harrisburg, PA: The Stackpole Co. 1976.

109.   MacDonald Fraser, George. *Steel Bonnets: The Story of the Anglo-Scottish Border Reivers.* London: Barrie & Jackson. 1999.

110.   Hanson, Victor Davis. *The Western Way of War: Infantry Battle in Classical Greece.* Oxford: Oxford University Press. 1989. Page 26.

111.   Ibid. Page 9.

112.   Hanson, Victor Davis. *Carnage and Culture: Landmark Battles in the Rise of Western Power.* New York: Doubleday. 2001.

113.   Hanson, Victor Davis. *The Western Way of War: Infantry Battle in Classical Greece.* Oxford: Oxford University Press. 1989. Page 46.

114.   Ibid. Page 181.

115.   Krentz. Peter. Casualties in hoplite battles. *Greek, Roman and Byzantine Studies* 26:13–20. 1985.

116.   Hanson, Victor Davis. *The Western Way of War: Infantry Battle in Classical Greece.* Oxford: Oxford University Press. 1989. Page 203.

117.   Newby, P. H. *Warrior Pharaohs: The Rise and Fall of the Egyptian Empire.* London: Faber & Faber. 1980. Page 27.

118.   Riley-Smith, J. *The Oxford History of the Crusades.* Oxford: Oxford University Press. 1999. Page 240.

119.   Plutarch. *Lak Apophthemata.* Page 56.

120.   *Plutarch on Sparta.* Trans Richard Talbott. London: Penguin. 1988.

121.   Hodkinson, Stephen. Inheritance, marriage, and demography: perspectives upon the success and decline of classical Sparta. In *Sparta; New Perspectives.* S. Hodkinson, A Powell, Eds. London: Duckworth. 1999.

122.   Manchester, William. *Goodbye Darkness: A Memoir of the Pacific War.* Boston: Little, Brown. 1980.

123.   Ricks, Tomas E. *Making the Corps.* New York: Simon & Schuster. 1997. Page 88.

124.   Shepher, Joseph. Mate selection among second generation kibbutz adolescents and adults: incest avoidance and negative imprinting. *Archives of Sexual Behavior* 1:293–307. 1971.

125.   Blake, Joseph. A. Death by hand grenade: altruistic suicide in combat. *Suicide and Life-Threatening Behavior* 8:46–59. 1978.

126.   Gray, G. *The Warriors: Reflections on Men in Battle.* Lincoln, NE; London: University of Nebraska Press. 1959.

127.   Mason, Philip. *A Matter of Honour: An Account of the Indian Army, its Officers and Men.* London: Jonathan Cape. 1974.

128.   Robertson, Dudley. *George Mallory.* London: Faber and Faber. 2000.

129.   Keegan, John. *The Face of War: A Study of Agincourt, Waterloos and the Somme.* London: Jonathan Cape. 1976. Page 307.

130. Clinch, Nicolas. *A Walk in the Sky: Climbing Hidden Peak*. New York: The American Alpine Club Inc. 1982.

131. de Wall, Frans. *Our Inner Ape. A Leading Primatologist Explains Why We Are Who We Are*. New York: Riverside Books. 2005.

132. Fick, N. *One Bullet Away: The Making of a Marine Officer*. Boston: Houghton Mifflin Company. 2005. Page 106.

133. Parker, John. *The Gurkhas: The Inside Story of the World's Most Feared Soldiers*. London: Headline Book Publishing. 1999. Page 147.

134. Leckie, Robert. *Challenge for the Pacific; The Bloody Six Month Battle for Guadalcanal*. New York: Doubleday. 1968.

135. Ambrose, Stephen E. *Band of Brothers: E Company, 506th Regiment, 101st Airborne from Normandy to Hitler's Eagle's Nest*. New York: Simon & Schuster. 1992. Page 76.

136. Finkel, M. Playing War. *The New York Times Magazine*. December 24, 2000. Pages 30–50.

137. Beah, Ishmael. *A Long Way Gone: memoirs of a boy soldier by Ishmael Beah*. New York: Farrar, Straus & Giroux. 2007.

138. PBS *Hitler Youth*. October 26, 2000.

139. Fuller, G. The Demographic Backdrop of Ethnic Conflict: The Challenge of Ethnic Conflict to National and International Order in the 1990s. *Geographic Perspectives*. 151–156. 1995.

140. Mesquida, Christian.G., and Wiener, Neil I. Male age composition and the severity of conflicts. *Politics and the Life Sciences* 18:181–189. 1999.

141. Urdal, H. Population Pressure and Domestic Conflict: Assessing the Role of "Youth Bulges" in the Onset of Conflict. Fourth European International Relations Conference. University of Kent. 2001.

142. Warrick, J. CIA chief sees unrest growing with population growth. *Washington Post*. May 1, 2008.

143. Cincotta, Richard P., Engelman, Robert, Anastasion, Daniele. *The Security Demographic: Population and Civil Conflict After the Cold War*. Washington: Population Action International. 2003.

144. Watson, J. L. Self Defense Corps: Violence and Bachelor Subcultures in South China: Two Case Studies. *Proceedings of the Second Conference on Sinology, Academia Sinica*. Taipei: Academia Sinica. 1989.

145. Hudson, V. M., and den Boer, A. A Surplus of Men, A Deficit of Peace: Security and Sex Ratios in Asia's Largest States. *International Security*. 26:5–38. 2002.

146. Hudson, Valerie M., and den Boer, Andrea M. *Bare Branches: Security Implications of Asia's Surplus Women*. Cambridge, MA: MIT Press. 2004.

147. Worsnop, J. A Reevaluation of the "Problem" of Surplus Women in Nineteenth Century England. *Womens' International Studies Forum* 13:21–31. 1990.

148. Hudson, Valerie M., and den Boer, Andrea M. *Bare Branches: Security Implications of Asia's Surplus Women*. Cambridge, MA: MIT Press. 2004.

149. Donohue, J. L., and Levitt, S. D. The Impact of Legalized Abortion on Crime. *The Berkeley Law and Economics Working Papers*. 2001. http://www.bepress.com/blewp/default/vol2000/iss2/art7
   Donohue, John J. and Levitt, Steven D. The Impact of Legalized Abortion on Crime. *Quarterly Journal of Economics*. 379–420. 2000.
   Levitt, Steven, and Dubner, Stephen J. (2005). *Freakonomics: A Rogue Economist Explores the Hidden Side of Everything*. New York: HarperCollins. 2005.

150. Tietze, C. Two years' experience with a liberal abortion law: its impact on fertility trends in New York City. *Family Planning Perspectives* 5:36. 1973.

151. David, Henry.P., Dytrych, Z., Matejcek, Z., Schuller, V. *Born Unwanted: Developmental Effects of Denied Abortion*. Prague: Avicenum Czechoslovak Medical Press. 1988.

152. Slotow, R., van Dyke, G., Poole, J., Page, B., Klocke, A. Older bull elephants control young males. *Nature* 408:425–6. 2000.

153. Haddock, Vicki. Murderous otters prey on seal pups: serial killers elude marine experts. *San Francisco Chronicle*. December 10, 2000. Page 19.

154. Sapolsky, R. M., and Share, L. J. A pacific culture among wild baboons: its emergence and transmission. PloS 2 (4): e106.

155. Lewis Bernard. *The Assassins: A Radical Sect of Islam*. London: Phoenix. 2003. Page 130

156. Potts, D. Malcolm, and Potts, William T. W. *Queen Victoria's Gene: Haemophilia and the Royal Family*. Stroud, Gloucester: Alan Sutton Publishing. 1995. Page 77.

157. U.S. Congressional Budget Office. August 2007.

158. Sheftall, M. G. *Blossoms in the Wind: Human Legacies of the Kamikaze*. New York: Penguin. 2006. Page 214.

159. Ohnuki-Tieney, Emiko. *Kamikaze Diaries: Reflections of Japanese Student Soldiers*. Chicago: University of Chicago Press. 2006.

160. Sheftall, M. G. *Blossoms in the Wind: Human Legacies of the Kamikaze*. New York: Penguin. 2006. Page 221.

161. Shakur, Sanyika aka Monster Kody Scott. *Monster: The Autobiography of an L.A. Gang Member*. New York: Penguin Books. 1993. Page 11.

162. Alonso, Alex. *Steetgangs.Com. Magazine*. Update December 22, 2002.

163. Shakur, Sanyika. aka Monster Kody Scott. *Monster: The Autobiography of an L.A. Gang Member*. New York: Penguin Books. 1993. Page 11.

164. Terraine, John. *Right of the Line: The Royal Airforce in the European War 1939–1945*. London: Hodder & Stoughton Ltd. 1985. Page 208.

165. http://en.wikipedia.org/wiki/World_II_casualties. Accessed March 21, 2008.

166. Perlman, D. The next big one: worse than '06? *San Francisco Chronicle*. March 21, 2008. Page 1.

167. Gulf of Tonkin measure voted in haste and confusion in 1964. *The New York Times*. June 25, 1970.

168. Jenkins inflamed public opinion by displaying his ear—suitably pickled—in the House of Commons, London. Britain invaded Florida and the "War of Jenkins' Ear" lasted until 1743.

169. Shakur, Sanyika aka Monster Kody Scott. *Monster: The Autobiography of an L.A. Gang Member*. New York: Penguin Books. 1993. Page 79.

170. Hrdy, Sarah. B. *Mother Nature: A History of Mothers, Infants, and Natural Selection*. New York: Pantheon Books. 1999.

171. Hoffman, B. All you need is love: How the terrorists stopped terrorism. *The Atlantic* 288:34–7. 2001.

172. Washington, Ebonya, and Leonhardy, David. Children, the littlest politicians. *New York Times*. February 19, 2006.
    Washington, Ebonya. Female socialization: how daughters affect their legislator fathers' voting on women's issues. *The American Economic Review* 98:311–32. 2008.

173. Garcia-Moreno, Claude, Heise, L., Jensen, H.A.F.M., Ellsberg, M., Watts, C. Violence against women. *Science* 310:1282–3. 2005.

174. Hrdy, Sarah. B. *Mother Nature: A History of Mothers, Infants, and Natural Selection*. New York: Pantheon Books. 1999. Page 37.
    Hrdy, Sarah B. *The Langurs of Abu: Female and Male Strategies of Reproduction*. Cambridge, MA: Harvard University Press. 1977.

175. Gowaty, Patricia A. *Feminism and Evolutionary Biology: Boundaries, Intersections and Frontiers*. New York: Chapman and Hall. 1997.
    Gowaty, Patricia A. Birds face sexual discrimination. *Nature* 385:486–7. 1997.
    Pusey, Anne, and Goodall, Jane. The Influence of Dominance Rank on the Reproductive Success of Female Chimpanzees. *Science* 211:1171–93. 1997.

176. Tutin, Caroline E. G. Mating patterns and reproductive strategies in a community of chimpanzees. *Behavioral Ecology and Sociobiology* 6:29–38. 1979.

177. Potts, Malcolm, and Short, Roger. *Ever Since Adam and Eve: The Evolution of Human Sexuality*. Cambridge, Cambridge University Press. 1999.

178. The *Demographic and Health Surveys,* now conducted in a great many countries, include a module on domestic violence. It is common across nearly all cultures, from Latin America to

Africa (e.g. 2003 Kenyan DHS) and to Asia (e.g. data from India). http://www.measuredhs/aboutdhs.

179. Susman, R. L. *The Pygmy Chimpanzee.* New York: Plenum Press. 1984.

   Kano, T. *The Last Ape: Pygmy Chimpanzee Behavior and Ecology.* Stanford, CA: Stanford University Press. 1992.

   De Waal, Frans, and Lanting, F. *Bonobos: The Forgotten Ape.* Berkeley: University of California Press. 1997.

180. DeBartolo, Anthony. "Newest" apes are teaching us about ourselves. *Chicago Tribune.* June 11, 1998.

181. Hrdy, S. B. *Mother Nature: A History of Mothers, Infants, and Natural Selection.* New York: Pantheon Books. 1999.

182. Whyte, Martin K. *The Status of Women in Preindustrial Society.* Princeton: Princeton University Press. 1978.

183. Shepard, D. J. The Elusive Warrior Maiden Tradition: Bearing Weapons in Anglo-Saxon Society. In *Ancient War.* Ed. A. Harding. Trowbridge, Wiltshire: Sutton Publishing. 1999. Pages 219–43.

184. Costello, John. *Love, Sex and War: Changing Values 1939–45.* London: Collins. 1985. Page 54.

185. Ibid. Page 47.

186. Regan, T. Some women soldiers say they carry knives to protect themselves from other US soldiers. *Christian Science Monitor.* March 19, 2007.

187. Fisher I. Like mother, like daughter, Eritrean women wage war. *The New York Times.* August 26, 1999. Pages A1, A8.

188. Takeste, Assefaw. Personal communication. 2007.

189. Gray, G. (1959). *The Warriors: Reflections on Men in Battle.* Lincoln, NE and London: University of Nebraska Press. 1959. Page 61.

190. Fagan, Brian. Battles at the bottom of the world. In *The Archaeology of War.* Intro. Mark Rose. New York: Hatherleigh Press. 2005. Pages 23–2.

191. Vayda, A. P. *Maori Warfare.* Wellington: The Polynesian Society Incorporated. 1960.

192. Sheftall, M. G. *Blossoms in the Wind: Human Legacies of the Kamikaze.* New York: Penguin. 2006. Only Americans were admitted to the Recreation and Entertainment Association, and the 8 cent entrance fee included a bottle of beer.

193. Richie, A. *Faust's Metropolis: A History of Berlin.* New York: Carroll & Graf. 1998.

194. Edgerton, Robert B. *Warrior Women: The Amazons of Dahomey and the Nature of War.* Bolder, CO: Westview Press. 2000.

   Alpern, Stanley B. *The Amazons of Black Sparta: The Women Warriors of Dahomey.* New York: New York University Press.1998.

195. See Campbell, Kenneth, and Wood, Jim. Fertility in Traditional Societies. In *Natural Human Fertility.* M. Potts, P. Diggory, S. Teper, Eds. London: Macmillan Press. 1988. Pages 39–69.

196. National Institute of Population Research and Training and ORC Macro 2007. *Bangladesh Demographic and Health Survey 2007 Preliminary Report.* Dhaka, Bangladesh and Calverton, MD: National Institute of Population Research and ORC Macro. 2007.

197. Huxley, Thomas Henry. *Evolution and Ethics and Other Essays.* 1894. Page 111.

198. Friend, T. "Iceman" was murdered, science sleuths say. *USA Today.* 2003. See same story at www.sciscoop.com/story/2003/8/12/7419/29586.

199. Benenson, A. Murder or War? In *The Archaeology of War.* Intro. Mark Rose. New York: Hatherleigh Press. 2005. Page 7.

200. Keith, A. *A New Theory of Human Evolution.* London: Watts. 1948.

201. Wright, Quincy. *A Study of War.* Vol 1. Chicago: University of Chicago Press. 1942. (Abridged edition published 1964).

   Turney-High, H. H. *Primitive War: Its Practice and Concepts.* Colombia: University of South Carolina Press. 1971.

202. The Taung child, the first Australopithecine skull Dart found, is now thought to show evidence of attack by a large eagle. Berger, L., and Clark, R. Eagle involvement in the accumulation of the Taung child fauns. *Ethology and Sociobiology* 12:315–33. 1995.

203. Dart, R. The predatory transition from ape to man. *International Anthropological and Linguistic Review* 1:201–17. 1953.

204. Roper, M. K. A Survey of Evidence for Intra-Human Killing in the Pleistocene. *Current Anthropology* 10: 427–59. 1969.

205. Pennisi, E. Was Lucy's a fighting family? Look at her legs. *Science* 311:330. 2006.

206. Weidenreich, F. The Skull of Sindathropus Pekinensis. *Palaeotologica Sinica NS.* 10. 1943.

207. Boaz, N., Chiochon, Q. Xu, and Liu, J. Large mammalian carnivores as a taphonomic factor in the bone accumulations at Zhoukoudian. *Acta Anthroplogica Sinica* (Suppl) 19:224–34. 2000.

208. Hart, Donna, and Sussman, Robert.W. *Man the Hunted: Primates, Predators and Human Evolution.* New York: Westview. 2005.

209. Tutin, Caroline, McGrew, W., and Baldwin, P. Social organization of savana dwelling chimpanzees, *Pan troglodytes verus,* at Mt. Assirik, Senegal. *Journal of Medical Primatology* 20:357–60. 1983.

210. Arsuaga, J. L. *The Neanderthal's Necklace: In Search of the First Thinkers.* New York: Four Walls, Eight Windows. 2002.

211. White, T. D. Cut-Marks on the Bodo Eranium: A Case of Prehistoric Defleshing. *American Journal of Physical Anthropology* 69:503–11. 1968.
     White, T. D. Once We Were Cannibals. *Scientific American* 265:58–65. 2001.

212. LeBlanc, Steven. A. *Prehistoric Warfare in the American Southwest.* Salt Lake City: University of Utah Press. 1999.
     LeBlanc, Steven A. and Register, K. E. *Constant Battles: The Myth of the Peaceful Noble Savage.* New York: St. Martin's Press. 2003.

213. Weiss, Elizabeth. Kennewicks Man's Funeral; The Burying of Scientific Evidence. *Politics and the Life Sciences* 20 (1):13–18. 2001.
     Holdren, C. Court battle ends: bones still off limits. *Science* 305:591. 2004.

214. Bachechi, L., Fabbri, P .E., Mallegni, F. An Arrow-Caused Lesion in a Late Upper Paleolithic Human Pelvis. *Current Anthropology* 38:135– 40. 1997.

215. Wendorf, F. Site 117: A Nubian final Paleolithic graveyard near Jebal Sahaba, Sudan. In *The Prehistory of Nubia.* Ed. F. Wendorf. Vol. 2: 954–95. 1968.

216. Frayer, David. Ofnet: evidence for a Mesolithic massacre. In *Troubled Times: Violence and Warfare in the Past.* Martin, Debra, and Frayer, David W., Eds. Amsterdam: Gordon and Breach. Pages 181–215. 1998.

217. Teschler-Nicola, M., Gerold, F., et al. Evidence of Genocide 7000 BP—Neolithic Paradigm and Geo-climate Reality. *Coll. Anthropology.* 23:437–50. 1999.

218. Latkoczy, C., Prohaska, T., et al. Investigation of Sr Isotope Ratios in Prehistoric Human Bones and Teeth Using Laser Ablation ICP-MS and ICP-MS after Rb/Sr Separation. *Journal of Analytical Atomic Spectroscopy* 17:887–91. 2002.

219. Lawler, A. Murder in Mesopotamia? *Science.* 317:1164–65. 2007.

220. Wilford, John N. A 1,200-year-old murder mystery in Guatemala. *New York Times.* November 17, 2005.

221. Blick, J. Genocidal warfare in tribal societies as a result of European-induced culture conflict. *Man,* New Series. 23:654–70. 1988.

222. LeBlanc, S. A. *Prehistoric Warfare in the American Southwest.* Salt Lake City: University of Utah Press. 1999.

223. Zimmerman, L. J., Whitten, R. G. Prehistoric Bones Tell a Grim Tale of Indian V Indian + Possible South-Dakota Massacre Site. *Smithsonian* 11:100–108. 1980.
     Zimmerman, Larry J. The Crow Creek massacre. www,usd/anth/crow/crow1.html.
     Willey, P. *Prehistoric Warfare on the Great Plains: skeletal analysis of the Crow Creek massacre victims.* New York: Garland. 1990.

224. Benenson, Alexander. Unearthing a violent past. In *The Archaeology of War.* Intro. Mark Rose. New York: Hatherleigh Press. 2005. Pages 11–13.

225. Owsley, D. H., Berryman, D., et al. Demographic and osteological evidence of warfare at the Larsen site, South Dakota. Plains Anthropology Memoirs. 13:119–31. 1977.

226. Krober, C. B., Fontana, B. L. *Massacre at Gila: An Account of the Last Major Battle Between American Indians, with Reflections on the Origins of War*. Tucson: University of Arizona Press. 1986.

227. Sandin, Benedict. *The Sea-Dayaks of Borneo before White Rajah Rule*. London: Macmillan. 1967.

228. Caesar (100 BC–44 BC). *The Conquest of Gaul*. London: Penguin Books. 1982.

229. Strabo (64 BC–AD 24) *Geography*. Cambridge, MA, Horace White: Harvard University Press. 1932.

230. Brunaux, J.-L. (1996). *Les religions Galoises*. Paris: Editions France. 1996.

     Brunaux, J.-L. Pleasing the Gods; Terrifying the Enemy: Gallic Blood Rites. *Archaeology* 54:57. 2001.

231. Morgan, J. *The Life and Adventures on William Buckley: Thirty-two Years a Wanderer Amongst the Aborigines of the Unexplored Country Round Port Philip*. Canberra: Australian National University. 1852 (Republished 1979).

232. Darwin, Charles. *Voyage of the Beagle*. London: Henry Colburn. 1839. Penguin Books edition, 1989. Page 178.

233. Read, K. Morality and the Concept of the Person Among the Gahuku-Gama, Eastern Highlands, New Guinea. *Oceania* 25:233–82. 1955.

     Read, Kenneth. *The High Valley*. New York: Charles Scribner's Sons. 1965.

     Diamond, Jared. *Guns, Germs, and Steel*. New York: W.W. Norton and Company. 1997.

234. Feil, D. K. *The Evolution of Highland Papua New Guinea Societies*. Cambridge: Cambridge University Press. 1987.

235. Langness, Lewis. *The Life History in Anthropological Science*. New York: Holt Rinehart & Winston. 1965.

236. Chagnon, Napoleon A. Yanomamö social organization and warfare. In *The Anthropology of Armed Conflict and Aggression*. Fried, M., Harris, M., Murphy, R. New York: Natural History Press. 1968.

     Chagnon, Napoleon A. *Yanomamö: The Fierce People*. New York: Holt, Rinehart & Winston. 1968.

     Biocca, E. *Yanomama: The Story of Helena Valero, a Girl Kidnapped by Amazonian Indians as told to Ettore Biocca*. New York: Kondasha America, Inc. 1996.

237. Chagnon, Napoleon A. Life histories, blood revenge, and warfare in a tribal population. *Science* 239:985–92. 1988.

238. Goodall, Jane. *Chimpanzees of Gombe: Patterns of Behavior*. Cambridge, MA: The Belknap Press of Harvard University. 1986. Page 330.

239. Bowden, M. *Black Hawk Down: A Story of Modern War*. New York: Atlantic Monthly Press. 1994.

240. Schlenger, W. E., Caddell, J. M, Ebert, L., Jordan, B. K., Rouke, K. K., Wilson, D. Psychological reactions to terrorist attacks: Findings from a National Study of American's Reactions to September 11. *Journal of the American Medical Association* 5:581–8. 2002.

241. Catalano, Ralph, Kesssell, E., McConnell, W., Pirkle, E. Psychiatric emergencies following the attacks of Spetember 11, 2001. *Psychiatric Services* 55:163–6. 2004.

242. Holman, Alison, Cohen Silver, Roxane, Poulin, Michael, Andersen, Judith, Gil-Rivers, Virginia., McIntosh, Daniel N. Acute stress and cardiovascular health: a 3-year study following the September 11th attacks. *Archives of General Psychiatry* 65:73–80. 2008.

     Tierney J. Living in fear and paying a high cost in heart risk. *New York Times*. January 15, 2008. Page D1.

243. Rosenthal, A. M. *Thirty-eight Witnesses: The Kitty Genovese Case*. Berkeley: University of California Press. 1999.

244. McNamara, R. S., and VanDeMark, B. (1995). *In retrospect: the tragedy and lessons of Vietnam*. New York: Times Books. 1995. Page 414.

245. Morton, Tom. ABC Radio National: *Torn Curtain—The Secret History of the Cold War*. June 11, 2006.

246. Builder, Carl H. *The Masks of War: American Military Styles in Strategy and Analysis*. Baltimore: Johns Hopkins Press. 1989.

247. Stewart, K. Mohave Warfare. In *The Californian Indians: A Source Book*. Ed. M. A. Whipple. Berkeley, University of California Press. 1962.

248. Eckhart, W., Wilkinson, D. *Civilization, Empires and Wars: A Quantitative History of War*. Jefferson, North Carolina: McFarland & Company. 1992.

249. Heider, K. G. *Grand Valley Dani: Peaceful Warriors*. New York: Holt Rinehart and Winston, 1979. Page 106.

250. In eight hours on the blue Aegean Sea at Salamis, on a September day in 480 BC, 40,000 Persian sailors and soldiers were drowned, but this was probably less than 70 per 100,000 of the population of the Persian Empire.

251. Badsey, S. (Ed). 2000. *Atlas of World War II Battle Plans: Before and After*. Oxford: Helicon. Operation Barbarossa opened on a 1,000-mile front on June 22, 1940 with 3,360,000 men, 3,600 guns, and 2,500 aircraft.

252. Wright, Quincy. *A Study of War*. Chicago: Chicago University Press. 1942. (2nd edition 1965).

253. Talbott, Strobe (trans). *Khrushchev Remembers; The Last Testament*. New York: Little Brown. 1974.
McNamara, Robert S. *Blundering into Disaster: Surviving in the First Century of the Nuclear Age*. New York: Pantheon Books, 1986.
McNamara, Robert S. (with Brian Van-DeMark). *In Retrosepct: The Tragedy and Lessons of Vietnam*. New York: Times Books. 1995.

254. Rees, Sir Martin. *Our Final Hour: A Scientist's Warning: How terror, error, and environmental disaster threaten humankind's future in this century—on earth and beyond*. New York: Basic Books. 2003. Page 26.

255. Schlesinger, A. H. Jr. *New York Times*. October 1, 2002.

256. Chagnon, Napoleon. *Yanomamö: The Fierce People*. Fort Worth: Harcourt Brace Jovanovich. 1992.

257. Faulkner, R. O. Egyptian Military Organisation. *Journal of Egyptian Archaeology* 39:32–47. 1953.

258. Lloyd, A. B. (Ed.) *Battle in Antiquity*. London: Dickerson. 1966. Page 233.

259. Vayda, A. P. *Maori Warfare*. Wellington: The Polynesia Society Incorporated.
Clarke, George. *Notes on Early Life in New Zealand*. Hobart: Watch. 1903. Page 10.

260. Dunnigan, James F. *A Comprehensive Guide to Modern Warfare*. New York: William Morrow. 1982. Page 331.
van Creveld, Martin, *Technology and War: From 2000 BC to the Present*. Toronto: Maxwell Macmillian Canada. 1991.

261. George Roux, George. *Ancient Iraq*. New York: Penguin. 1986.

262. Newby, P. H. *Warrior Pharaohs: The Rise and Fall of the Egyptian Empire*. London: Faber and Faber. 1980.

263. Dupuy, R. Ernest, and Dupuy, Trevor. *Encyclopedia of Military History from 3500 BC to Present*. New York: Harper and Row. 1970. Page 286.

264. Dunnigan, James F. *A Comprehensive Guide to Modern Warfare*. New York: William Morrow. 1982. Page 215.

265. Boehm, Christopher. *Hierarchy in the Forest: The Evolution of Egalitarian Behavior*. Cambridge, MA: Harvard University Press. 1999. Page 61.

266. Gabriel, Richard & Karen Metz. *Sumer to Rome: The Military Capabilities of Ancient Armies*. New York: Greenwood Press. 1991.

267. Boehm, Christopher. *Hierarchy in the Forest: The Evolution of Egalitarian Behavior*. Cambridge, MA: Harvard University Press. 1999. Page 18.

268. Wrangham, Richard. Is military incompetence adaptive? *Evolution and Human Behavior* 20:3–17. 1999.

269. Johnson, Dominic, Wrangham, Richard, Rosen, Stephen P. Is military incompetence adaptive? An empirical test of risk-taking behavior in modern warfare. *Evolution and Human Behavior* 23:245–64. 2002.

270. Regan, Geoffrey. *Someone Had Blundered: A Historical Survey of Military Incompetence*. London: Batsford. 1987. Page 232.

271. Ibid. Page 263.

272. Lukas, John. *Five Days In London, May 1940*. New Haven: Yale University Press. 1999.
273. Caesar, Julius. *The Conquest of Gaul* IV.1 (Trans. S. A. Handforth). London: Penguin Books. Page 88.
274. Hanson, Victor Davis. *Carnage and Culture: Landmark Battles in the Rise of Western Power*. New York: Doubleday. 2001.
275. Holmes, Richard. Battle: The Experience of Modern Combat. In *The Oxford History of Modern War*. Ed. Charles Townshend. Oxford: Oxford University Press. 2000. Page 198.
    Holmes, Clive. *The Gentry in England and Wales, 1500–1700*. Stanford: Stanford University Press. 1994.
276. Madden, David. *Beyond the Battlefield: The Ordinary Life and Extraordinary Times of the Civil War Soldier*. New York: Touchstone. 2000.
277. Beevor, Antony. *Stalingrad: The Fateful Siege: 1942–1943*. London: Viking. 1998. Page 170.
278. Holper, J. J. Kin term usage in the Federalist: Evolutionary foundations of public rhetoric. *Politics and the Life Sciences* 15:265–72. 1996.
279. Ellis, John. *Eye-Deep in Hell: Trench Warfare in World War I*. Baltimore: Johns Hopkins Press. 1989. Page 84.
280. Miles, Maria. *Patriarchy and Accumulation on a World Scale: Women in the International Division of Labour*. London: Zen Books. 1986.
281. Gabriel, R. A., Metz, K. S. *From Sumer to Rome: The Military Capabilities of Ancient Armies*. Westport, NY: Greenwood Press. 1991. Page 87.
282. Ephesians 6:5.
283. Jopling, Carol. *The Coppers of the Northwest Coast Indians: Their Origin, Development and Possible Antecedents*. Philadelphia: American Philosophical Society. 1989.
284. Meillassoux, C. *L'esclavage en Afrique précoloniale*. Paris: Maspero. 1975.
285. *Exodus*. 21:2–6
286. Levy, Barry S., and Sidel, Victor W. (Eds.) *War and Public Health*. Washington: American Public Health Association. 2000.
287. Adams, Gordon. Iraq's sticker shock. *Foreign Policy* March/April 2007. Page 34.
288. Survey of the defence industry. *The Economist*. July, 20, 2002.
289. United Nations Development Program. *Human Development Report*. New York: Oxford University Press. 1994.
290. Finn, Mark V., Geary, David C., Ward, Carol V. Ecological dominance, social competition, and coalitionary arms races: Why human evolved extraordinary intelligence. *Evolution and Human Behavior* 26:10–46. 2005.
291. Carneiro, Robert. A theory of the origin of the state. *Science*. August 21, 1970. Pages 734–7.
292. Samuel. 17:45
293. Newby, P. H. *Warrior Pharaohs: The Rise and Fall of the Egyptian Empire*. London: Faber and Faber. 1980. Page 65.
294. Ekelund, R. B., Hebert, R. F., Tollison R. D., Anderson, G. M., Davidson, A. M. Sacred Trust: The Medieval Church as an Economic Firm. New York: Oxford University Press. 1996.
295. Woodward, Bob. *Plan of Attack*. New York: Simon and Schuster. 2004. Page 282.
296. *New York Times*. March 11, 2003. Page A28.
297. Charles Darwin. 1871. *The Descent of Man and Selection in Relation to Sex*. New York: The Modern Library. Page 470.
298. Donald Brown. *Human Universals*. New York: McGraw Hill. 1991.
299. Hardy, Sir Alister. *The Divine Flame*. London: Collins. 1966. Page 174.
300. Lee A. Kilpatrick. *Attachment, Evolution, and the Psychology of Religion*. New York: The Guildford Press. 2005.
301. Wilson, Edward O. *Consilience: The Unity of Knowledge*. New York: Alfred Knopf. 1998.
302. Wilson, David Sloan. *Darwin's Cathedral: Evolution, Religion and the Nature of Society*. Chicago: University of Chicago Press. 2002.
303. Media & Society Research Group. Restrictions on Civil Liberties, Views of Islam, and Muslim Americans. Quoted in *The Atlantic Monthly*. April 2005. Page 44.

304. Plante, Thomas G. and Thoresen, Carl T. *Spirit, Science, and Health: How the Spiritual Mind Fuels Physical Wellness*. Westport, CT: Prager. 2007.

Steffen, P. R., Hinderliter, A. L., Blumenthal, J. A., Sherwood, A. Religious coping, ethnicity, and ambulatory blood pressure. *Psychosomatic Medicine* 63: 523–30. 2001.

Oman, D., Kurata, J. H., Strawbridge, W. J., Cohen, R. D. Religious attendance and cause of death over 31 years. *International Journal of Psychiatry in Medicine* 32:69–89. 2002.

305. Hummer, R. A., Rogers, R. G., Nam, C. B., Ellison, C. G. Religious involvement and {U.S.} adult mortality. *Demography* 36, 273–85. 1999.

306. Yagoub, Yousif Al-Kandari. Religiosity and its relation to blood pressure among selected Kuwait-is. *Journal of Biosocial Science* 35:463–73. 2003.

307. Waller, N., Kojetin, B., Bouchard,, T., Lykken, D., Tellegen, A. Genetic and environmental influences on religious interests, attitudes, and values: A study of twins reared apart and together. *Psychological Science* 1: 138–42. 1990.

308. Rodinson, Maxime (Trans. Anne Carter). *Mohammed*. London: Penguin Books. 1971. Page 167.

309. Dupuy, R. Ernest and Dupuy, Trevor. *Encyclopedia of Military History from 3500 BC to Present*. New York: Harper and Row. 1970. Page 848.

310. Carrasco, David. *City of Sacrifice: The Aztec Empire and the Role Of Violence in Civilization*. Boston: Beacon Press. 1999.

311. Duran, Diego. *The History of the Indies of New Spain*, c. 1581 (Trans. Doris Heyden). Oklahoma City: University of Oklahoma Press. 1994.

312. Soisson, Pierrs & Janine Soisson. *Life of the Aztecs in Ancient Mexico*. Geneva: Editions Minerva SA. 1978.

313. Hassig, Ross, *Aztec Warfare: Imperial Expansion and Political Control*. Norman: University of Oklahoma Press. 1989. Page 124.

314. Soisson, Pierre. *Life of the Aztecs in Ancient Mexico*. Geneva: Editions Minerva SA. 1978.

315. Benjamin, Daniel, and Simon, Steven. *The Age of Sacred Terror*. New York: Random House. 2002.

316. Lewis, C. S. *The Screwtape Letters*. London: Collins. 1942.

317. Lewis, C. S. *A Grief Observed*. London: Collins. 1961.

318. Potts, Malcolm. Grief has to be. *The Lancet*. 343:279. 1994.

319. Rhodes, Richard. *The Making of the Atomic Bomb*. New York: Simon and Schuster. 1986. Page 704.

320. de Waal, Frans. *Peacekeeping among Primates*. Boston, MA: Harvard 1989. Page 10.

321. Goodall, Jane. *Chimpanzees of Gombe: Patterns of Behavior*. Cambridge, MA: The Belknap Press of Harvard University. 1986. Page 556. Quoting A. Krotlandt.

322. Gabriel, R. A., Metz, K. S. *From Sumer to Rome: The Military Capabilities of Ancient Armies*. Westport, NY: Greenwood Press. 1991.

323. Edward Mewan, E., Miller, R., Bergson, C. Early bow design and construction. *Scientific American* June.1991.

324. Buchan, J. *Montrose*. London: Nelson 1928.

325. Chapman, John. The Origins of Warfare in the Prehistory of Central and Eastern Europe. In *Ancient Warfare*. John Carman and Anthony Harding (Eds). Stroud, Gloucestershire: Sutton. Pages 101–142. 1999.

326. Ritchie, W. F., and Ritchie, J. N. G. *Celtic Warriors*. Princess Risborough: Shire Archaeology. 1997.

327. Frolich H. *Die Militarmedicin Homers*. Stuggart 1879. Gabriel, R.A., Metz, K.S. *From Sumer to Rome: The Military Capabilities of Ancient Armies*. Westport, NY: Greenwood Press. 1991 Quoted Page 9.

328. Richie, W. F. & J. N. G. Ritchie. 1997. *Celtic Warriors*. Princess Risborough: Shire Archaeology. 1997.

329. Postgate, J. N. *Taxation and Conscription in the Assyrian Empire*. Rome: Biblical Institute Press. 1974.

330. Brocon, G. I. *The Big Bang: A History of Explosives*. Stroud, Gloucestershire: Sutton. 1998.

331. Davis, Victor Hanson. *Carnage and Culture: Landmark Battles in the Rise of Western Power.* New York: Random House. 2001.Page 64.

332. Fornander, Abraham. *An Account of The Polynesian Race Its Origins and Migrations and the Ancient History of the Hawaiian People to the Times of Kamehameha 1*. London: Turner & Co, 1880.

333. van Creveld, M. *Technology and War: from 2000 BC to the Present*. London: Collier Macmillan.1981.

334. Ibid. Page 38.

335. McCallum, H. D., McCallum F.T. *The Wire That Fenced the West*. Norman: University of Oklahoma Press. 1965.

336. Cross, Wilbur. 1991. *Zeppelins of Word War I*. London: Paragon House. Page 19.

337. *Kolnische Zeitung,* January 21, 1915.

338. Wyatt, R. J. *Death From the Skies: The Zeppelin Raids over Norfolk, 19 January, 1915*. Norwich: Gliddon Books. 1990.

339. Jones, H. A. *The Air War*. Oxford: Oxford University Press. 1937. Appendix.

340. Hastings, Max. *Bomber Command,* London: Michael Joseph. 1979. Page 19.

341. Blair, Clay. *Hitler's U-Boat War: The Hunters*, 1939–1942. New York: Random House. 1996.

342. Hastings, Max. *Bomber Command*. London: Michael Joseph. 1979. Page 106.

343. Ibid. Page 94.

344. Lee, Raymond. *The London Observer*. London: Hutchinson. 1972. Page 372.

345. Taylor, A. J.P. *The Second World War*. London: Hamish Hamilton. 1975. Page 129.

346. Bishop Bell said, "I desire to challenge the government on the policy which directs the bombing of the enemy towns on the present scale. . . . It is of immense importance that we who are the liberators of Europe should so use power that it is always under the control of war."

347. Terraine, John. *Right of the Line; The Royal Air Force in the European War 1939–1945*. London: Hodder & Stoughton.1985. 45,000 people were rendered homeless and nearly 400 killed. The RAF lost under four percent of their force, fewer than in many smaller raids. Page 425.

348. Caidin, Martin. *The Night Hamburg Died: Allied Bombs Burned a City to Death*. New York: Ballantine Books. 1960.

349. Sir Arthur Harris. *Bomber Offensive*. London: Collins. 1947. Page 267.

350. Terraine, John. *Right of the Line; The Royal Air Force in the European War 1939–1945*. London: Hodder & Stoughton.1985. Page 682.

351. Fourth Supplement of *The London Gazette* No. 36235 of Friday, November 5, 1943.

352. Hinchcliffe, Peter. *The Other Battle: Luftwaffe Night Aces versus Bomber Command*. London: Zenith Press. 1996.

353. Hilling, John. *Strike Hard: A Bomber Airfield at War—RAF Downham Market and its Squadrons, 1942–1946*. Stroud, Gloucestershire: Alan Sutton Publishing Limited. 1996. Page 89.

354. Sherry, Michael S. *The Rise of American Air Power: The Creation of Armageddon*. New Haven: Yale University Press. 1987. Page 406.

355. Ohnuki-Tierney, Emiko. *Kamikaze Diaries: Reflections of Japanese Student Soldiers*. Chicago: University of Chicago Press. 2006.

356. Davis, Victor Hanson. *Carnage and Culture: Landmark Battles in the Rise of Western Power.* New York: Random House. 2001. Page 84.

357. Ibid. Page 104.

358. Rossabi, M. All the Khan's horses. *Natural History* October 1994. Pages 49–56.

359. Nicolle, D, Shpakovsky V. *Kalka River 1223: Genghiz Khan's Mongols invade Russia*. Wellingborough Northants: Osprey Books. 2001.

360. Norwich University. Master of Arts in Military History on Line. http:www,historyguy.com/GulfWar.html (accessed March 16, 2008).

361. Fornander, Abraham. *Ancient History of the Hawaiian People to the Times of Kamehameha 1*. Honolulu: Mutual Publishing. 1996 (Reissue).

362. Vayda, A. P. *Maori Warfare*. Wellington: The Polynesian Society Incorporated. 1960.

363. Reuter, Christopher and Ragg-Kirby. Helene. *My Life as a Weapon: Modern History of Suicide Bombing*. Princeton: Princeton University Press. 2004. Page 86.

364. Rhodes, Richard. *The Making of the Atomic Bomb*. New York: Simon and Schuster. 1986. Page 725.

365. Broad, W. J. From the start, the space race was an arms race. *New York Times*. September 25, 2007. Page D10.

366. Sobel, Dava. *Galileo's Daughter: A Historical Memoir of Science, Faith and Love*. New York: Walker Publishing. 1999.

367. Johnson, David. *V-1 V-2 Hitler's Vengeance on London*. Chelsea: Scarborough House. 1981.

368. Ellis, John. *The Social History of the Machine Gun*. Manchester: Ayer Publishing. 1981. Page 56.

369. Howard, Martin. *Wellington's Doctors: the British Army Medical Services in the Napoleonic Wars*. Staplehurst: UK Spellmount. 2002.

370. Isaiah 37:36.

371. McNeil, William. *Plagues and People*. London: Doubleday. 1977. Page 94.

372. Lobell, J. A. Digging Napoleon's dead. In *The Archaeology of War*. Intro. Mark Rose. New York: Hatherleigh Press. 2005. Pages 114–7.

373. Mann, Charles C. *1491: Discovering what Americas were like before Columbus*. New York: Knopf. 2005.

374. Caesar. *The Conquest of Gaul*. Trans S.A. Handford. London: Penguin Books. 1951. VI. 43.

375. Deuteronomy 20:13–14.

376. Exodus 21:24–25.

377. Luke 6:29.

378. Tertullian. *On Ideology*. Trans S. Thewall. 27.

379. Aristotle. *Politics* 1,7,12552, 3,12555b.

380. St. Ambrose. *On the Christian Faith* Bk II, XVI, 136, 142, pages 241–2.

381. Christopher, P. *The Ethics of War and Peace: An Introduction to Legal and Moral Issues*. New Jersey: Prentice Hall. 1999. Page 48.

382. Russell, F. H. *The Just War in the Middle Ages*. Cambridge: Cambridge University Press. 1975. Page 308.

383. Nef, John U. *War and Human Progress on the Rise of Industrial Civilization*, Cambridge, MA: Harvard University Press. 1952.

384. Wedgewod, C.V. *The Thirty years War*. New Haven: Princeton. 1939. Pages 512–3.

385. Grotius, Hugo. *The Law of War and Peace*. New York: Boobs-Merrill Co. 1962.

386. *Encyclopedia American* 17: 143–144.

387. Massie, Robert K. *Dreadnought: Britain, Germany, and the Coming of the Great War*. New York: Random House. 1991. Page 431.

388. Krugman, Paul. King of pain. *The New York Times* September 18, 2006. Page A29.

389. Friedman, Leon (Ed.). *The Law of War: A Documentary History*. New York: Random House. 1972. Vol. 1. Page 3.

390. Cicero. *De Officiis* Bk 1. XI. Page 35.

391. Hartigan, Richard Shelly. *The Forgotten Victim: A History of the Civilian*. Chicago: Precedent. 1982. Pages 651–66.

392. Jehl, Douglas, and Schmitt, Eric. In Abuse, a Portrayal of Ill-Prepared, Overwhelmed GI.'s *The New York Times*. May 9, 2004. Page 1.

393. Goldhagen, Daniel J. *Hitler's Willing Executioners: Ordinary Germans and the Holocaust*. New York: Knopf. 1996.

394. Argell, N. *The Public Mind*. London: Noel Douglas. 1926. Page 75.

395. Kiernan, Ben, *The Pol Pot Regime: Race, Power, and Genocide in Cambodia under the Khmer Rouge. 1975–79*. New Haven: Yale University Press. 1996.

396. Slack, Jack. *Lieutenant Calley: His Own Story*. New York City: The Viking Press. 1970. Pages 108–9.

397. *New York Times Magazine*. April 26, 2001. Pages 50–133.

398. Vistica, Gregory L. One Awful Night in Thanh Phong. *The New York Times Magazine*. April 25, 2001.

399. *Basic Facts About the United Nations.* New York: UN. 1989.

400. Http://www.endevil.com/guncontrol.html.

401. Koestler, Arthur, *The Ghost in the Machine.* London: Penguin. 1967.

402. van Lawick, Hugo, and Goodall, Jane. *Innocent Killers.* London: Collins. 1970.

403. Augustine (trans. George G. Leckie). *Concerning the Teacher.* 1938. Appleton-Century-Crofts. Inc. Quoted in Freemantle, Anne. *The Age of Belief: The Mediaeval Philosophers.* Boston: Houghton Mifflin. 1954. Page 41.

404. Mian, Mohammad Sharif. *A History of Muslim Philosophy.* Vol. 1. Wiesbaden: Harrassowitz. 1963.

405. Choi, J-Y, and Bowles, S. The coevolution of parochial altruism and war. *Science* 316:636–40. 2007.

   Arrow, H. The sharp end of altruism. *Science* 318:581–2. 2007.

406. Mavany. *I Live Across a Danger.* Bangkok: SVITA. 1980.

407. Burford, Bill. *Among the Thugs.* New York: W.W. Norton & Company. 1991. Page 239.

408. Burdor, S. (Ed) *Complete Josephus.* London: Hurst. Book V, Chapter 13.

409. Talbot Rice, T. The Crucible of Peoples: Eastern Europe and the Rise of the Slavs. In *The Dawn of European Civilization: The Dark Ages.* New York: McGraw Hill Book Co. Inc. 1965. Page 143.

410. Bray, B. *Montaillou: The Promised Land of Error* (Trans. LeRoy Ladurie). New York: Vintage Books. 1975.

411. Kirkpatrick, S. *The Conquest of Paradise: Christopher Columbus and Columbian Legacy.* New York: Alfred Knopf. 1991. Page 32.

412. Ibid. Page 157.

413. Mason, Capt. John. *A Brief History of the Pequot War: Especially of the Memorable taking of their Fort at Mistick in Connecticut in 1637.* Boston: S. Kneeland & T. Green. 1736. Page 30.

414. Chang, Iris. *The Rape of Nanking: The Forgotten Holocaust of World War II.* New York: Basic Books. 1997. Page 85.

415. Fornander, Abraham. *An Account of The Polynesian Race Its Origins and Migrations and the Ancient History of the Hawaiian People to the Times of Kamehameha 1.* London: Turner & Co, 1880. Page 123.

416. White, John. *The Ancient History of the Maori.* Wellington: Government Printer. Vol. IV. 1889.

   Vayda, A. P. *Maori Warfare.* Wellington: The Polynesian Society Incorporated. 1960. Page 93.

417. Bullock, Alan. *Hitler and Stalin: Parallel Lives.* London: Harper Collins. 1991. Page 551.

418. Conquest, Robert. *The Great Terror: A Reassessment.* Oxford: Oxford University Press. 1990. Page 223.

419. Goldhagen, Daniel Jonah. *Hitler's Willing Executioners: Ordinary Germans and the Holocaust.* New York: Alfred Knopf. 1994. Page 5.

420. Luther, Martin. *Von den Jueden und iren Lügen.* 1543. Quoted by Roul Hiberg in *The Destruction of the European Jews.* New York: New Viewpoints. 1973. Page 9.

421. http://en.wikipedia.org/wiki/Animal rights in_Germany (accessed March 25, 2008).

422. Deuteronomy 20:13.

423. Solzhenitsyn, Alexander S. *The Gulag Archipelago 1918–1956* (Trans. T. P. Whitney). New York: Harper and Row. 1973.

424. Zimbardo, P. G. The SPE: What it was, where it came from, and what came out of it. In *Obedience to Authority: Current Perspectives on the Milgram Paradigm.* T. Bliss, Ed. Mahwah, NJ: Lawrence Erlbaum Associates. 2000. Pages 198–210.

425. Declaration on Atomic Bomb by President Truman and Prime Ministers Attlee [Great Britain] and King [Canada]. The White House, Washington.

426. Harris, Robert, Paxman, Jeremy. *A Higher Form of Killing: The Secret Story of Chemical and Biological Warfare.* New York: Hill & Wang. 1982. Page 10.

427. Ibid. Page 43.

428. Rhodes, Richard H. *The Making of the Atomic Bomb.* New York: Simon & Schuster. 1986.

429. Powers, Thomas. *Heisenberg's War: The Secret History of the German Bomb.* New York: Alfred Knopf. 1993.

430. Trevisanato, Siro. The 'Hittite plague', an epidemic of tularemia and the first record of biological warfare. *Journal of Medical Hypotheses* 69:1371–4. 2005.

431. Enserink, M. How devastating would a smallpox attack really be? *Science* 296:1592–5. 2003.

432. Stone R. Smallpox: WHO puts off destruction of U.S., Russian caches. *Science* 295:598–9. 2002.

433. Finkel, E. Engineered mouse virus spurs bioweapon fears. *Science* 291:585. 2001.

434. Hutchinson, S. J. *Population and Food Supply*. Cambridge: Cambridge University Press.1969.

435. Smith, M. O. Scalping in the Archaic Period: Evidence from the Western Tennessee Valley. *Southeastern Archaeology* 14. 1995.

436. LeBlanc, Steven A. *Prehistoric Warfare in the American Southwest*. Salt Lake City: University of Utah Press. 1999.

437. Turner, A. K. Genetic and hormonal influences on male violence. In *Male Violence*. J. Archer (Ed.). New York: Routledge. 1995. 233–252.

438. Arens, William. *The Man-Eating Myth: Anthropology and Anthropophagy*. New York: Oxford University Press. 1979.

439. Martar, R. A., Leonard, B. L., Billman, B. R., Lambert, P. M., Martar, J. E. Biochemical evidence of cannibalism at a prehistoric Puebloan site in southwestern Colorado. *Nature* 407:74–8. 2000.

440. Miguel, E., Satyanath, S., Sergenti, E. Economic shocks and civil conflict: An instrumental variables approach. *Journal of Political Economy* 112:725–53. 2004.

    Miguel, E. Poverty and violence: an overview of recent research and implications for foreign aid. In *Too Poor for Peace? Global Poverty, Conflict, and Security in the 21st Century*. L. Brainard, D. Chollet (Eds). Washington. DC: Brookings Institution Press. 2007. Pages 50–59.

441. Diamond, Jared. Easter Island's End. *Discover Magazine*. 1995.

442. Ramphal, S., Sinding, S. W. (Eds). *Population Growth and Environmental Issues*. Westport, CT: Praeger. 1996.

443. Peluso, N. L., Watts, M. *Violent Environments*. Ithaca: Cornell University Press. 2001

444. Kaplan, R. D. The coming anarchy: how scarcity, crime, overpopulation, and disease are rapidly destroying the social fabric of our planet. *Atlantic Monthly* February 1994. Pages 44–76.

    Kaplan, R. D. *The Ends of the Earth: A Journey at the Dawn of the 21st Century*. New York: Random House. 1996.

445. May, John. Demographic pressure and population policies in Rwanda, 1962–1994. *Population and Societies* 319. 1996.

446. Fairhead, J. International dimensions of conflict over natural and environmental resources. In *Violent Environments*. N. L. Peluso and M. Watts (Eds). Ithaca: Cornell University Press. 2001.

447. Turchin, P., and Korotayev, A. *Population Dynamics and Internal Warfare: a Reconstruction*. In press.

448. Homer-Dixon, Thomas. *Environment, Scarcity, and Violence*. Princeton; Princeton University Press. 1999.

449. Benedick, Richard. Human Population and Environment Stress in the Twenty-first Century. *Environmental Change and Security Project Report*. 6:5–18. 2000.

450. Renner, M. The Anatomy of Resource Wars. *State of the World Library*. Washington, DC. Worldwatch Paper. 2002.

    Gleditsch, N. P. Armed Conflict and the Environment: A Critique of the Literature. *Journal of Peace Research* 35:381–400. 1998.

451. Myers, N., and Kent, J. New consumers: The influence of affluence on the environment. *Proceedings of the National Academy of Sciences U.S.* 100: 4963–8. 2003.

452. Cincotta, Richard P., Engelman, Robert, Anastasion, Daniele. *The Security Demographic: Population and Civil Conflict After the Cold War.* Washington: Population Action International. 2003.

453. Benjamin, Daniel, and Simon, Steven. *The Age of Sacred Terror*. New York: Random House. 2002.

454. Reuter, Christopher. *My Life is a Weapon*. Princeton and Oxford: Princeton University Press. 2004. Page 8.

455. Noonan, John. *Contraception: A History of Its Treatment by the Catholic Theologians and Canonists*. Cambridge, MA: Harvard University Press. 1968.

Pagels, Elaine. *Adam, Eve, and the Serpent*. New York: Random House. 1988. Page 99.

Haskins, Susan. *Mary Magdalen: Myth and Metaphor*. New York: Harper Collins. 1994. Pages 72–4.

456. Ranke-Heinemann, Uta. *Eunuchs for Heaven: The Catholic Church and Sexuality* (Trans. John Brownjohn). London: Andre Deutsch. 1990. Page 138.

457. Noonan, John. *Contraception: A History of Its Treatment by the Catholic Theologians and Canonists*. Cambridge, MA: Harvard University Press. 1968.

458. Fildes, Valerie. *Breasts, Bottles and Babies: A History of Infant Feeding*. Edinburgh: Edinburgh University Press. 1986. Page 175.

459. Stopes-Roe, Harry V., and Scott, Ian. *Marie Stopes and Birth Control*. London: Priory Press. 1974.

Soloway, Richard A, *Birth Control and the Population Question in England, 1877–1930*. Chapel Hill: University of North Carolina Press. 1982.

460. Mohr, James C. *Abortion in America: The Origins and Evolution of National Policy*. New York: Oxford University Press. 1978.

461. Potts, Malcolm. Sex and the Birth Rate. *Population and Development Review* 23:1–40. 1997.

462. Campbell, Martha M., Salin-Hodoglugil, Nuriye S., Potts, Malcolm. Barriers to Fertility Regulation. *Studies in Family Planning* 37:87–98. 2006.

463. Sanger, Margaret. *Margaret Sanger: Pioneering Advocate for Birth Control*. New York: Cooper Press. 1999. Facsimile of 1938 edition.

464. Gray, Madeline. *Margaret Sanger: A Biography of the Champion of Birth Control*. New York: Richard Marek Publishers. 1979.

465. Asbell, Bernard. *The Pill: A Biography of the Drug that Changed the World*. New York: Random House. 1995.

466. Temkin, Elizabeth. Nurses and the prevention of war: public health nurses and the peace movement in World I. In *War and Public Health*. Barry S. Levy and Victor Sidel, Eds. Washington: American Public Health Association. 2000. Pages 350–9.

467. The Alan Guttmacher Institute. *Sharing Responsibility: Women, Society and Abortion Worldwide*. New York: The Alan Guttmacher Instutute. 1999.

468. Wills, Garry. *Papal Sin: Structures of Deceit*. New York: Doubleday. 2000. Page 94.

469. All Party Group on Population, Development and Reproductive Health. *Return of the Population Growth Factor: Its Impact on the Millennium Development Goals*. London: Portcullis House, House of Commons. 2006. Page 16.

470. Ibid. Page 20.

471. Davis, Kinsley. Population policies: Will current program succeed? *Science* 158:730–9. 1967

472. Connelly, Matthew. *Fatal Misconceptions: The Struggle to Control World Population*. Cambridge, MA: The Belknap Press of Harvard University Press. 2008. Page 320.

473. Thomas D'Agnes. *From Condoms to Cabbages: An Authorized Biography of Mechai Viravaidya*. Bangkok: Post Books. 2001.

474. Anon. *From Battlefront to Community: The Story of Gonoshasthaya Kendra*. Dhaka, Bangladesh: Gonoshasthaya Kendra. 2006.

475. Vahidnia, Farnaz. Case study: fertility decline in Iran. *Population and Environment* 28:259–66. 2007.

476. Sissmann, Sarah, and Barbier, Christophe. L'effect Soleilland: Une affaire de femmes. *L'Express*. August 30, 2004.

Corliss, Richard. Shades of Gray. *Time*. January 15, 1990.

477. Harry S. Truman Library Oral History. http://www.trumanlibrary.org/oralhist/draperw.htm (accessed March 26, 2008).

478. Smelser, Neil J., and Mitchell, Faith (Eds). *Discouraging Terrorism: Some Implications of 9/11*. Washington: National Academies Press. 2002.

479. Mumford, Stephen D. *The Life and Death of NSSM 200: How the Destruction of Political Will Doomed a US Population Policy.* Research Triangle Park, North Carolina: Center for research on Population and Security. 1996.

480. *The 9/11 Commission Report: Final Report of the national Commission on Terrorist Attacks upon the United States.* New York: W.W Norton & Co. 2004. Page 54.

481. Campbell, Martha M. Schools of thought: An analysis of interest groups influential in international population policy. *Population and the Environment* 19: 487–512. 1998.
Campbell, Martha. Why the silence on population? *Population and the Environment* 29:237–46. 2007.
Potts, Malcolm, and Campbell, Martha. Reverse gear: Cairo's dependence on a disappearing paradigm. *Journal of Reproduction and Contraception* 16:179–86. 2005.

482. Speidel, J. Joseph, Weiss, Deborah, Ethgelston, Sally A., Gilbert, Sarah M. Family planning and reproductive health: the link to environmental preservation. *Population and Environment* 28: 247–58. 2007. See also: The donor landscape http://www.packard.org/assets/files/population/programs%20review/pop_rev_speidel_030606.pdf (Accessed July 2006).

483. *The Return of the Population Factor: Impact on the Millennium Development Goals.* Report of hearings in U.K. Parliament, report November 2006. Page 29.

484. Krugman, Paul. March of Folly. *International Herald Tribune.* July 18, 2006.

485. Harris, Sam. *The End of Faith: Religion, Terror and the Future of Reason.* New York: Norton & Co. 2004. Page 100.

486. Belgium SABENA and Swiss.

487. No civil airliner has been high jacked since this simple change was instituted.

488. Wright, Lawrence. *The Looming Tower: Al Qaeda and the Road to 9/11.* New York: Alfred A. Knopf. 2006.

489. Alexander, D. A., and Klein, S. The psychological aspects of terrorism from denial to hyperbole. *Journal of the Royal Society of Medicine.* 98:557–62. 2005.

490. http://teaching.arts.usyd.edu.au/history/hsty3080/3rdYr3080/Callous%20Bystanders/isolation-ism.html

491. Naylor, S. *Not a Good Day to Die: The Untold Story of Operation Anaconda.* New York: Berkeley Caliber Books. 2005. Page 56.

492. Ibid. Page 272.

493. Godges, J. Afghanistan on the edge. *Rand Review.* Summer 2007. Pages 14–22.

494. Yingling, P. A Failure in Generalship. *Armed Forces Journal.* April 27, 2007.

495. Webb, M. *Illusions of Security: Global Surveillance and Democracy in the Post 9/11 World.* San Francisco: City Lights. 2006.

496. Johnson, Dominic, McDermott, Rose E., Barrett, Emily S., Cowden, Jonathan, Wrangham, Richard, McIntyre, Matthew H., Rosen, S. P. Overconfidence in wargames: experimental evidence on expectations, aggression, gender and testosterone. *Proceedings of the Royal Society B: Biological Sciences* 273:2513–20. 2006.

497. Woods, Kevin, Lacey, James, Murray, William. Saddam's Delusions: The View From the Inside. *Foreign Affairs.* May/June 2006. Pages 2–26.

498. Phillips, Kevin. *American Theocracy: Perils and Politics of Radical Religion, Oil, and Borrowed Money in the 21st Century.* New York: Viking. 2006.

499. Quotation from Richard Kerr, Deputy Director CIA under George H. W. Bush. In Joseph S. Nye, Jr. Transformational Leadership and U.S. Grand Strategy. *Foreign Affairs.* July/August 2006. Pages 139–48.

500. Machiavelli, Niccolò (Trans. Daniel Donno). *The Prince.* 1531. New York: Bantam Classic. Page 31.

501. Phillip Gordon. Air Power Won't Do It. *Washington Post.* July 25, 2006. Page A15.

502. Poll by the Jerusalem Media and Communications Center, quoted in *The Economist,* July 22, 2006. Page 31.

503. Sudan: catastrophe looms. *The Economist.* September 9, 2006. Page 13.

504. Bridgland F. Riding with the Janjaweed. *Sunday Herald* (Scotland). July 8, 2004. http://www.sundayherald.com/43939

505. *Strategic Survey 2003/4: An Evaluation and Forecast of World Affairs.* International Institute of Strategic Studies. May 2004.

506. U.S. State Department and U.S. Department of Defense, Foreign Military Assistance Act. *Report to Congress.* http://www.fas.org/asmp/profiles/index.html.

507. Langewiesche, Lawrence. The Wrath of Khan. *The Atlantic.* November 2005.

508. Sobhani, S. *The Pragmatic Entente.* New York: Prager. 1989. Page 146.
      Hiro, D. *The Longest war: the Iran-Iraq Military Conflict.* New York: Routledge. 1991. Page 118.

509. Kaplan, F. All it touched off was a debate. *New York Times.* September 18, 2005. Page 5.

510. The long, long half-life. *The Economist.* June 8, 2006. Pages 21–3.

511. Knightley, Phillip. *The First Casualty. From the Crimea to the Falklands: The War Correspondent as Hero, Propagandist and Myth Maker.* London: Pan Books. 1989.

512. Kronblut, A., and Stolberg, S. G. In speech, Bush warns of risks in quitting Iraq. *New York Times.* September 1, 2006.
      Rich, Frank., Rumsfeld, Donald. Dance with the Nazis (Op-ed). *New York Times.* September 3, 2006.

513. Human Development Report 2007/2008. United Nations Development Program. New York: Palgrave Macmillan, 2007. Page 25.

514. *EarthPulse: The Essential Visual Report on Global Trends.* Washington, DC: National Geographic Society. 2008.

515. Gourevitch, Philip. *We Wish to Inform You That Tomorrow We Will Be Killed With Our Families: Stories From Rwanda.* New York: Farrar Straus and Giroux. 1998.

516. Crigin, K., and Curiel, A. Prime Numbers. *Foreign Affairs.* September/October 2006. Pages 34–5.

517. Thun, M. J., Apicella, L. F., Henley, S. J. Smoking vs other risk factors as causes of smoking-attributable deaths. *Journal of the American Medical Association.* 284:706–12. 2000.

518. Herbert, B. Ike saw it coming. *New York Times.* February 27, 2002. Page A23.

519. Nichols, John. *The Nation Magazine.* May 17, 2004. Speech in Greenboro, North Carolina, Oct 10, 2000.
      Blumenthal, Sidney. The religious warrior of Abu Ghraib: an evangelical U.S. general played a pivotal role in Iraqi Prison reform. *The Guardian* (London). May 20, 2004.
      Cooper, Richard T. General casts war in religious terms: The top soldier assigned to track down Bin Laden and Hussein is an evangelical Christian who speaks publicly of "the army of God." *Los Angeles Times.* October 16, 2003.

521. President Bush and some relatives of the fallen lean on each other. *The New York Times.* November 10, 2007.

522. Deuteronomy 20:16–17.

523. Barrett, David, Kurian, George, Johnson, Todd (Eds). *World Christian Encyclopedia* (2nd edition). New York: Oxford University Press. 2001. 2 vols.

524. Harris, Sam. *The End of Faith: Religion, Terror, and the Future of Reason.* New York: W. W. Norton. 2004. Page 35.

525. Shah, Timothy, and Toft, Monica. Why God is winning. *Foreign Policy* July/August 2006. Pages 39–43.

526. Wilson, Edward O. *Sociobiology.* Cambridge, MA: Harvard University Press. 1975.

527. Haidt, J. The new synthesis in moral psychology. *Science* 316: 998–1001. 2007.

528. Nowak, M. A., and Sigmund, K. Evolution of indirect reciprocity. *Nature* 437: 499–502. 2005.

529. Haidt, J. The new synthesis in moral psychology. *Science* 316: 998–1001. 2007.
      Also: Bowles, Samuel. Group Competition, Reproductive Leveling, and the Evolution of Human Altruism. *Science* 314:1569–72. 2006.

530. Greene, Joshua D., Sommerville, R. Brian, Nystrom, Leigh E., Darley, John M., Cohen, Jonathan D. An fMRI Investigation of Emotional Engagement in Moral Judgment. *Science* 293:2105–8. 2001.

531. Market forces. *The Economist.* August 5, 2006. Page 37.

532. Goodall, Jane. *Reason for Hope: A Spiritual Journey.* New York: Warner Books. 1999. Page 293.

533. Ibid. Page 117.
534. Churchill, Winston. Speech, September 11, 1940.
535. Boothby, J. C., and Halperin, J. Mozambique child soldier life outcome study: lessons learned in rehabilitation and reintegration efforts. *Global Public Health* 1:87–107. 2006.
536. Liddell Hart, B.H. *Strategy*. London: Meridian. 1967.
537. Pinker, S. *The Blank Slate: The Modern Denial of Human Nature*. London: Penguin Books. 2002.
538. Adams, D. The Seville Statement on Violence: A progress report. *Journal of Peace Research* 26:113–21. 1989.
539. For example, in 1641 by Massachusetts.
540. Woolman, John. *Some Considerations on Keeping Negroes*. 1754. Northampton, MA: Gehenna Press. 1970 (reprint).
541. Reardon, Betty. *Women and Peace: Feministic Vision of Global Security*. Albany, NY: State University of New York Press. 1993.
542. Potts, Malcolm. Why can't a man be more like a woman? Sex, Power and Politics *Obstetrics and Gynecology* 106:1065–70. 2005.
543. Fearon, Kate. *Women's Work: The Story of the Northern Ireland Women's Coalition*. Belfast: The Blackstaff Press. 1999.
544. information@huntalternatives.org.
545. Fearon, James D., and Laitin, David D. Ethnicity, insurgency and civil war. *American Political Science Review* 97:36–58. 2003.
    Rice, S.E. Poverty breeds insecurity. In *Too Poor for Peace: Global Conflict, and Security in the 21st Century*. L. Brainard, D. Chollet (Eds). Washington, DC: Brookings Institution Press. 2007.
    A Prime Minister's Strategy Unit Report to the Government of the UK . Investing in Prevention: An International Strategy to Manage Risks of Instability and Improve Crisis Response. February 2005.
546. Potts. M. Sex and the birth rate: human biology, demographic change, and access to fertility regulation methods. *Population and Development Review* 23:1–39. 1997.
547. Abu Bakr Naji. *The Management of Savagery: The Most Critical Stage through which the Umma Will Pass* (Trans. W McCants). Combating Terrorism Center, May 23, 2006. www.ctc.usma.edu/Management of_Savergy.pdf.
548. Campbell, M., Sahin, Hodoglugil, Potts, M. Barriers to fertility regulation: a review of the literature. *Studies in Family Planning* 37:87–98. 2006.
549. Kaplan, Robert D. *Soldiers of God: With Islamic Warriors in Afghanistan and Pakistan*. New York: Vintage Books. 1990. Page 49.
550. Stockholm International Peaces Institute (SIPI). www.sipri.org/contents/milap/milex/mex_database1.html (Accessed April 10, 2008).
    Library Congress Country Studies. http://memory.loc.gov/cgi-bin/query2/r?frd/cstdy:@field(DOCID+il0184 (Accessed August 12, 2006).
551. Out of their silos. *The Economist*. June 10, 2006. Page 36.
552. Tian, Jingdong, Hui Gong, Nijing Sheng, Xiaochuan Zhou, Erdogan Gulari, Xiaolian Gao, and George Church. Accurate multiplex gene synthesis from programmable DNA microchips. *Nature* 432:1050– 54. 2004.
553. Wilson, E. O. *Consilience; The Unity of Knowledge*. New York: Alfred A Knopf. 1998. Page 264.
554. Abraham Lincoln, Second Inaugural Address, 1865.

Index

About the Authors

MALCOLM POTTS, MB, BChir, PhD, FRCOG, is the Bixby Professor at the University of California, Berkeley. A graduate of Cambridge University and trained as an obstetrician and research biologist, his profession has taken him all over the world. Potts led a medical team into Bangladesh immediately after the War of Liberation in 1972, and he has worked in many other war-torn places including Vietnam and Cambodia, Afghanistan, Egypt, the Gaza Strip, Liberia, and Angola. His most recent books are *Queen Victoria's Gene* and *Ever Since Adam and Eve: The Evolution of Human Sexuality*.

THOMAS HAYDEN is a freelance journalist who writes frequently about science, medicine and culture. Formerly a staff writer at *Newsweek* and *U.S. News & World Report*, his articles and reviews have appeared in more than twenty publications, including *National Geographic*, *Nature*, and *The Washington Post*. He is coauthor of *On Call in Hell: A Doctor's Iraq War Story*, a 2007 national bestseller. He lives in San Francisco with his wife and fellow writer, Erika Check Hayden.